Modern Technology in Vegetable Production

NIPA® GENX ELECTRONIC RESOURCES & SOLUTIONS P. LTD.
New Delhi-110 034

Modern Technology in Vegetable Production

P. Hazra
Professor
Department of Vegetable Crops
Bidhan Chandra Krishi Viswavidyalaya
Mohanpur-741252, West Bengal, India

A. Chattopadhyay
Reader (Dept. of Vegetable Crops) and Incharge
AICRP on Vegetable Crops,
Directorate of Research, Bidhan Chandra Krishi Viswavidyalaya
Kalyani-741235, West Bengal, India

K. Karmakar
Reader (Dept. of Agricultural Entomology) and Incharge
AINP on Agricultural Acarology,
Directorate of Research, Bidhan Chandra Krishi Viswavidyalaya
Kalyani-741235, West Bengal, India

&

S. Dutta
Lecturer (Dept. of Plant Pathology)
AICRP on Vegetable Crops,
Directorate of Research, Bidhan Chandra Krishi Viswavidyalaya
Kalyani-741235, West Bengal, India

NIPA® GENX ELECTRONIC
RESOURCES & SOLUTIONS P. LTD.

101,103, Vikas Surya Plaza, CU Block
L.S.C. Market, Pitam Pura, New Delhi-110 034
Ph : +91 11 27341616, 27341717, 27341718
E-mail: newindiapublishingagency@gmail.com
Web: www.nipabooks.com

For customer assistance, please contact
Phone: + 91-11-27 34 17 17
Fax: + 91-11- 27 34 16 16
E-Mail: feedbacks@nipabooks.com

ISBN: 978-81-19235-12-4

Composed and Designed by NIPA®.

Dedicated to Our Parents

P. Hazra, A. Chattopadhyay
K. Karmakar and S. Dutta

Preface

Vegetables play major role in the National developmental programmes to alleviate hunger and malnutrition. In India tremendous progress has been made in the recent past in the vegetable production and India emerged second largest producer of vegetables in the World after China. The total production of vegetables has increased from 23.4 million tones in 1961-62 to the tune of 125 million tones in 2007-08. This remarkable transformation from subsistence vegetable farming in sixties to the present commercial production has happened mainly due to huge increase in population and resultant increase in market demand mainly from the high and middle income segment of the population. However, considering the demand to meet the requirement of 300 g vegetables per capita per day for the burgeoning population with annual growth rate of 1.93 per cent and for processing industries and export market, India need to produce 150 million tones of vegetables by 2011 mainly through the enhancement in productivity. These challenges call for greater technology support to maximize productivity against unfortunate diminishing soil fertility, natural resources particularly land and water and rising cost of vegetable production.

This book "Modern Technology in Vegetable Production" deals with classification of different vegetable crops, basic principles of different crop management practices and specific production technologies of 29 important vegetable crops. Unique feature of this book lay on 150 coloured photographs on four important aspects of vegetable production viz., nursery management, physiological disorder, disease and insect pests of different vegetable crops. This type of text book cum manual with extensive photographic documentation is the new addition in the teaching and demonstrative field of Vegetable science.

This text book cum manual has basically been prepared for the graduate and post graduate level students of Agricultural Universities of India and abroad. Technical personnels of the commercial vegetable farms, planners, extension and development officers and even nutritionists and dieticians will also get benefit from this book.

The authors appreciate their wives for their lasting patience during preparation of this book.

We thank all our students for helping in taking most of the photographs. We owe to everybody whose photographs, downloaded from the internet, have been used in this book.

We would very much appreciate receiving suggestions and criticism related to different aspects of this book.

We thank Messers New India Publishing Agency, New Delhi for publishing this edition of the book.

B.C.K.V., 2010 **Authors**

Contents

Dedication v

Preface vii

1. Classification of vegetable crops 1

2. Principles and practices of vegetable production management ... 11

1. Seedling management in the nursery 11
2. Water management 21
3. Plant nutrient management 27
4. Pollination management 44
5. Integrated pest management 48
6. Integrated disease management 53
7. Biological management of diseases 56
8. Weed management 60

3. Production technology of vegetable crops 71

1. Potato 71

Solanaceous fruit vegetables 87

2. Tomato 87
3. Brinjal 103
4. Chilli 114
5. Capsicum or Sweet pepper 128

Cole crops 140

6. Cauliflower 140
7. Cabbage 155
8. Broccoli *or* sprouting broccoli 168

Cucurbits **178**
9. Watermelon 178
10. Muskmelon 190
11. Pumpkin 201
12. Bottle gourd 212
13. Bitter gourd 224
14. Cucumber 236
15. Ridge gourd 249
Root crops **256**
16. Radish 256
17. Carrot 266
Bulb crops **275**
18. Onion 275
19. Garlic 287
Greens and Pot herbs. **298**
20. Amaranthus 298
21. Palak 305
Peas and Beans **310**
22. Garden pea 310
23. French bean 318
24. Vegetable cowpea 328
25. Hyacinth bean 340
Tropical Tuber Crops **350**
26. Elephant foot yam 350
27. Sweet potato 358
Okra **370**
28. Okra 370
Pointed gourd **380**
29. Pointed gourd 380
Appendix **395**
Colour Plates **411**
Index **427**

1
Classification of Vegetable Crops

Introduction

The book has been framed as a text book cum vegetable production manual for both under graduate and post graduate students of Agriculture/ Horticulture as well as commercial vegetable production professionals. This book has been oriented on different management practices of vegetable culture in open field condition viz., Classification of vegetable crops, Management of seedlings in the nursery, Water managent, Plant nutrient management, Pollination management, Integrated pest management, Biological management of diseases, Weed management and Production technology of different vegetable crops.

Classification of vegetable crops

Vegetables are defined as herbaceous plant origin material with generally low calorific value providing micronutrients, vitamins, antioxidants, fibre and induce alkaline reaction inside human body. Vegetable crops are not only botanically different but also their cultural and climatic adaptation shows wide variability. It is, therefore, necessary to classify the vegetable crops for their easy recognition. Vegetable crops can conveniently be classified "Botanically" for the purpose of vegetable breeding and according to "Method of culture" from the point of crop husbandry.

Botanical classification

The crops used for vegetable purpose in the world belong to 1200 species under 78 families and of them, more than 860 species under 59 families belong to dicotyledoneae and about 340 species under 19 families belong to monocotyledoneae. In the tropical and subtropical parts of the world, about 90 species of vegetable crops are cultivated, but hardly 20 of them are commercially important.

In the present botanical classification, over 140 different vegetable crops under 6 monocotyledoneae and 20 dicotyledoneae families have been included.

Name	**Botanical name**	**Chromosome number (2n)**
Monocotyledoneae		
1. Amarylidaceae (Alliaceae)		
Onion	*Allium cepa* L.	16
• Multiplier onion	*A. cepa* var. *aggregatum* L.	16
• Top onion	*A. cepa* var. *viviparum* (Metz.) Alef.	16
Garlic	*A. sativum* L.	16-32 (2x – 4x)
Leek	*A. ampeloprasum* L. var. *porrum*	32, 48 (4x, 6x)
Welsh onion	*A. fistulosum* L.	16
Shallot	*A. ascalonicum* L.	16
European chive	*A. schoenoprasum* L.	16
Chinese chive	*A. tuberosum* Rottler ex Spreng	32
Kurrat	*A. kurrat*	32 (4x)
2. Araceae		
Taro	*Colocasia esculenta* (L.) Scott	-
• Eddoe type	*C. esculenta* var. *antiquorum*	28,42 (2x, 3x)
• Dasheen type	*C. esculenta* var. *esculenta*	42 (3x)
Giant taro	*Alocasia macrorrhiza* (L.) Scott, *A. indica* (Roxb.) Scott.	26,28
Swamp taro	*Colocasia esculenta var. stolonifera*	26
Giant Swamp taro	*Cyrtosperma chamisonis*	26,28
Tannia	*Xanthosoma sagittifolius* (L.) Scott.	26
Elephant foot yam	*Amorphophallus paeoniifolius* (Dennst.) Nicolson	26,28
3. Bambusaceae		
Bamboo	*Bambusa vulgaris* Schrader ex Wendland	
4. Dioscoreaceae		
Yam	*Dioscorea* spp.	30-80 (3x-8x)
• Greater yam	*D. alata* L.	40 (4x)

• Lesser yam	*D. esculenta* (Lour.) Burkill	40 (4x)
• White yam	*D. rotundata* (L). Poir.	40 (4x)
5. Gramineae (Poaceae)		
Sweet corn	*Zea mays* L. var. *rugosa*	20
6. Liliaceae		
Asparagus	*Asparagus officinalis* L.	20
7. Nelumbonaceae		
Lotus	*Nelumbo nucifera* Gaertn.	
Dicotyledoneae *1. Aizoaceae*		
New Zealand spinach	*Tetragonia expansa* (Murr.)	
2. Amaranthaceae		
Leafy amaranth	*Amaranthus tricolor* L., *A. blitum* L., *A. viridis* L., *A. tristis* L., etc.	32 or 34
	A. dubius L.	64 (4x)
3. Basellaceae		
Indian spinach	*Basella rubra* L. var. *alba* (green type)	24
(Malabar night shade)	*B. rubra* L. var. *rubra* (red type)	24
4. Chenopodiaceae		
Garden beet	*Beta vulgaris* L. subsp. *vulgaris*	18
Palak	*B. vulgaris* L. var. *bengalensis* Hort.	18
Chard	*B. vulgaris* L. var. *cicla*	18
Spinach	*Spinacea oleracea* L.	12
French spinach (orach)	*Artiplex hortensis* L.	12
Pigweed	*Chenopodium album* L.	36 (4x)
5. Compositae (Asteraceae)		
Lettuce	*Lactuca sativa* L.	
• Head type	*L. sativa* var. *capitata* L.	18
• Leaf type	*L. sativa* var. *crispa* L.	18

• Cos type	*L. sativa* var. *longifolia* L.	18
• Asparagus type	*L. sativa* var. *asparagina* L.	18
Chicory	*Cichorium intybus* L.	18
Endive	*Cichorium endivia* L.	18
Globe artichoke	*Cynara scolymus* L.	34
Jerusalem artichoke	*Helianthus tuberosus*	102 (6x)
Terragon	*Artemisia dracunculus* L.	18
Salsify	*Tragapogon porrifolius* L.	-
Black salsify	*Scorzonera hispanica* L.	-
Spanish salsify	*Scolymus hispanica* L.	-
Dandelion	*Taraxacum officinale* Weber ex Wiggers	-
6. Convolvulaceae		
Sweet potato	*Ipomoea batatas* (L.) Lam.	90 (6x)
Water spinach	*Ipomoea aquatica* Forsk	-
7. Cruciferae (Brassicaceae)		
Cabbage	*Brassica oleracea* L.var. *capitata* L.*f. alba* DC	18
Red cabbage	*B. oleacea* L. var. *capitata f. rubra* (L.) Thell	18
Savoy cabbage	*B. oleacea* var. *sabauda* L.	18
Cauliflower	*B. oleacea* L. var. *botrytis* L.	18
Brussels sprouts	*B. oleracea* L. var. *gemifera* DC.	18
Sprouting broccoli	*B. oleracea* L. var. *italica* Plenck	18
Knolkhol (Kholrabi)	*B. oleracea* var. *gongylodes* L.	18
Kale/collard	*B. oleracea* var. *acephala*	18
Chinese cabbage	*B. chinensis* L. (Pak-choi), *B. chinensis* L. var. *pekinensis* (Pet-sai)	20
Chinese kale	*B. alboglabra* Bailey	-
Mibuna Turnip	*B. japonica* *B. rapa* L. var. *glabra* Kitamura	- 20
Rutabaga (swede)	*B. napus* var. *napobrassica* Peterm.	38(amphi-diploid)
Leaf mustard	*B. juncea* (L.) Czern. & Coss. var. *cuneifolia Roxb.*	36 (4x)

Curled mustard	*B. juncea* var. *crispifolia*	36 (4x)
White mustard	*Sinapis alba* L.	24
Radish	*Ranphanus sativus* L.	18
• European radish	*R. sativus* L. var. *radicula* DC	18
• Japanese & Chinese winter radish	*R. sativus* L. var. *raphanistroides* Makins	18
• Black or Spanish radish	*R. sativus* L var. *niger* (Mill.) Pers.	18
• Oilseed radish	*R. sativus* L var. *caudatus* (L) Hooker et Anderson	18
Horse radish	*Armoracia rusticana* Gaertn.Mey. & Schr ed.	32 (4x)
Water cress (brahmi sag)	*Rorippa nasturtium-aquaticum* (Syn. *Nasturtium officinale R. Br.)*	32
Garden cress	*Lepidium sativum* L.	16,32
Upland cress	*Barbarea vulgaris* R. Br.	-
Sea kale	*Crambe maritime*	-
Catran	*Crambe tatarica* Busch	-
Tuberous rooted Chinese mustard	*Sinapis juncea* var. *napiformis* Paill & Bois.	-
8. Cucurbitaceae		
Cucumber	*Cucumis sativus* L.	14
Muskmelon	*Cucumis melo* L.	24
• Snap melon	*C. melo* L. var. *momordica*	24
• Long melon	*C. melo* L. var. *utilissimus* Duth.& Full.	24
• Snake melon	*C. melo* L. var. *flexuosus* Naud	24
• Netted melon	*C. melo* L. var. *reticutalus*	24
• Pickling melon	*C. melo* L. var. *conomon* Mak.	24
• Cantaloupe	*C. melo L.* var. *cantalupensis* Naud.	24
• Winter melon	*C. melo* L. var. *inodorus*	24
• Mango melon	*C. melo* L. var. *dudain*	24
• Small gourd	*C. melo* L. var. *agretis* Naud.	24

Gherkin	*Cucumis anguria* L.	24
Watermelon	*Citrullus lanatus* (Thumb.) Matsum & Nakai	22
Round melon	*Praecitrullus fistulosus (*Stocks) Pang.	24
Pumpkin	*Cucurbita moschata* (Duch.) Poir.	40
Summer squash	*C. pepo* L.	40
Winter squash	*C. maxima* Duch.	40
Buffalow gourd	*C. foetidissima* Bouche	40
Bottle gourd	*Lagenaria siceraria* (Molina) Stadl.	22
Bitter gourd	*Momordica charantia*	22
Balsam apple	*M. balsamina* L.	22
Giant Spine gourd or sweet gourd	*M. cochinchinensis* (Lour.) Spreng	28
Teasle gourd	*M. subangulata* ssp. *renigera* (G. Don) de Wilde	56
Spine gourd	*M. dioica* Roxb. Ex Willd.	28
Ridge gourd	*Luffa acutangula* (L.) Roxb.	26
Sponge gourd	*L. cylindrica* Roem.	26
Pointed gourd	*Trichosanthes dioica* Roxb.	22
Snake gourd	*T. anguina* L.	22
Wax gourd	*Benincasa hispida* (Thunb.) Cogn.	24
Ivy gourd	*Coccina grandis* (L.) Voigt.	24
Chow Chow	*Sechium edule* (Jacq.) Swartz.	28
Caigua	*Cyclanthera pedata* (L.) Schrader	32
9. Euphorbiaceae		
Cassava	*Manihot esculenta* Crantz.	36
Chekurmanis	*Sauropus androgynus* (L.) Merr.	-
10. Labitae		
Chinese potato	*Solenostemon rotundifolius (Poir) Morton* (Syn. *Coleus parviflorus* Benth.)	-
Chinese artichoke	*Stachys siebeldi* Miq.	-
Hoary basil	*Ocimum americanum* L.	-
11. Leguminoceae (Fabaceae)		
Garden pea & snow pea	*Pisum sativum* L. sub sp. *hortense*	14

French bean	*Phaseolus vulgaris* L.	22
Lima bean	*P. lunatus* L.	22
Runner bean	*P. coccineus* L.	22
Tepary bean	*P. acutifolius* Grey	22
Thicket bean	*P. polystachyus*	22
Adzuki bean	*P. angularis* (Wild.) Wright	22
Sword bean	*Canavalia gladiata* (Gucav.) DC.	22,24
Gotani bean	*C. plagosperma*	22
Jack bean	*C. ensiformis* (L.) D.C.	22
Valvet bean	*Mucuna deeringiana* (Bor.) Merr.	22
Cluster bean	*Cyamopsis tetragonolobus* (L.) Taub.	14
Winged bean	*Psophocarpus tetragonolobus* (L.) Taub.	18,22
Broad bean	*Vicia faba* L.	12
Hyacinth bean	*Lablab perpureus* (L.) Sweet.	22,24
• Vegetable type	*Lablab perpureus* (L.) var. *typicus*	22,24
• Pulse type	*Lablab perpureus* (L.) var. *lignosus*	22,24
Horse gram	*Dolichos uniflorus* Lam.	22,24
Cowpea	*Vigna unguiculata*	22
• Small podded vegetable and pulse type	*Vigna unguiculata* sub sp. *unguiculata*	22
• Long podded vegetable type	*V. unguiculata* sub sp. *sesquipedalis*	22
Soybean	*Glycine max* (L.) Merrill	40
Yam bean	*Pachyrhizus erosus* Urbn.	22
African yam bean	*Sphenotylis stenocarpa*	-
African-locust bean	*Parkia clappertonia*	-
Ground bean	*Voandzeia subterranea* (L.) Thou.	-
Marama bean	*Tylosema esculentum* (Burchell) Schreiber	-
Agathi	*Sesbania grandiflora* Poit.	24
Fenugreek	*Trigonella foenum-graceum* L.	16

12. Malvaceae		
Okra	*Abelmoschus esculentus* (L.) Moench	126,130,132 (Allopolyploid)
13. Martynaceae		
Martynia	*Proboscidea jussieeui* (Keller)	-
14. Moringaceae		
Drumstick	*Moringa oleifera* Lamk.	28
15. Polygonaceae		
Rhubarb	*Rheum rhaponticum* L., *Rheum rhabarbarum* L.	22, 44 (4x)
Sorrel	*Rumex acetosa* L.	-
Patience dock	*R. patientia* L.	-
Bladder dock	*R. vesicarius* L.	-
Buck wheat	*Fagopyrum tatariocum* (L.) Gaertn.	16
16. Portulaceceae		
Pusley	*Portulaca oleracea* L.	54
Ceylon spinach	*Talinum triagulare* Willd.	-
17. Rutaceae		
Curry leaf	*Murraya koenigii* (L.) Spreng	18
18. Solanaceae		
Potato	*Solanum tuberosum* L.	48 (auto 4x)
Brinjal	*Solanum melongena* L.	24
• Fruit round or egg-shaped	*S. melongena* var.*esculentum*	24
• Fruit long & slender	*S. melongena* var. *serpentinum*	24
• Plant dwarf	*S. melongena* var. *depressum*	24
Tomato	*Lycopersicon esculentum* Mill.	24
• Common	*L. esculentum* var. *commune Bailey*	24
• Large-leafed	*L. esculentum* var. *grandifolium* Bailey	24
• Upright	*L. esculentum* var. *validum* Bailey	24
• Cherry	*L. esculentum* var. *cerasiforme* Bailey	24

• Pear	*L. esculentum* var. *pyriforme* bailey	24
Currant tomato	*L. pimpinellifolium* (Juslen) Mill.	24
Sweet pepper & Chilli	*Capsicum annuum* L.	24
Tabasco pepper	*C. frutescens* L.	24
Husk tomato	*Phesalis pubescens* L.	24
Tomatillo	*P. ixocarpa* Brot.	24
Tree tomato	*Cyphomandra betacea*	24
19. Tiliaceae		
Jute	*Corchorus olitorius* L.	14
20. Umbelliferae (Apiaceae)		
Carrot	*Daucus carota* L.	18
Celery	*Apium graveolens* L. var. *dulce* (Mill.) Pers.	22
• Celeriac	*A. graveolens* L. var. *rapaceum* (Mill.) Gaud.	22
• Leaf celery	*A. graveolens* L. var. *secalinum*	22
Parsley	*Petroselinum crispum* (Mill.) Nym. Ex. Hill.	22
• Turnip-rooted parsley	*P. crispum* var. *tuberosum*	22
Parsnip	*Pastinaca sativa* L.	22
Turnip-rooted chervil	*Chaerophyllum bulbosum* L.	-
Skirret	*Sium sisarum* L.	-
Chervil	*Anthriscus cerefolium*	-
Dill	*Anethum graveolens* L.	22
Coriander	*Coriandrum sativum* L.	22

Classification based on method of culture

This classification is the best for the vegetable growers because the vegetable crops requiring the same cultural practices (but not same season for cultivation) are grouped together though they may be divergent botanically. However, the crop species in the groups like cole crops, bulb crops, cucurbits, peas and beans belong to the same family. In this book the crops have been listed following this classification.

Group 1	:	**Potato**
Group 2	:	**Solaneceous fruits** (Tomato, brinjal, chilli, sweet pepper)
Group 3	:	**Cole crops** (Cabbage, cauliflower, sprouting broccoli, Brussels sprouts, knolkhol, Chinese cabbage).
Group 4	:	**Cucurbits** (Pumpkin, summer squash, winter squash, muskmelon, watermelon, cucumber, bottle gourd, ash gourd, ridge gourd, snake gourd, sponge gourd, bitter gourd).
Group 5	:	**Root crops** (Radish, carrot, beet, turnip, horse radish, rutabaga, salsify, turnip-rooted chervil, Spanish salsify, black salsify)
Group 6	:	**Bulb crops** (Onion, garlic, leek, shallot, welsh onion, chive).
Group 7	:	**Salad crops** (Lettuce, celery, endive, chicory, parsley, sorrel, water cress, garden cress, and upland cress).
Group 8	:	**Greens and pot herbs** (Amaranth, palak, spinach, New Zealand spinach, coriander, Indian spinach, French spinach, dandelion, pigweed, tarragon, parsley, Ceylon spinach, fenugreek).
Group 9	:	**Peas and beans** (Pea, French bean, cowpea, hyacinth bean, asparagus bean, cluster bean, winged bean, lima bean, velvet bean).
Group 10	:	**Tuber crops other than potato** (Taro, giant taro, tannia, yam, cassava, elephant-foot yam, Chinese potato).
Group 11	:	**Sweet potato**
Group 12		**Okra**
Group 13	:	**Pointed gourd**
Group 14	:	**Chayote (chow chow)**
Group 15	:	**Tropical perennial vegetable crops** (Drumstick, agathi, chekurmanis, curry leaf)
Group 16	:	**Temperate perennial vegetable crops** (Asparagus, rhubarb, seakale, globe artichoke)

2

Principles and Practices of Vegetable Production Management

1. Seedling management in the nursery

Seedlings of many seed propagated vegetable crops like, tomato, brinjal, chillies, sweet pepper, cauliflower, cabbage, knolkhol, broccoli, Brussels sprout, kale, endive, chicory, celery, lettuce, parsley, onion, etc. are first raised in the nursery beds before transplanted in the main field although, many of these crops like, tomato, sweet pepper, onion, cabbage, cauliflower, broccoli, etc. are also direct seeded using precision seed drills in the USA, Japan, the Netherlands, Italy and other European countries. Many seed propagated vegetable crops viz., okra, peas, beans, radish, carrot, beet, turnip and all the cucurbits (e.g. bitter gourd, bottle gourd, pumpkin, cucumber, watermelon, muskmelon, snake gourd, sponge gourd, ridge gourd, etc.) are conventionally grown through direct seeding in the main field.

Transplant groups

1. Plants easy to transplant: They form new roots rapidly, put forth shoots rather slowly and absorb water readily e.g. beet, broccoli, Brussels sprouts, cabbage, cauliflower, Swiss chard, lettuce and tomato.

2. Plants transplanted with moderate ease: This group replace roots and grow shoots a bit slower than the previous group and do not absorb water very efficiently e.g. celery, brinjal, onion, pepper, leek and cauliflower.

3. Plants transplanted with difficulty, requiring special care: They are very poor or slow to both root replacement and enhanced shoot growth. They take long time to cope up root injury and do not survive very ordinary transplanting. They require special care in handling e.g. cucumber, sweet corn, legumes, pumpkin, squash and melons.

A number of vegetable crops are also propagated and grown by different vegetative plant parts like, underground stem tuber (potato), underground root tuber and above ground vines (sweet potato), corm, corm pieces and cormel (elephant-foot yam), cormel (taro), bulb (onion), clove i.e., bulblet, bulbil (garlic), fleshy root and vine cuttings (pointed gourd), fleshy root (teasle gourd), stem cutting (cassava), limb cutting (drumstick), root tuber(Jerusalem artichoke), root suckers or offshoots and division of old crown (globe artichoke), pieces of stem tuber and vine cuttings (yam) and division of crowns (rhubarb).

Followings are different advantages of raising seedlings in the nursery.

1. Uniform, vigorous and healthy planting materials can be raised for better growth, development and yield.
2. Economy of seed rate which cut down the cost of cultivation particularly for hybrids.
3. Seedlings can be raised in off-season vegetable forcing.
4. Favourable growing conditions can be provided to the seedlings in nursery bed through convenient protection against biotic and abiotic stresses.
5. Efficient time management for the preparation of main field.
6. Improvement in land use efficiency by increasing the cropping intensity.
7. Facilities for intercropping and relay cropping.
8. Easy adjustment of date of planting according to pest and disease vulnerability and market demand.
9. Maintainance of desired plant density in the field.
10. Efficient management of weed, pest and disease.

Different features of a nursery

Nursery site: The nursery site should be located away from shade at higher topography of land to facilitate proper drainage. The area should preferably be fenced or protected from pet and wild animals. Irrigation, transport, electricity, communication facilities, availability of basic material are some important considerations for establishing a commercial nursery.

Nursery soil: Success of seedling nursery depends mainly upon the quality of soil. Before selecting a nursery, the soil analysis must be done to avoid any risk. The soil should be light to medium in texture with high porosity, good fertility, sufficient water holding capacity and good drainage. A soil pH of 6.5-7.5 is most suitable for vegetable nursery. Well drained sandy loam soil is the best for raising the seedlings. If the nursery soil is heavy, sand @ 2 kg/sqm and if the soil is light, farm yard manure or compost @ 2.5 kg/sqm need to be added to raise good seedlings. Farm yard manure or compost also replenishes the macro and micronutrients in the growing medium. Well rotten and powdered FYM or compost is also used as organic mulch to improve the water holding capacity of soil.

Growing media for the seedlings

Seedling growing media are one of the most important factors in raising vegetable nursery because growth and development of nursery plants and roots proliferation depends on it. Growing media are divided into soil, soil-less group or mixture of both. The growing media are soil, sand, coco peat, compost, vermicompost, peatmoss, perlite, pumice, sawdust, ashes, etc. However, under open field condition, farmers generally use soil, sand and well rotten farm yard manure as they are cheap, easily available, retain sufficient water, air and allow sufficient drainage. Growing media should be easily available and must have the properties like, neutral in reaction, sufficient water and air retentive capacity and allow proper drainage, free from weed seeds, nematodes, fungi and bacterial inoculum and capable of being sterilized. Different growing media are hereunder mentioned.

Vermicompost: Vermicompost is a useful growing medium and it can be prepared easily by rearing of earthworms which provides all the required qualities like firmness, good water and air holding capacity and better nutrition and support to the seedlings.

Saw dust: This is a by-product of wood and timber industry and generally free from harmful substances. It is available at a very cheap rate everywhere. It has very good moisture retention capacity, free from microorganism inoculum and neutral in reaction.

Peat: Peat consists of the remains of aquatic, marsh, or swamp vegetation, formed as a result of partial decomposition under anaerobic or semi-anaerobic conditions. It is dark brown to black in colour, low in pH with a low moisture-holding capacity but contain 2.0 to 3.5 percent nitrogen.

Sphagnum moss: Commercial sphagnum moss is the dehydrated young residue or living portions of plants in the genus Sphagnum, such as S. papillosum, S. capillaceum, and S. palustre. It is relatively sterile, light in weight and has a very high water holding capacity, being able to absorb water 10 to 20 times of its weight. This material is generally shredded, either by hand or mechanically, before it is used in a propagation or growing medium. It has a pH value of about 3.5 to 4.0 and contains small quantity of minerals.

Perlite: Perlite is a grey white aluminium silicate mineral from lava flows. It is light in weight, highly expandable (4-20 times to its original volume) when heated approximately at 957°C. The surface holds moisture and nutrients to make them available to the seedlings. Since it is sterile in nature, it is free from the diseases causing inoculums and nematodes. It has low cation exchange capacity and does not influence the pH of the media. High porosity of perlite helps to control water holding capacity and aeration of substrates. A coarse perlite is 1.2-2.0 mm in diameter with pH of 7.5.

Vermiculite: Vermiculite is a micaceous mineral that expands markedly when heated. Chemically it is hydrated magnesium aluminium/iron silicate. When expanded, vermiculite is very light in weight, 90 to 150 kg per cubic meter, neutral in reaction with good buffering properties, and insoluble in water. It is able to absorb large quantities of water (80 percent porosity). It has a relatively high cation exchange capacity (CEC 100-150 meq/100 g) and thus can hold nutrients in reserve and later on release them for the use of seedlings. It contains 9-12% magnesium and 5-8% potassium for the growing plants.

Pumice: Chemically pumice is mostly silicon dioxide and aluminium oxide, with small amounts of iron, calcium, magnesium and sodium in the oxide form. Pumice is screened to different size grades but is not heat treated. It increases aeration and drainage in rooting mixture and can be used alone or mixed with peat moss.

Rock wool: Rock wool is a man made mineral fibre. It has a high percentage of pore space, which holds both the nutrient solution and air for the growing plants. The rock wool contains 40-50% easily available water and 40-45% air capacity by volume. It acts as a support medium for the root system and as a means for holding nutrient solution and air in close proximity to the roots. It is sterile, easily managed and transported, completely consistent in performance and has a pH of about 7.0-7.5.

Coco peat: Coco peat is biodegradable and eco-friendly growing medium prepared from natural coir fibre. It is completely organic and soil free potting medium, a mixture of corky cellular material and short fibre which can replace soil and peatmoss. It has uniform texture and consistency, free from pathogen, weed seeds and toxins. It has a high water holding capacity, almost five times of its volume and retains and releases nutrients over a long period which provides better aeration and enhances strong and healthy root growth.

Soil treatment

For raising healthy seedlings, nursery soil must be treated by physical, chemical or biological means to kill the soil borne disease causing pathogens particularly, *Pythium, Rhizoctonia, Fusarium, Sclerotinia, Botrytis, Phytophthora, Phomopsis, Sclerotium, Alternaria* and *Pseudomonas solanacearum*, nematodes (particularly *Meloidogyne* spp.) and insect pests. Different species of *Pythium, Sclerotinia* and *Phytophthora* are more likely to cause damping off in cool wet soils, where as different species of *Rhizoctonia,*

Fusarium and *Sclerotium* cause damping off in warmer and drier conditions which are most common disease in the nursery. Symptom manifestation of bacterial wilt in different vegetable crops particularly tomato and brinjal caused by *Pseudomonas solanacearum* mostly occur in the main field but very often the bacteria harbour the seedling in the nursery which is carried forward in the main field. Similarly infestation of nematode in the seedbed is carried forward to the main field. Besides, different fungi and bacteria cause some specific disease in the seedling stage viz., early blight of tomato (*Alternaria solani*), leaf blight of tomato (*Alternaria alternata* f.sp. *lycopersici*), collar rot of tomato, brinjal, chilli (*Sclerotium rolfsii*), bacterial spot of tomato (*Xanthomonas campestris* pv. *vesicatoria*), anthracnose of chilli and capsicum (*Colletotrichum orbaculare*), purple blotch of onion (*Alternaria porri*), etc. Nursery soil may be treated by different methods as mentioned below.

Soil solarization: The whole nursery area after being irrigated lightly is covered with transparent polyethylene sheet of 400 gauge for about 5-6 weeks during the summer months (from mid-April to June) when air temperature may raise upto 40-48°C and soil temperature inside the polyethylene reaches upto 48-56°C which is sufficient to kill most of the harmful pathogens, nematodes and weed seeds. After 5-6 weeks, the polyethylene sheet is removed and the area is used for the preparation of nursery beds for sowing the seeds.

Steam treatment: The nursery area is first covered with polyethylene sheet in such a manner that air can not pass in or out and then steam is passed inside for at least 4-6 hours continuously to kill the harmful pathogen, nematodes and insects pest. The soil is then ploughed and prepared for seed sowing.

Formalin treatment: Formalin solution is prepared by adding 250 ml commercial grade formaldehyde (40%) per 10 litres of water and soil is drenched @ 4.5 litres per square meter so as to saturate top soil surface upto a depth of 15-20 cm 15-20 days before seed sowing and the drenched area is covered with 400 gauge polyethylene sheet so that fumes of formalin penetrate inside the soil to kill the pathogens. The polyethylene cover is removed after 24 hours and the bed kept open for 15-20 days prior to seed sowing.

Soil treatment with fungicides: Fungicides like captan or thiram can also be used @ 5 gram per square meter for dry treatment or 0.25-0.30% solution @ 4-5 litres per square meter as soil drenching.

Soil treatment with insecticides: Granular insecticides like, Furadon, Phorate 10 G, etc. are applied to the soil @ 4-5 gram per square meter and then thoroughly mixed to a depth of 15 to 20 cm to kill insects, their eggs and nematodes.

Soil treatment with bio-agents: Different species of Trichoderma spp., Pseudomonas fluorescence and Aspergillus niger are very much effective biological agents to controls majority of the soil borne pathogens and these bio-agents are added @ 10-25 gram per square meter and well mixed in the soil and seeds should be sown 2 days after the application of bio-agents. However, success of biological control depends on the activity of the spores, spore counts in the formulation and sufficient organic matter in the beds for rapid multiplication of the bio-agents.

Seed treatment

Most of the vegetable seeds are treated with captan or thiram or carboxin-thiram @ 2.5-3 g/kg or carbendazim @ 2.0 g/kg of seed to disinfect the seed surface. Hot water treatment of seeds of cabbage and cauliflower at 50°C for 30 minutes just before sowing prevents seed borne black rot disease caused by the bacteria *Xanthomonas campestris* pv. *campestris*. Biological control of the soil borne damping off diseases caused by *Pythium* sp., *Rhizoctonia* sp., *Fusarium* sp., etc can be done through seed treatment with different antagonistic fungi like, *Trichoderma* spp., *Gliocladum virens*, *Penicillium* spp., and bacteria like, *Bacillus subtilis* and *Pseudomonas fluorescens*.

The use of coated or pelleted seeds has now become quite common in vegetable crops especially those having small seeds like, tomato, brinjal, chilli, sweet pepper, lettuce, etc. The pelleted seeds are often coated with pesticides for better protection against pest and disease causing pathogens. In the developed countries enhanced seeds are also available which are developed by osmoconditioning or priming. In this process, seeds are allowed to imbibe water in a temperature controlled dilute aerated solution of an organic or inorganic osmoticum to ensure faster and uniform germination.

Preparation of nursery bed

Nursery bed should generally be raised 15-20 cm from ground level and the soil must be fine and have optimum moisture for uniform and high percentage of germination. The length of the bed may be kept 3 to 5 meter, and width is restricted to a maximum of 1 meter only to facilitate intercultural operations. The beds should preferably be prepared in the east and west direction and line should be made north to south direction on the beds to explore and capture maximum radiation energy especially during winter.

Seed sowing in the nursery

Proper seed sowing in the nursery beds is one of the important factors to get healthy and vigorous seedlings. After the seed bed preparation, seeds are sown in the nursery bed either by broadcasting or in lines depending upon the nature and season of crop and availability of time and labour, although line sowing is the best. About 1 cm deep lines parallel to the width at a distance of 5.0 cm apart are opened and the seeds are sown or placed at a distance of about 1.0-2.0cm in the line. After sowing, the seed bed should be covered with 0.5 cm thick mixture of farm yard manure: sand: soil in the ratio of 1:1:2 for better moisture conservation and germination. Spraying the seed covering mixture first with 0.25% Dursban and then after two days with 0.25% captan or 0.1% carbendazim will protect seedlings from termite, ant and damping off disease. The seedlings in the nursery should be protected from different sucking insects particularly white fly which transmit different virus for 25 to 30 days by spraying imidacloprid (0.3 ml/litre of water) at weekly interval. Mulching the seedbeds with paddy straw or sugar cane trash in hot weather with intermittent sprinkling of water and polyethylene mulch during winter helps in quick and uniform germination of seeds. The time to remove the mulch from the seedbed is extremely important. The seed beds should be observed daily and as and when the white thread like structure is seen above the ground, the mulch should be removed carefully in the evening hours to avoid any damage to the emerging seedlings. During the seedling growth, if temperature reaches above 30°C then the beds should be covered by green or black coloured 50% or 60% shade nets about 1 meter above the ground with suitable support to protect the seedlings from direct scorching sunlight. The newly introduced translucent blue shade net gives blue light, which is excellent for plant growth. The nursery beds require light irrigation with the help of rose can from the seed sowing until the seedlings become ready to transplant. Young roots of seedlings absorb water primarily by osmotic forces. Anaerobic condition caused by excessive soil moisture retard seed germination. It is important to remove weak, diseased, insect pests damaged and dense plants from the nursery beds keeping 5 to 1.0 cm distance from plant to plant for proper growth of the seedlings.

Pricking out of seedlings

It is a method of producing healthy and hardened seedlings by pricking out the young seedlings when first true leaves have come out and transplanting them to another seed bed or seed box or grown individually in small plastic thumb pots of about 5 cm diameter. The seedlings should be uprooted carefully with the help of V-notched wooden stick for second transplanting and spaced about 5 cm within a row which are 7-8 cm apart, both in second seed beds or seed boxes. The soil mixture generally used to grow the pricked out seedlings is slightly richer than the previous one for first seed bed. Seedlings become ready for planting in the main field generally three weeks after first transplanting.

Seedling raising under poly house

The direct sown crops like, okra, peas, beans, radish, carrot, beet, turnip , bitter gourd, bottle gourd, pumpkin, cucumber, watermelon, muskmelon, snake gourd, sponge gourd, ridge gourd, etc. cannot be transplanted after uprooting because of the formation of an external layer of corky cell "suberine" in the existing cells of the roots. Since suberine is impervious to water which results in the death of the seedlings due to water deficit. Poor root regeneration also hinder plant establishment after transplanting in these crops. But these difficult to transplant vegetable crops can be transplanted without disturbing the roots in the growing media to minimize root damage. Generally nursery raised plants of these crops are economic in the adverse situation when there is too low temperature for seed germination or the field preparation is late due to some standing crop in the field. It is now a common practice to sow seeds of different cucurbits like, bottle gourd, bitter gourd, muskmelon, watermelon, cucumber etc in winter months(December – February) in small polybags (15 × 10cm size of 200-300 gauges) filled with soil, farm yard manure /compost and sand mixture (1:1:1) to get early crop. Seeds having hard or leathery seed coat such as bottle gourd, bitter gourd, pumpkin, watermelon, etc. may be soaked in water overnight before sowing for better germination. The polybags are kept in polyhouse where temperature is raised above 20°C during day time in winter months. Simple, improvised structure with bamboo poles or iron pipes with a covering of polyethylene sheet of 200 gauge can be used in raising the vegetable seedlings under unfavorable environmental condition. Micro sprinklers are the best source to irrigate such poly house grown vegetable seedlings. The micro sprinklers connected with water supply and sensors automatically sprinkle water at regular intervals. Seedlings of different cucurbits become ready within 28-30 days under the poly houses especially during winter months (December-February) while under open condition it takes 50-60 days. As soon as field conditions are favourable for planting, the seedlings are transplanted in well prepared mounds either by removing the polyethylene or by cutting lower portion of the polyethylene so that roots may penetrate into the soil. In this way the crops can be advanced by 30- 45 days compared to normal direct seeding in the main field.

Protection of seedlings

Different vegetable crops like, tomato, chilli, sweet pepper, etc are very susceptible to virus particularly leaf curl virus which are usually deposited inside the plant at the seedling stage by the insect vector primarily white fly (*Bemisia tabaci*). Fruit and shoot borer also attacks brinjal even in the seedling stage. Seedling of these vegetable crops should be raised inside the fine mosquito net (40-50 mesh) coupled with spraying of insecticides outside the mosquito net to keep away the infestation of the insect vectors to produce virus free seedlings. It is better to spray the combination of fungicides (metalaxyl-mancozeb@ 2.0 g / litre or 0.2% chlorothalonil or 0.25% mancozeb + 0.1% carbendazim), insecticides (metasystox 1.5 ml/litre or imidachloprid 0.3 ml/litre) and sticker (triton or sandovit @ 0.5 ml/litre of solution) to control the fungal pathogen and sucking insect pests like, white fly, jassids, mite, thrips, etc. Red pumpkin beetle which is the most prominent insect pest of the seedlings of cucubits can effectively be controlled by spraying carbaryl 50 WP @ 2 g/ litre at 2-3 leaf stage.

Hardening of seedlings

Hardening of seedlings is done for better field stand by different ways i) lowering the available soil moisture to 20% by withholding watering to the plant 5-7 days before transplanting, ii) lowering the temperature which retards the growth, iii) by application of 4000 ppm NaCI with irrigation water, iv) by spraying of 2000 ppm of cycocel + 0.25% $ZnSO_4$ and v) by spraying of 0.2% muriate of potash.

Step-wise seedling raising in field nursery

- Use healthy and improved seed from a reliable seed source. If the seed is not already treated, mix with 2-3 g thiram/captan per kilogram of seed.
- Disinfect the nursery area by soil solarization. Cover the nursery beds with a 400 gauge polyethylene sheet for 5-6 hours. After removing the sheet, stir the soil with a disinfected rake.
- Prepare fine seed beds about 20 cm high and 1-1.5 m wide.
- Sow seeds in lines about 5 cm apart, about 1-2 cm deep. Adjust the planting depth according to the seed size.
- Cover the seed with a fine layer of compost or farm yard manure. Be sure that compost or farm yard manure is fully decomposed and free from pathogens. Mulch the nursery beds with paddy straw or dry leaves.
- Water the nursery beds with a rose can twice a day. After seedlings sprout, water once a day.
- Remove the mulch immediately after seedlings emerge from the soil.
- Protect nursery beds from excessive rain and stagnant water by providing rain shelters. Allow for good air circulation to protect seedlings from damping off and foliar diseases. Protect seedlings from insects and insect-transmitted diseases, especially viruses, by covering nursery beds with 50-mesh nylon net.
- Spray the seedlings with mancozeb + carbendazim mixture @ 2.5 g/ litre of water if damping off occurs.
- Harden the seedlings by withholding water 4-5 days before transplanting. Hardening is facilitated by the application of potassium chloride (KCl) at 1 ml per 250-300 ml of water, or the application of a 10% sucrose solution 7-10 days before transplanting.
- Drench the nursery beds with water one day before transplanting.
- Transplant 4- to 6-week-old seedlings, preferably in the afternoon hours.

Step-wise seedling rising in plug trays

- Use healthy and improved seed from a reliable seed source. If the seed is not already treated, mix with 2-3 g thiram/captan per kilogram of seed.
- Use clean trays. Disinfect the trays with 1% solution of chlorine bleach. Wash the trays thoroughly with clean water before use.
- Use the correct cell size for the crop. Cucurbitaceous crops require a larger cell size than other vegetable crops.
- Prepare growing media by mixing cocoa peat, perlite, and vermiculite in 3:1:1 ratio. A mixture of peat moss, compost, and coconut coir in equal proportions can also be used.
- Adjust the pH of the mixture between 5.5 and 6.5 with lime to improve micronutrient availability. Avoid mixing soil into the media to minimize the occurrence of soil-borne diseases.
- Fill the plugs tightly with the pre-moistened media.
- Sow one seed in the center of each cell, about 0.5-1 cm deep, depending upon the seed size. Cover the seeds with the same mixture.
- Water the trays thoroughly to moisten the entire plug. This promotes root growth to the bottom of the plug. Avoid watering late in the afternoon to minimize occurrence of foliar diseases.
- Apply water-soluble fertilizer with irrigation water at 50-75 g per 100 liters of water at weekly intervals. Excessive use of fertilizers makes seedlings leggy.
- Protect the trays from excessive rain by providing rain shelters. Allow good air circulation to protect seedlings from damping off and foliar diseases. Protect seedlings from insects and insect-transmitted diseases, especially viruses, by covering the sides of nursery area with 50-mesh nylon net.
- Spray the seedlings with mancozeb + carbendazim mixture @ 2.5 g/ litre of water if damping off occurs.
- Harden the seedlings by withholding water 4-5 days before transplanting. If the seedlings are being raised in protected conditions, move them outside for several days to acclimatize before transplanting.
- Drench the plugs with water one day before transplanting.
- Transplant 4- to 6-week-old seedlings, preferably in the afternoon hours.

2. Water management

Water is the most vital component of the living cells and the basis of plant feeding and in fact, both literally and poetically is the blood of civilization. Water is also characterized with extraordinary physical properties: it has high specific heat, biggest latent heat vaporization, very high thermal conductivity and greatest surface tension. Large amounts of heat are tied up with water because of very high specific heat, which contribute to both climatic stability and to heat budgets of the plants and other organisms.

Integral relationship of plant and water

Water molecules are integral part of living systems being the solvent for metabolites and structural component of proteins and nucleic acids. In fact, biological macromolecules cannot exist in their complicated secondary folding or spiraled helixes without the support of hydrogen bonding within their water matrix. The water matrix is also critical for metabolic activity because its properties of hydration permit ready reactivity between molecules in solution as well as between enzymes and their substrates, and the biochemical processes are sensitive to tissue hydration. A reduction of only 20-25 per cent below maximum hydration is sufficient to cause metabolic malfunctioning of many enzymes. It is the substrate for photosynthesis and a product of respiration. Water also helps to buffer plant tissue against fluctuations in the external environment. Plant growth including leaf expansion, organ enlargement, reproductive development and stomatal function are all clearly associated with moisture status. Hormonal changes (abscisic acid and cytokinins) in plant body occur with brief reduction in water potential. Water is also the major force in shaping the climatic patterns.

Transpiration lowers water potential at evaporative sites within leaves, and this effect is immediately translated to the root system through creating tension in the plants' vascular system. This demand of water must be satisfied continuously to maintain leaf water potential. The internal water balance of plants depends on the relative rates of water absorption and water loss. Despite stomatal resistance, the steep gradient in vapour pressure from leaf to air favours the transpiration rate well in excess of CO_2 fixation. Transpiration rate commonly ranges between 500 to 2500 mg H_2O/dm^2-h whereas CO_2 fixation rates generally between 5 to 25 mg CO_2/dm^2-h. This physiological phenomenon amply justifies high water requirement to maintain high dry matter productivity. For this reason, actively growing plants as the case of all the vegetable crops need to maintain liquid phase continuity from soil water through its vascular system and all the way to evaporative sites in leaves. Vegetables contain large amount of water (more than 85%) and the product quality like, tenderness, succulence, crispness and flavour is very much related to water supply at proper stages. In fact, texture of the vegetables is determined by the combination of tissue structure, cell wall properties and turgor pressure.

Development of moisture stress in foliage results different physiological consequences *viz.* (1) decrease in stomatal opening, (2) reduction in photosynthesis and transpiration, (3) dehydration of protoplasm, (4) reduction in cell division and enlargement, (5) increase in respiration initially which later decrease due to less availability of respiratory substrate, (6) decrease in total dry matter production and growth, (7) hastening in maturity ,(8) accumulation of sugar .particularly during later part of growth, etc. Virtually every facets of the plant's physiology is dislocated to some degree during temporary moisture stress which is recovered subsequently upon adequate moisture supply later and under severe moisture stress such recovery of essential physiological functions is not operative causing permanent wilting of the plant.

Consumptive use and water use efficiency

Soil water is lost by evaporation from the soil surface, water surface and from the surfaces of leaves of the plant and transpiration from plant body. In the field, these two processes cannot be separated so far as plant water requirements are concerned. The water lost in these two ways is called evapotranspiration which is nearly equal in amount to the consumptive use or net water requirement of the crop. Of the total absorbed water, hardly 0.1 to 0.5 per cent water is actually incorporated into the plant tissues and the remaining water is ward off from the plant body by transpiration. So, consumptive use is the quantity of water lost by evapotranspiration plus the quantity that is actually incorporated in the plant tissues. Water loss from mature plants is about 50 to 90 per cent water which is lost from a free water surface on the same area. Evaporation from free water surface is estimated by using "Evaporation pan". Potential evapotranspiration is indirectly estimated from pan evaporation data after necessary adjustment.

Water requirement of a crop can be calculated from consumptive use or evapotranspiration of a crop which is influenced by temperature, irrigation practices, length of growing season, precipitation and other factors. The volume of water transpired by crop plants depends on availability of water, absorption of water by the roots, temperature and humidity of the atmosphere, wind movement, intensity and duration of sunlight, stage of development of plant, type of foliage, nature of the leaves and kind of crop. Variations in consumptive use rate occur from day to day because of changes in climate. It is increased in hot, dry and windy day and is decreased in cool day. Consumptive use also varies with relative growth of the crop. The phenology of the growing plant can be characterized by vegetative, flowering, fruiting, and other distinctive characteristics in vegetative stage like, curding in cauliflower, heading in cabbage and head lettuce, bulbing in onion and garlic, tuberization in potato and sweet potato, etc. During the vegetative stage consumptive use continues to increase till the end of this stage of growth and flowering occurs near and during the peak of consumptive use. The fruiting stage is accompanied by a decrease in consumptive use until the transpiration essentially ceases during the latter part of the development of seeds inside the fruit.

Stomata are the compulsory passage way for CO_2 and H_2O gas exchange between plant and the atmosphere. Therefore, in order to fix carbon, the plant loses water to the atmosphere, the ratio being variable depending on species and growing conditions. This ratio is called water use efficiency *i.e.,* the quantity of water transpired per unit carbon fixed in dry matter. Water use efficiency increases as the amount of water required per unit of dry matter production decreases. "Transpiration ratio" is used to express the water use efficiency. Transpiration ratio refers to the quantity of water (in gram) transpired by the plant to accumulate 1 g of dry matter. This ratio ranges from 200 to 1000 (i.e., 200: 1 to 1000: 1) depending on the crop species, cultivation conditions and vegetation period. Vegetable crops can be categorized into four groups according to such index of water exchange. Leafy vegetables require much more water to produce 1 g dry matter than pumpkin. Vegetable crops showing high transpiration ratio cannot endure water stress, and any shortage of water during the period of growth and development severely affects yield. On the other hand, vegetable crops characterized by very low transpiration ratio can somewhat endure water stress condition and can give satisfactory yield even in moisture-deficient condition. Vegetable crops can be classified into four groups according to the water requirement which is expressed as the units of absorbed water required for the production of one unit of dry matter.

Classification of vegetable crops according to their water requirement

High	**Moderate**	**Low**	**Very low**
Palak	Onion	Pea	Watermelon
Amaranth	Cucumber	French bean	Muskmelon
Lettuce	Chilli	Cowpea	Pumpkin
Celery	Brinjal	Broad bean	Wax gourd
Other leafy greens	Tomato	Cluster bean	
Sweet pepper	Carrot	Winged bean	
Cabbage	Potato	Hyacinth bean	
Cauliflower			
Asparagus			
Broccoli			
Rhubarb			
Brussels sprouts			
Horse radish			
Radish			
Beet			
Turnip			
Ridge gourd			
Rutabaga			
Green onion			

Source: Vegetable Growing in Home Gardens of Tropical and Subtropical Areas (Konokov and Vemulapalli, 1988)

Scheduling of irrigation

Frequency of irrigation and amount of water to be applied depend on a number of factors such as, depth of root system, water use efficiency, stages of growth, soil type, prevailing weather condition and actual consumptive use of the crop however, it is a rather complex consideration.

Irrigation requirement depends on the stages of growth of the particular vegetable crop. When the plants are young although they require and use less water, need highly moist soil because the root system of the young plants is weak, sparsely distributed and located in the upper 15 to 20 cm layer of the soil that gets quickly dried. The vegetable crops utilize and transpire more water in the later stages of growth with more area and size of the leaves. Although most plants have critical stages of growth during which water stress markedly reduce yield. However, all the leafy vegetable crops, cauliflower, radish, carrot, beet, celery, etc. require optimum soil moisture level constantly during the whole vegetation period for proper growth and yield.

Over-irrigation also causes other harms like, bursting of heads in cabbage, branching of roots in carrot, fruit cracking in tomato, fruit rot in pointed gourd, etc. Abundant irrigation during ripening of fruits causes an increase in the water content of fruits and subsequently reduces their sweetness and storability. It is thus suggested to stop irrigation before fruit ripening of watermelon, muskmelon, pumpkin, winter squash, etc. In potato, onion, elephant foot yam, sweet potato, etc. irrigation should also be stopped 15 days before the harvest which greatly extends their storability.

Irrigation needs and critical soil moisture period, and impact of water deficit of vegetable crops

Crop	SMT(bars)	ASM	Critical stage of growth	Major defects apart from low yield caused by water deficit
Potato	-0.30	65%	Stolon formation, tuberization and tuber enlargement	Poor and mis-shaped tuber
Tomato	-0.35	60%	Flowering and fruit development and after each harvest	Blossom-end rot and fruit cracking
Brinjal	-0.45	50%	Flowering and fruit development and after each harvest	Blossom-end rot, misshapen fruit, poor fruit colour
Chilli	-0.45	50%	Tenth leaf to flowering and fruit development and after periodical harvests	Low fruit set, shriveled fruits

Sweet pepper	-0.35	60%	Flowering and fruit development and after each harvest	Poor fruit size and blossom-end rot
Cauliflower	-0.25	70%	Curd formation and enlargement	Ricey curd, buttoning
Cabbage	-0.35	60%	Head formation and enlargement	Growth cracks and tough leaf
Broccoli	-0.25	70%	Heads development	Strong flavour, ricey curd and premature bolting
Onion	-0.35	60%	Bulb formation and bulb enlargement	Poor-sized bulb and early bolting
Garlic	-0.35	60%	Bulb formation and bulb enlargement	Poor-sized cloves
Carrot, beet, turnip	-0.40	55%	Root enlargement	Growth cracks, misshapen roots
Cucumber	-0.45	50%	Flowering and fruit development	Pointed and cracked fruits with bitterness
Pumpkin and all gourds	-0.50	55%	Flower bud development and early fruit development	Misshaped, small and cracked fruits
Watermelon	-0.60	40%	Flowering and fruit development	Blossom-end rot
Muskmelon	-0.60	40%	Flowering and fruit development	Poor fruit growth and quality
Okra	-0.55	45%	Flowering and pod development	Toughness and fibre development in pods
Peas	-0.60	40%	Flowering and pod filling	Poor pod filling and shriveled grains
French bean, hyacinth bean	-0.45	50%	Pod setting and pod enlargement	Poor pod filling with stringy pods
Cowpea	-0.45	50%	Prior to flowering and after pod set	Poor pod filling with fibrous pods
Leafy vegetables	-0.30	70%	Entire crop duration	Toughness in leaves
Celery	-0.30	70%	Entire crop duration	Toughness in leaves
Head lettuce	-0.35	60%	Head formation	Poor head with tough leaves
Sweet potato	-0.70	35%	Tuber initiation, early bulking and late bulking stages	Poor and misshaped tuber
Elephant foot yam	-0.60	40%	Shoot emergence	Low bulking of corm

SMT: Soil moisture tension; ASM: Available soil moisture.

The depth, extent and configuration of the root system are also important in judging water absorptive potential of the particular crop. Roots of the shallow-rooted vegetable crops are confined to the surface soil which remove water much more rapidly and dry the soil. So, shallow rooted vegetable crops require frequent irrigations. On the other hand, crops with extensive root system continue to extract water from greater soil depths. Thus, interval between irrigation for the deep-rooted vegetable crops can be extended. So, knowledge of rooting depths is, therefore, important for scheduling the time and quantity of irrigation. Rooting depths vary greatly among the vegetable crops. The vegetable crops can be classified into 5 groups according to their rooting depth, *viz.,* very shallow-rooted, shallow-rooted, moderately deep-rooted, deep-rooted and very deep-rooted.

Rooting depth of different vegetable crops

Very shallow rooted (15-30cm)	Shallow-rooted (30-60 cm)	Moderately deep-rooted (60-90 cm)	Deep-rooted (90-120 cm)	Very deep-rooted (120-180 cm)
Onion	Broccoli	French bean	Pea	Artichoke
Lettuce	Brussels sprouts	Beet	Chilli	Asparagus
Small radish	Cabbage	Carrot	Rutabaga	Lima bean
	Cauliflower	Chard	Summer squash	Parsnip
	Celery	Cucumber	Turnip	Pumpkin
	Chinese cabbage	Brinjal		Winter squash
	Endive	Muskmelon		Sweet potato
	Garlic			Tomato
	Leek			
	Parsley			
	Potato			
	Palak			
	Spinach			
	Radish			
	Cowpea			

However, the relationship between rooting depth and water requirement is not always direct. Some shallow-rooted crops, like cabbage, cauliflower, spinach and onion require relatively lesser quantity of water as compared to the crops with deep root system like, tomato, muskmelon, etc., which is quite obvious. On the other hand, water requirement of some shallow-rooted crops, like potato, celery, etc., is much higher than some very deep-rooted crops like sweet potato, artichoke, lima bean, etc. Similarly, water requirement of very shallow-rooted crops like onion and lettuce is same as that of the moderately deep to deep-rooted crops like pole bean, beet, carrot, cucumber, pea, chilli, etc. In most of the vegetable crops, a great proportion of the total root weight is frequently found to occur near the upper surface of the soil. Hence, the deep-rooted vegetable crops often extract water mostly from the upper layers of the soil. Water and nutrient absorption also decreases rapidly with the depth of roots. So, moisture extraction depths of shallow- and deep-rooted vegetable crops are sometimes equated. In such situations, water requirement of the shallow- and deep rooted vegetable crops differs with the parameters other than the rooting depths, like water use efficiency, duration of the crop, consumptive use, etc. Generally, shallow-rooted crops grown in sandy soil require moderate and frequent irrigation while deep-rooted crops in clayey soils need heavy irrigation at longer intervals.

3. Plant nutrient management

Plants obtain from soil, besides water, different inorganic substances that are essential in the general metabolism of the plants. The supply and absorption of inorganic elements needed for growth and development is defined as nutrition, and these elements are called plant nutrients. An essential nutrient element may be regarded as one which is required for normal life processes and whose functions cannot be substituted by other elements.

Carbon (C), Hydrogen (H), Oxygen (O), Nitrogen (N), Phosphorus (P), Potassium (K), Calcium (Ca), Magnesium (Mg), Sulphur (S), Iron (Fe), Manganese (Mn), Copper (Cu), Zinc (Zn), Molybdenum (Mo), Boron (B) and Chlorine (Cl) are known to be essential for higher plants. Sodium (Na), Silicon (Si) and Cobalt (Co) have been found to be essential in some higher plants and hence, cannot be regarded as essential for all the higher plants. Growth and yield get restricted; any if of these elements falls short hence, balanced nutrition demands supply of all the essential elements. Contrast to the essential elements, there are some, like Aluminium (AI), Nickel (Ni), Selenium (Se) and Fluorine (Fl) which are toxic to the plants.

Classification of plant nutrients

1. Based on the quantity used by the plant

These classifications of plant nutrients are based on the nutrient element content in the plant body.

Macronutrients: Those nutrients which are used by the crop in higher quantities *e.g.*, N, P, K, Ca, Mg, and S (Na and Si may also be regarded as macronutrients for the plant which requires those nutrient elements). The elements carbon, hydrogen and oxygen are very much needed for plant body but these are not essentially supplied, rather crop plants get these essential elements from nature itself as C and O from CO_2 and O_2 from the atmosphere and H from H_2O in soil or precipitation.

- Primary nutrients: Those nutrients which are used in considerable quantities by all the crops e.g., N, P, K.

- Secondary nutrients: Those nutrients which are used in moderate quantities by the crops and influence soil pH. e.g., Ca, Mg, S.

Micronutrients: Those nutrients which are essential but used by the crop in relatively smaller quantities by the crops e.g., Fe, Mn, Cu, Zn, Mo, B and Cl (CO is also micronutrient for the plant which needs it).

However, these classifications of plant nutrients are somewhat arbitrary. For example, the Fe and Mn contents in plant tissues are sometimes found as high as that of S or Mg. Again, comparatively higher concentration of Cl is present in many plants but this nutrient element is needed in minute quantities for its specific role in photosynthesis.

2. Based on physiological point of view

Group 1: This group contains C, H, O, N and S which are the major constituents of the organic materials of the plant body. Carbon is taken up in the form of CO_2 from the atmosphere at the time of photosynthesis, and possibly in the form of HCO_3^- ion from the soil solution. Hydrogen is taken up in the form of water which is reduced to H+ ion (photolysis) at the time of photosynthesis. Nitrogen is taken up in the form of NO_3^- or NH_4^{++} ions from the soil solution. The leguminous plants may get elemental N_2 gas from the soil air through the symbiotic nitrogen fixation by *Rhizobium* bacteria. Certain free living bacteria *(Azotobacter; Clostridium butyricum),* blue-green algae and fungi are able to fix elemental N_2 from soil air into their body tissues which in turn enrich the soil with nitrogen which is called non-symbiotic nitrogen fixation. Sulphur is not only taken up in the form of $SO_4^=$ or SO_3^- ion from the soil solution but can also be absorbed as SO_2 gas from the atmosphere.

Group 2: This group contains P, B and Si. Phosphorus is taken up in the form of $H_2PO_4^-$, $HPO_4^=$ or $PO_4^=$ ions, boron in the form of boric acid (H_3BO_3) or borate ($B_4O_7^=$) ions, and silicon in the form of silicate ions from the soil solution.

Group 3: This group contains K, Na, Mg, Ca, Mn and Cl and these nutrient elements are taken up in the form of their ions from the soil solution.

Group 4: This group contains Fe, Cu, Zn and Mo. These elements, with the exception of Mo, are taken up as chelate complexes (an organic compound which holds tightly certain cation attached to it) from the soil solution in addition to their ionic forms. Molybdenum is taken up in the form of molybdate (MoO_4^-) ions from the soil solution.

Nutrient requirements of vegetable crops

Nutrient requirement is basically a genetic characteristic of the crop plant, and this requirement may vary with the cultivars of the particular crop. Nutrient removal by high yielding cultivars and hybrids is much higher than what are removed by the low yielding cultivars of the same crop. Again, nutrient removal by any crop depends on the nutrient availability and their absorption which is influenced by soil pH, moisture and soil temperature. In general, vegetable crops exert tremendous pressure on the soil for nutrients due to their high productivity. When the crop is harvested it marks the removal of plant nutrients from the soil. There is no effective way to reduce the nutrient demand because nutrient requirement at a given yield level is quite stable. Nutrient removal is perhaps the most critical factor when sustainability of a farming system is considered. If the nutrients removed from the field are not replaced, the farming system will not remain sustainable due to decline in soil productivity. Some idea of the optimum fertility levels needed for vegetable production can be obtained from the quantities of nutrients removed by the crops.

Removal of primary nutrients from the soil by some vegetable crops

Crop	Nutrient removal (kg/100 kg yield)		
	Nitrogen	Phosphorus	Potassium
Potato	0.74	0.28	1.39
Tomato	0.33	0.11	0.44
Brinjal	0.29	0.08	0.50
Cabbage	0.53	0.16	0.50
Cauliflower	0.40	0.16	0.50
Knolkhol	0.50	0.42	0.85
Carrot	0.42	0.18	0.67
Radish	0.60	0.30	0.60
Beet	0.24	0.16	0.45
Onion	0.30	0.13	0.40
Leek	0.34	0.13	0.80
Cucumber	0.18	0.13	0.30
Pumpkin	0.18	0.14	0.32
Muskmelon	0.36	0.11	0.64
Bottle gourd	0.15	0.16	0.16
Pea	1.25	0.45	0.90
Beans	0.87	0.27	1.07
Okra	0.30	0.13	0.45
Celery	0.67	0.27	1.01
Lettuce	0.30	0.12	0.53
Spinach	0.50	0.15	0.30
Asparagus	2.40	1.20	3.00
Cassava	0.38	0.18	0.88
Sweetpotato	0.48	0.19	0.85
Elephant foot yam	0.34	0.08	0.49
Yam	0.57	0.14	0.61

Vegetable crops may be categorized into four groups based on the quantity of nutrient removal *viz.*, very high removal (asparagus, peas, beans, potato, celery, etc.), high removal (cabbage, cauliflower, knolkhol, radish, sweet potato, yam, spinach, carrot, etc.), moderate removal (onion, garlic, tomato, brinjal, beet, leek, muskmelon, okra, cassava, elephant foot yam, lettuce, etc.) and low removal (cucumber, pumpkin, bottle gourd, etc).

From the pattern of primary nutrient removal by different vegetable crops, it appears that removal of phosphorus is comparatively low while that of potassium is luxuriantly high. This does not mean that application of very high potassium will give good yield in all the vegetable crops. Again, plants require more phosphorus than its removal for giving optimum yield. On the other hand, they require as much sulphur as phosphorus. Nutrient requirements depend on the specific demands of roots, shoots, leaves and fruit development of the crop, although all the vegetable crops are very responsive to nitrogen application. In general, the vegetable crops whose above ground vegetative parts are consumed (cabbage, cauliflower, knolkhol, amaranth, palak, spinach and other leafy vegetables, lettuce, celery and other salad crops, etc.) require more nitrogen; the bulb, root and tuberous vegetable crops (carrot, radish, beet, turnip, onion, garlic, leek, shallot, potato, sweet potato, elephant-foot yam, yam, taro, etc.) require more potash; and fruit vegetable crops (tomato, brinjal, chilli, okra, pumpkin, bottle gourd, bitter gourd, ridge gourd, summer squash, cucumber, gherkin, watermelon, muskmelon, etc.) require more phosphorus. Peas and beans require high nitrogen but most of the required nitrogen is supplied through symbiotic nitrogen fixation by *Rhizobium* bacteria in their root nodules. In these crops, a low dose of nitrogen should be applied as basal to stimulate early growth. Cabbage, cauliflower, radish, onion, turnip require comparatively more sulphur as the compounds which impart characteristic odour to these vegetable crops contain sulphur. Removal of sulphur from soil per 100 kg yield by some vegetable viz., cabbage, cauliflower, tomato, onion, radish and cassava is 0.11 kg, 0.08 kg, 0.06 kg, 0.09 kg, 0.20 kg and 0.03 kg, respectively.

Some common disorders/problems of vegetables associated with imbalanced nutrition

<table>
<tr><th>Disorder/other problems</th><th>Due to</th><th>Remedy</th></tr>
<tr><td>Blossom end rot of tomato, sweet pepper and watermelon</td><td rowspan="3">Low pH and calcium deficiency in tomato, sweet pepper, watermelon and onion; In cauliflower and cabbage club root is incited by the fungus Plasmodiophora brassicae but aggravated by low pH and calcium deficiency.</td><td rowspan="3">Soil application of ground lime stone ($CaCO_3$) 10-14 days before planting</td></tr>
<tr><td>Club-root of cabbage and cauliflower</td></tr>
<tr><td>Damping off of onion</td></tr>
<tr><td>Black rot in cabbage and cauliflower</td><td>Incited by the bacteria Xanthomonas campestris pv. campestris but aggravated by disproportionately higher application of nitrogen</td><td>Application of P and K in proportionately higher quantity</td></tr>
</table>

Browning of cauliflower and broccoli; hollow stem and head of cabbage, broccoli and cauliflower; black heart of beet; internal browning of turnip; stem rosetting of tomato; fruit cracking in tomato; cracked stem in celery	Deficiency of boron	Soil application of borax with basal NPK fertilizer
Root rot of cucumber	Deficiencies of manganese and zinc	Soil or foliar application of sulfates of manganese and zinc
Interveinal chlorosis and necrosis in older leaves in tomato, brinjal and chilli; interveinal chlorosis and puckering of older leaves in cabbage; interveinal yellow spots followed by chlorosis in leaves in okra; die back of leaves in onion	Deficiency of magnesium	Soil or foliar application of magnesium sulfate
Root splitting in radish	High ammonium in soil	Application of N through non ammonium fertilizer or organic matter.
Whiptail of cauliflower; upwardly rolled leaves with diffused marginal and interveinal yellow mottling in potato; purple young leaves in tomato; discoloured leaf and death of growing tip in radish; yellow or white veins in the leaves of pea; pale green leaves with interveinal mottling and scorching in bean; deep blue leaves with conspicuous yellow and green mottling in onion	Deficiency of molybdenum	Application of ammonium molybdate
Susceptibility of brinjal to leaf hopper	Disproportionately higher application of nitrogen	Application of P and K fertilizer in proportionately higher quantity

Evaluation of soil fertility

• Plant tissue nutrient analysis: The basic principle behind this technique is that the nutrient concentration of plants is related to the amount of nutrient element available in soil. General range of nutrient content in fully developed leaves of vegetable crops is given below. If the nutrient level in the tissues falls below the critical concentration, the soil may be deficient in that element for optimum plant growth and deficiency symptoms in plant may occur.

Sufficient and critical level of nutrients in the leaf tissue of vegetable crops (general level)

Nutrient		Per cent or ppm of dry weight of leaf	
		Sufficient level	Critical level
Nitrogen	(%)	1.50-4.0	1.00-1.50
Phosphorus	(%)	0.25-0.80	0.20-0.25
Potassium	(%)	2.00-9.00	1.50-2.00
Calcium	(%)	0.35-2.00	0.35-0.80
Magnesium	(%)	0.25-1.00	0.20-0.25
Sulphur	(%)	0.16-0.50	0.10-0.16
Manganese	(ppm)	30-200	30-50
Iron	(ppm)	50-250	50-80
Zinc	(ppm)	30-100	20-30
Boron	(ppm)	30-60	20-30
Copper	(ppm)	8-20	4-8
Molybdenum	(ppm)	0.5-5.0	0.2-0.5

Different plant parts of the same plant contain different concentrations of the same nutrient and nutrient concentration again varies with the stage of the crop. So, leaf samples for tissue analyses should be selected on the basis of specific physiological age *i.e.*, developmental stage. It is important that the samples must be free from diseases, insect damage and physical or chemical injury. Leaf near the fruit should not be sampled as the nutrients, it might have contained, are often translocated to the fruits.

• Soil analysis: It is a rapid chemical method and is important for predicting the fertility status of the soil. Fertilizer recommendations can be advocated judiciously if the results of soil tests are available. Soil tests and their interpretations are based on the soil samples. It is, therefore, important that the soil samples should be properly collected and should represent the area under soil test. After scrapping the surface litter, uniform core of a thin slice of soil from the surface to plough depth (15-22 cm) from 15 to 20 spots of each sampled field should be taken. When soil chemical analyses (soil tests) are employed to determine which nutrients are deficient and by how much, nutrient balances in the soil can be achieved by fertilizer application. Ranges of nutrient content commonly found in mineral soils are given below.

Range in nutrient content commonly found in soils

Nutrient	Normal range	
	Per cent	ppm
Nitrogen	0.02-0.50	200-5000
Phosphorus	0.01-0.20	100-2000
Potassium	0.17-3.30	1700-33000
Calcium	0.07-3.60	700-36000
Magnesium	0.12-1.50	1200-15000
Sulphur	0.01-0.20	100-2000
Iron	0.50-5.00	5000-50000
Manganese	0.02-1.00	200-10000
Zinc	0.001-0.025	10-250
Boron	0.0005-0.015	5-150
Copper	0.0005-0.015	5-150
Chlorine	0.001-0.10	10-1000
Cobalt	0.0001-0.005	1-50
Molybdenum	0.00002-0.0005	0.2-5

Source: The Nature and Properties of Soil (N C. Brady).

Percentage or ppm of nutrient element can be converted to kg of nutrient per hectare *e.g.,* the soil containing 0.1 per cent potassium means 1 g potassium per kg of soil. One hectare furrow slice, about 15 cm deep of upland mineral soil weighs about 2,240,000 kg. So, the conversion of nutrient is 2,240 kg of potassium per hectare.

Based on the soil tests, soils can be classified into three fertility level categories *viz.,* high, medium and low. Fertilizer recommendations should be based on this soil fertility level.

General rating chart of the soil test data for N, P and K

Fertility level	Organic C (%)	Available N (kg/ha)	Available P_20_5 (kg/ha)	Available K_20 (kg/ha)
High	> 0.75	> 450	> 90	> 340
Medium	0.50-0.75	280-450	45-90	150-340
Low	< 0.50	< 280	< 45	< 150

Management of nutrients in soil

Organic matter

Soil organic matter consists of decomposed plant and animal residues and numerous dead microorganisms particularly which are nitrogen fixers (both free living and symbiotic) which are very active and important portion of soil. After active decomposition, the organic matter residues are collectively called humus. Beneficial effect of soil organic matter can be summarized as:

1. It is the nitrogen reservoir and source of 90-95 per cent nitrogen in unfertilized soil.
2. It supplies large portions of soil phosphorus and sulphur. It is also the source of other macro- and micronutrients.
3. It supplies cementing substances for desirable aggregate formation of soil.
4. It loosens up the soil to provide better aeration and water movement mainly due to soil aggregation which increases soil pores and water holding capacity of the soil is thereby increased.
5. The large available surface of humus has many cation exchange sites that absorb nutrients for eventual plant use.
6. Organic matter acts as a chelate and make bond to different micro-nutrient elements like Fe, Zn, Cu, and Mn. The soluble chelate helps to mobilize these micronutrients in soils and increase their availability to plants.
7. It supplies carbon to many microbes that perfom other beneficial functions in soils and growth substances produced by the micro-organisms in soil help plant growth in many ways.
8. It reduces soil erosion, shades the soil, and keeps the soil cooler in very hot weather and warmer in winter.

For maximum benefit, organic matter must be readily decomposable and continuously replenished with fresh residues. In cultivable soil, organic matter content ranges from less than 1 per cent to 15 per cent. The majority of the tropical and subtropical soils like ours have low humus content which is the consequence of its quick year round mineralization under the influence of tillage and favourable weather condition. Moreover, vegetable crops are generally cultivated intensively, and the crop residues are not sufficient to replace the organic matter lost from the soil. The soil organic matter can be replenished by adding farmyard manure, oilcakes, compost, green manure and other bulky organic residues.

Mineral values of some commonly used organic resources

Organic resource	Nutrient content (%)		
	N	P_2O_5	K_2O
FYM	0.5-1.0	0.3-0.5	0.4-1.0
Compost (urban)	1.2-2.0	1.0	1.5
Compost (rural)	0.4-0.8	0.3-0.6	0.7-1.0
Slurry from gobar gas plant	1.6-1.8	1.1-2.0	0.8-1.2
Dhaincha (green manure)	0.6	-	-
Cowpea (green manure)	0.7	0.1	0.6
Mustard cake	5.1	1.8	1 .1
Neem cake	5.2	1.0	1.4
Groundnut cake	7.0	1.5	1-3
Linseed cake	5.5	1.4	1.2
Mahua cake	2.5	0.8	1.8
Castor cake	5.5	1.9	1.0
Sludge	2.5	1.0	0.4
Activated sludge	4.0-7.0	2.1-4.2	0.5-0.7
Vermi compost	1.0	0.5	1.5

Different manures and oilcakes should be applied and ploughed under the soil well in advance of seeding or planting (15 to 20 days) because of the slow availability of nutrients from the manures.

Composting is the process of reducing plant and animal refuse to a quickly utilizable form for improving and maintaining soil fertility. Raw vegetable residues are rich in cellulose and other readily decomposable carbohydrates, and have a high carbon-nitrogen ratio of 40: 1 or more. Direct application of such undecomposed, low nitrogen organic matter as manure brings about a temporary deficiency of nitrates in soil by stimulating the growth of microorganisms, which in turn compete with the crop for nitrogen, phosphorus and other nutrient elements. For this reason, the organic wastes are allowed to decompose before application and thereby the carbon nitrogen ratio is brought down to 10-12: 1 during the process of composting.

When green manuring crops are turned under the soil for decomposition, various nutrient elements in the plant body are released and returned to the soil. A fairly good quantity of organic matter is also added in this process. Leguminous green manuring crops like cowpea, dhaincha (*Sesbania* sp.), rice bean, clover, etc. are preferred to non-legume crops because legumes supply substantial quantity of nitrogen to the soil. Two important considerations with respect to green manuring are: (i) the green manuring crops should be turned under before they are matured because succulent plant materials decompose quickly, and (ii) the green manuring plants should be turned under well in advance of planting the successive crop, otherwise (a) the micro-organisms that decompose the plant materials consume soil nitrates and may depress the early crop growth, and (b) nutrients that are released from the plant materials in the course of decomposition will not be available for early crop growth. Balanced fertilization with P, K, Cu, Mo and Fe to the leguminous green manuring crops is essential for most efficient nodule activity and nitrogen fixation.

Most soils under continuous tillage for production of crops like vegetables lose organic matter at the rate of approximately 1-2 per cent per year. In this situation, returning of the crop residues judiciously is beneficial because it helps in aggregation of soil particles, improving water infiltration and aeration, in reserving plant nutrients and also acts as energy reservoir for the micro-organisms.

Macro-nutrient through fertilizer

In natural ecosystems, the minerals absorbed by the crops return to the soil after organic matter decomposition, and soil fertility is more or less maintained through nutrient cycling. In cultivated ecosystems like vegetable cultivation, however, all harvested biomass (product and plant residue) withdrawn from the field contains nutrients that no longer return to the soil. Hence, maintenance of soil fertility and crop yield depend on counterbalancing of fertilizer inputs. Chemical fertilizers are inorganic or synthetic organic materials of concentrated nature. They contain one or more plant nutrients in easily soluble and quickly available forms, and being concentrated, fertilizers have the advantage of smaller bulk with resultant economy and ease in storage, transport and handling.

Nitrogen

It is most often the limiting nutrient element in plant growth and the vegetable crops require more nitrogen than any other nutrients. Nitrogen is highly mobile and easily lost from the soil. In cultivable land, crops can utilize only 50-60 per cent of the available nitrogen. When the nitrogen sources are applied to the soil, they all are eventually converted to NH_4^{++} and NO_3^- forms of nitrogen. Plant roots take up these two inorganic forms of nitrogen almost exclusively. Nitrate, if left unused, is lost through downward leaching as negatively charged NO_3^- is naturally repelled by soil particles and remains free to move with soil water, but positively charged NH_4^{++} is held by soil particles and generally, are not subjected to leaching. However, both NO_3^- and NH_4^{++} forms may be consumed by micro-organisms or converted to gaseous nitrogen forms (N_2 or NH_3) and lost to atmosphere. Nitrogen use efficiency in vegetable crops can be increased by adopting the following measures:

1. Split application of nitrogen fertilizers at the time of peak requirements of the crop which decrease nitrate leaching and thereby increase nitrogen use efficiency. In general, half the dose of nitrogen fertilizers is given as basal at the time of final land preparation and the rest half is applied in different splits. However, in leguminous vegetable crops, entire nitrogen fertilizer is applied as basal dose. In other fruit vegetable crops, like tomato, brinjal, chilli, okra, etc. split applications after periodical harvests prove beneficial.
2. Nitrogen fertilizer should be applied 5-10 cm deep in the soil. When nitrogen fertilizers are applied deep in the soil, plants can take 30-35 per cent more nitrogen than when applied on the soil surface.
3. Optimum soil moisture increases nitrogen use efficiency.
4. Nitrogen fertilizers should not be applied in large quantities near the root zone because of the danger of salt damage.
5. Nitrogen use efficiency often depends on the interaction among other nutrients, particularly phosphorus and potassium. In general, when adequate P and K fertilizers are applied to soil, the most efficient use of nitrogen results. So, when P and K fertilizers are reduced, the supply of N fertilizers should also be adjusted accordingly because the resulting root system would not be able to absorb nitrogen more than its capacity.

Phosphorus

Young plants assimilate phosphorus very rapidly and thereby need water soluble forms that migrate to the initial meagre root system. Plants generally accumulate as much as 75 per cent of their phosphorus need in the early stages of growth. This emphasizes the importance of ample supply of water soluble phosphorus in the soil zone of the germinating seeds. Most soluble phosphorus ($HPO_4^{=}$, $H_2PO_4^{-}$) become fixed (precipitate and form insoluble compounds) before plant can absorb them, and hardly 10-15 per cent of the applied phosphorus is utilized by the plants. So, leaching loss of phosphorus is practically nil, and it is lost primarily by erosion of surface soil. In general, phosphorus is more available at a pH of 6.0 to 6.5 than at higher or lower pH values. In acidic soils, low availability of phosphorus is common due to formation of complex compounds of phosphorus with aluminium and iron. Phosphorus use efficiency can be increased by adopting the following measures:

1. Maintenance of soil pH at 6.5 to 7.0.
2. Phosphatic fertilizers should be applied totally as basal near the active root zone of the plant.
3. It should not be applied very deep in the soil, rather shallow band placement (2.5-5 cm below and 5-6 cm away from seeds) in rows is very effective.
4. Late application of phosphorus is not effective for plant growth.
5. Adequate moisture content in soil and combined application of N and P fertilizers increase phosphorus availability to the plants.

6. Maintenance of organic matter in soil increases P availability to the plants.
7. In cold weather, more phosphorus fertilizer should be applied as absorption of P is less in cold weather.
8. Rock phosphate or basic slag should be applied 2-4 weeks before sowing or transplanting.

Potassium

It occurs in plant body only as a mobile ion K^+ rather than as an integral part of any specific compound. In vegetable crops, luxury consumption of potassium is a very common feature. Rapid uptake of this nutrient occurs in the early stages of growth and it usually occurs faster than either N or P. So, ample potassium should be made available to the seedlings. Potassium is less mobile than nitrogen fertilizers but more mobile than phosphates, and are generally applied in full dose at the time of sowing/transplanting. Potassium fixation embraces (a) immobilization into microbial bodies and (b) becoming entrapped in the clays. It is also leached substantially through the downward movement of water. For this reason, it often falls deficient in high rainfall areas. Potassium use efficiency can be increased by adopting the following measures:

1. Split application of potassic fertilizers, particularly in loose soils of high rainfall areas.
2. Maintenance of soil pH at 6.0-7.0 by liming which reduces leaching loss.
3. Returning of crop residue and application of manures which add much potassium.

General fertilizer guide (Kg nutrient per hectare) for some vegetable crops for the soil having moderate fertility status

Crop	Seed source					
	Improved open pollinated variety			Hybrid		
	N	P_2O_5	K_2O	N	P_2O_5	K_2O
Potato	180	120	150	-	-	-
Tomato	100	80	80	200	100	100
Brinjal	125	80	80	200	100	80
Chilli	100	60	60	150	80	80
Sweet pepper	120	80	60	200	150	100
Cauliflower	125	80	60	200	100	100
Cabbage	150	60	80	200	125	150
Onion	100	60	100	-	-	-
Garlic	100	65	100	-	-	-

Okra	125	75	60	200	100	100
Pumpkin	100	60	80	-	-	-
Bottle gourd	90	60	60	200	100	100
Bitter gourd	100	60	60	200	100	100
Cucumber	100	60	60	150	90	90
Pointed gourd	150	60	60			
Ridge gourd	100	60	60	150	90	90
Hyacinth bean	30	75	60	-	-	-
Cowpea	25	75	60	-	-	-
Radish, beet, carrot, turnip	100	60	80	-	-	-
Amaranth	90	50	50	-	-	-
Palak	100	60	40	-	-	-
Sweet potato	75	50	75	-	-	-
Elephant foot yam	150	100	150	-	-	-
Colocasia	120	60	80	-	-	-

Sulphur

Crops require as much sulphur as they do phosphorus and therefore, sulphur is called the fourth major nutrient. Some vegetable crops like cabbage, onion, turnip, etc. have high sulphur requirement. Sulphur content in the soil is being decreased gradually because (a) modern high analysis fertilizers are free of sulphur, (b) modern plant protection chemicals are sulphur-free compounds. Organic sources of sulphur are about 70-90 per cent and rainfall also dissolve atmospheric SO_2 evolved during burning of fuels and act as a source of sulphur. Leaching losses of sulphur from the soil are very much like nitrogen and caused by percolating water through light-textured soils. Soil acidity also affects the availability of sulphur due to fixation of sulphate ion on hydrated oxide. There could be sulphide toxicity due to reduction of sulphur to hydrogen sulphide (H_2S). Sulphur can be managed in the soil through the following measures:

1. Maintenance of soil pH to 6.0-7.0 by liming.
2. Application of organic matter to the soil.
3. Application of sulphur-containing fertilizers, like ammonium sulphate, superphosphate, potassium sulphate, etc. periodically on acid and sandy soil low in humus.

4. In acute deficiency, elemental sulphur may be applied before sowing/transplanting. Application of elemental sulphur may cause soil acidity, particularly in high rainfall areas where acidity is a problem. So, at the time of sulphur application in this type of soil, application of calcium in the soil must be ensured in the form of lime and other fertilizer sources like calcium ammonium nitrate, rockphosphate, etc.

Calcium

It is needed in large quantities for cell division. It is seldom deficient except in sandy and strongly acidic soils and the soils having high fixed phosphorus content. However, calcium is lost in a number of ways like, crop removal, leaching and soil erosion. Leaching of bases like calcium by percolating water coupled with nitrogen and sulphur fertilization contributes to natural soil acidity. Calcium can be managed in the soil by liming.

Magnesium

In light-textured soil, magnesium is leached through the percolating water. Magnesium can be managed in the soil by application of dolomite ($CaCO_3$. $MgCO_3$) which contains 12.0 per cent magnesium or magnesium sulphate containing 20.2 per cent magnesium.

Micronutrients through different chemicals

The main reasons of accelerated exhaustion of available micronutrients in the soil are (i) cultivation of high-yielding cultivars and hybrids which demands more macro and micronutrients, (ii) application of high dose of N, P and K fertilizers as it decreases the availability of micronutrients, (iii) inadequate application of organic matters to the soil, (iv) intensive cropping which creates micro-nutrient deficiency, (v) leaching loss of micronutrients particularly in light-textured soils of high rainfall areas, (vi) increased application of compound fertilizers because straight fertilizers may add some amount of micronutrients *e.g.,* superphosphate contains 26 ppm Cu, 50 ppm Zn, 65 ppm Mn, 9.5 ppm B and 3.3 ppm Mo. On the other hand, compound fertilizers do not contain such micronutrients, (vii) maximum application of phosphatic fertilizers decreases the availability of zinc, whereas application of ammonium sulphate or other sulphate compounds increases the availability of zinc, (viii) inadequate cultural practices decrease the availability of micronutrients because all micronutrient elements but molybdenum remain within few inches from the surface soil and better tilth increases the availability of micronutrients, and (ix) both moisture stress and excess condition influences the availability of micro nutrients. Moisture stress condition decrease the availability of Fe and B while excess moisture condition causes Zn deficiency.

Vegetable crops response very well to the application of micronutrients under Indian condition. Generally, micronutrients are applied in four different ways:

Soil application

Micronutrient containing chemical compounds can well be applied to soil mixed with other fertilizers. Care should be taken to avoid excess application of micronutrients because optimum dose depends on soil type, crop, etc. As for example, higher rates of manganese are required for both organic soils and sandy soil that are alkaline in reaction and high rate of zinc is applied when soil is neutral to alkaline in reaction.

Seedling root dipping

This method of application is not widely practiced. Generally, 0.2-0.3 per cent solution of zinc sulphate is used for root dipping.

Seed treatment

Seeds may be sown after treating with the chemical compounds containing Cu, Fe, Mo, Zn, B, Mn, etc. Seed treatment of pumpkin and squashes with 0.5% borax or manganese sulphate increases staminate flower production and thereby yield.

Soil status and guide for soil application of chemicals for secondary and micronutrients

Micronutrients	Critical limit (mg/g)	Recommended dose of chemicals
Zinc	0.6-1.0	Zinc sulphate ($ZnSO_4$) @ 25 kg/ha with organic manure
Boron	0.5	Borax ($Na_2B_4O_7.10H_2O$)@ 10 kg/ha with basal NPK fertilizer
Molybdenum	0.2	Ammonium molybdate ($(NH_4)_6Mo_6O_{24}.4\ H_2O$ @ 1.0 kg/ha with basal NPK fertilizer
Iron	2.5-4.5	Ferrous sulphate ($FeSO_4$) @ 50 kg/ha with organic manure
Manganese	1.0	Manganese sulphate ($MnSO_4$) @ 30 kg/ha with basal NPK fertilizer
Copper	0.2	Copper sulphate ($CuSO_4$) @ 5 kg / ha with basal NPK fertilizer
Sulphur	10-12	Elemental sulphur (S) @ 25-45 kg / ha

A general guide for application of micronutrients as foliar spray

Micronutrients	Concentration of the chemicals
Iron	0.4% ferrous sulphate + 0.2% lime*
Manganese	0.4%-0.6% manganese sulphate + 0.2%-0.3% lime*
Zinc	0.2%-0.6% zinc sulphate + 0.1-0.3% lime*
Copper	0.1-0.2% copper sulphate + 0.5% lime*
Boron	0.5% -0.6% borax
Molybdenum	0.05% sodium or ammonium molybdate

* Lime is added to neutralize the solution, otherwise leaves may get scorched. Minimum 500 litres of water must be used per hectare.

Foliar spray

Most of the micronutrients after foliar spray enter the plant body through leaves although time required for absorption is few hours to one day depending on the nutrient. Foliar application of micronutrients is widely used and at the same time suitable because of (i) convenience in application, (ii) requirement of small quantity, (iii) quick correction of deficiency, and (iv) avoidance of fixation of the micronutrients in soil. Lime should be added to neutralize the solution of ferrous sulphate, manganese sulphate, copper sulphate and zinc sulphate, otherwise foliage may get scorched. Foliar sprays may be given 2-4 times at an interval of 7 -10 days to correct the deficiency of particular micronutrient.

Integrated nutrient management

Indian population is at present more than 110 crore and expected to the tune of 130 crore by 2025. Such increase in population will magnify the demand for different food production including vegetables from the rapidly shrinking cultivated land area. To produce the food, agricultural operation needs to be intensified using huge amount of fertilizer and other inputs of production. Such intensification of agriculture has its attendant problems of salinization, acidification, desertification, nutrient depletion, loss of soil biodiversity, soil erosion etc. For sustaining productivity on long-term basis because of irreparable damages it causes on natural resource base and in some cases, because of intensive cropping particularly with vegetable crops, soils have become "talcum powder" like structureless materials losing retentivity for nutrients as well as water. Besides, use of enormous fertilizer inputs substantially enhanced the production cost, deteriorated soil and water quality and increased the greenhouse gas effect making the production system unsustainable. Such intensified production system has also been implicated in over-taking the resilience of agricultural soils and those natural processes that maintain its health and quality. Sometimes the stress exceeds the resilience capacity of the soil so that restoration becomes very expensive or even impossible. Since, the long-term performance of intensive agricultural system is important for poverty alleviation and sustaining food production in the face of continuous population growth, it is needed to redo the present way of cultivation with only inorganic fertilizer and other chemicals to save the natural resource base, particularly soil, for posterity.

When these environmental problems in intensive agriculture came into spotlight, organic farming was advocated as a solution for most of the problems encountered. However, the organic farming is not only a very low yielder but has many of the problems of conventional high intensive agriculture. Thus organic farming methods do not offer solution to many of the problems. To achieve sustainability in production system and maintain natural resource base in good stead, blending or integrating organic, inorganic, biological and mechanical inputs with ecological principles in mind may be an option in the concept of "integrated nutrient management" (INM) for vegetable crops.

Sustainable nutrition of crops is based on the principle of a balance between availability of nutrients and requirement by the crops. In such a balanced situation, involving fine tuning towards weather and soil conditions, emphasis is given to the best time of application of nutrients, the best organic and mineral composition, the best placement of the nutrients and to the environment adapted crops/varieties. IPN enhances crop production and maintains its consistency over long periods without disrupting ecological base of farming. Once IPN is adopted, it becomes possible to cut-off the use of chemical fertilizers, other nutrient supplying chemicals and plant protection chemicals over a period of time. IPN for sustainable nutrient management may be accomplished through (a) proper crop rotation, (b) soil protection in dry weather, (c) covering the soil by a live canopy or mulch, (d) application of nutrients only as per crop requirements, (e) simultaneous application of organic matters with mineral nutrients, (f) application of quick release nutrients only as top dressing in rows, (g) use of biofertilizers (h) integrated pest and disease management, (i) enrichment of organic matters by adding mineral additives and microbial inoculants during composting and (j) balanced application of nutrients.

Components of INM

Proper management of plant nutrients is necessary to ensure higher productivity, balanced nutrition and for checking wastage of costly fertilizer nutrients as well. The basic concept underlying the integrated nutrient management system, nevertheless, is the maintenance and possible improvement of soil fertility for sustained crop productivity on long-term basis and also to reduce fertilizer input loss and environmental degradation.

Different INM components

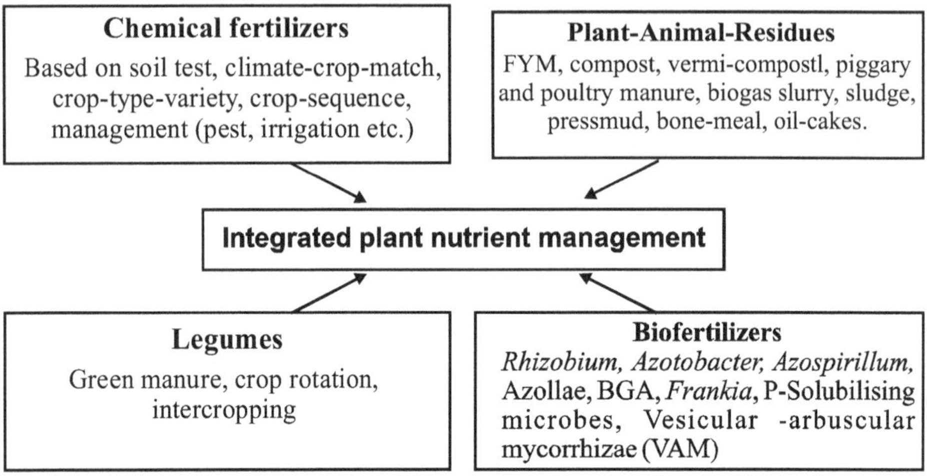

The judicious combination of organic and inorganic resources for balanced plant nutrient management is necessary since neither inorganic fertilizer nor the organic manures alone can bridge the gap between sustaining productivity and up-keeping soil health and resilience.

Biofertilizers are the cultures of microorganisms used for inoculating seed and soil or both under ideal condition to increase the availability of plant nutrients. They are cost-effective, eco-friendly and renewable sources of plant nutrients used to supplement chemical fertilizers in sustainable production system. Some common biofertilizers of agricultural importance are: *Azolla*, Blue green algae, *Rhizobium*, *Azotobactor*, *Acetobactor*, phosphate-solubilizer and sulphate-solubilizer. The free living N-fixers like *Azotobacter*, *Azospirillum* and the Phosphate solubilizers (*Bacillus polymyxa*, *Bacillus negatherium*, *Pseudomonas striata*, *Aspergillus awamonii*, *Penicillium digitatum*, etc.) are commonly used to vegetable crops through i) seed inoculation @ 1.5 kg biofertilizer dissolved in water with starch/molasses for one hectare, ii) seedling root soaking @ 1.5 kg biofertilizer dissolved in 35-40 litre of water for soaking seedlings required for one hectare for 30 minutes and iii) soil application @ 2 kg biofertilizer + 200-250 kg organic manure for one hectare during land preparation.

4. Pollination management

There are about 2,50,000 species of flowering plants on earth that require pollination, wind, gravity, water, birds, bats, and insects are the forces that accomplish pollination. Pollination refers to the transfer of pollen from the stamen (hermaphrodite/ staminate flower) to the pistil (hermaphrodite/ pistilate flower) because to produce a seed or fruit, pollen must move from the anther to a receptive stigma. Since both stamen and pistil are necessary for seed formation, they are called essential organs of a flower. Most flowers contain two other parts, the sepals and petals, which may help attract insect pollinators. A complete flower contains all four parts. If only one of the essential organs is present in a flower, it is called an incomplete flower. Vegetables may be grouped into five categories according to their manner of pollination:

Group 1

Vegetable crops with hermaphrodite sex form and predominantly self-pollinated i.e., the crops with flowers that can receive only their own pollens *eg.* Peas, French bean, hyacinth bean, cowpea, lima beans, cluster bean, chicory, endive, tomato, etc.

Group 2

Vegetable crops with hermaphrodite sex form and predominantly self-pollinated but fair amount of cross-pollination (both air-borne and insect-borne) happens in some environmental conditions. *eg.* brinjal, sweet pepper, okra, chilli, lettuce, etc.

Group 3

Vegetable crops with hermaphrodite sex form and cross-pollinated by insect-borne pollens and mostly form seed with the pollens from genetically different plants because of self-incompatibility. *eg.* cabbage, cauliflower, broccoli, Brussels sprouts, knolkhol, kale and radish,

Group 4

Vegetable crops with hermaphrodite/ monoecious/ andro-monoecious sex form and cross-pollinated by their own pollen (self-pollinated) or from pollens received from another plant (cross-pollinated) and can be divided as follows:

- Usually pollinated by air-borne pollens *eg.* garden beet, palak, corn, carrot, celery, Swiss chard, amaranth, etc.
- Usually pollinated by insect-borne pollens *eg.* onion, parsnip, muskmelon, watermelon, pumpkin, summer squash, winter squash, cucumber, gherkin, bottle gourd, bitter gourd, ridge gourd, sponge gourd, collards, mustard, rutabaga, parsley, etc.

Group 5

Vegetable crops with dioecious sex form and cross-pollinated by insect-borne pollens *eg.* asparagus, spinach, pointed gourd, teasle gourd, ivy gourd, etc.

The pollination process

Vegetable crops with hermaphrodite sex form (peas, French bean, hyacinth bean, cowpea, lima bean, cluster bean, chicory, endive, tomato, brinjal, sweet pepper, okra, chilli, lettuce, etc) have both male and female parts in the same flower which favours self pollination. However, proper pollination both in open and green house is needed for optimal fruit set and production in these crops. Blossoms of tomato, brinjal, etc. require slight movement for sufficient pollen from the stamens to fall onto the stigma of the flower. Honey bees cause sonication due to buzz foraging over the flowers. Bumble bees cause movement by hanging upside down on the flower, fastening their jaws onto the staminal tube, and then setting the flower into vibration by activating their flight muscles which is called "buzz pollination". Such bee mediated buzz pollination and sonication cause liberation of pollens from the apical split of the anthers to effect self-pollination. Growing of tomato, sweet pepper, brinjal, etc in green house rely either on manual pollination or electrically operated vibrator for liberation of pollen and subsequently fruit set, which can be very time consuming. Using bumble bees for pollination is an effective alternative and can completely replace manual pollination. Fair amount of cross-pollination occur in okra by bees mainly due to showy petals and mucilage that attracts the insects and in chilli due to very fine dusty pollens. Some vegetable crops with hermaphrodite flower viz., onion, carrot, etc. are predominantly cross-pollinated because of dichogamy (differential maturity of male and female parts in a flower), presence of nectar and strong smell in the flower. Another group of vegetable crops with hermaphrodite flower like, cabbage, cauliflower, broccoli, Brussels sprouts, radish, etc. are also chiefly cross-pollinated because of self-incompatibility (a genetically governed phenomenon when either pollens do not germinate or pollen tubes do not grow through the stylar tissue if genetic constitution of pollen and stylar tissue is same) and the flowers offer sugary liquid nectar as an added enticement for these pollinating insects.

Monoecious vegetable crops (watermelon, muskmelon, pumpkin, bottle gourd, bitter gourd, cucumber, ridge gourd, snake gourd, sponge gourd, corn, amaranth, etc.) have both staminate and pistilate flowers separately on the same plant. Andromonoecious vegetable crops (muskmelon) have staminate and hermaphrodite flowers on the same plant. Dioecious vegetable crops (asparagus, spinach, pointed gourd, teasle gourd, ivy gourd, etc.) plants have only one sex of flower per plant. Monoecious, andromonoecious and dioecious vegetable crops generally require cross-pollination.

Many cross pollinated vegetable crops like, watermelon, muskmelon, pumpkin, bottle gourd, bitter gourd, cucumber, ridge gourd, snake gourd, sponge gourd, etc. produce pollen that is heavy and sticky and not blown easily from flower to flower which generally require insects to transfer pollens. Many insects visit flowers to collect pollen and nectar as food. Nectar is usually offered deep within the flower near the base where petals originate from around the ovary. In seeking nectar or gathering pollen, the insect accidentally brushes against anthers so that pollen grains are transferred to the stigma. So, pollination is not a deliberate behavior but an accidental one performed by the insect as they collect food. Among insect pollinators, bees are especially efficient because they eat pollen and nectar exclusively, visit many flowers of the same species during a single trip, and have hairy bodies that easily pick up pollen grains.

Cross-pollination predominantly by air-borne pollens happens in some vegetable crops which produce light pollens that are easily blown about by wind and can drift in air for long time. Flowers of these crops are either very inconspicuous (palak, garden beet, amaranth, etc.) or have special pollen receiving parts in the pistilate flower (silk in pistilate flower of corn).

When the proper compatible pollen adheres to the stigma, it germinates and a pollen tube grows through the stigma and the style to the ovary. Fertilization takes place in the ovary when the nucleus of the pollen or male germ cell unites with the nucleus of the ovule or female germ cell. Ovules become seeds and the surrounding ovary develops into the fruit.

The honeybee is the most common and effective insect pollinator of most of the vegetable crops. Honeybee activity closely coincides with the period when the flower is open. Honeybee visitation begins an hour or two after sunrise and continues until mid-afternoon. If temperatures are very warm, bee activity may decline about noon. However, honeybee visitations are generally maximum upto 10 a.m. and then decline until 2.30 to 3.00 p.m. when activity almost ceased. Bumble bees are most active in the morning and in the afternoon at temperatures between 10-30°C although they function best at temperatures between 15-25°C. Generally the number of bee visitations is more important than the length of time that each bee stayed on the flower. It has been documented that well-shaped, fully expanded fruit of watermelon occur following eight bee visitations to a pistilate flower and fruit set is significantly reduced when only two to four bee visitations are made.

Management of bees for effective pollination

1. Vegetable farming and bee keeping should go concurrently. The honey bee is well adapted for pollination. Each colony contains thousands of individuals which are available to forage for the food (pollen and nectar) that is required to rear their replacements. Like other bees, their sense of smell, eyes, mouthparts, and numerous branched body hairs are ideally suited for finding food sources, sipping nectar, and collecting and distributing pollen. Honey bees visit a wide variety of flower types. In a single day, one bee makes 12 or more trips from the hive, visiting several thousand flowers. On each trip, it confines its visit to one plant species, collecting one kind of pollen. It has a complex communication system that enables the colony to find and collect food with maximum efficiency. These characteristics make honey bees most valuable agent for cross-pollinating crops. Bumble bee

pollination has some advantages because they are active at low temperatures of 10-12°C and windy and cloudy conditions and effective in greenhouses, high tunnels and in open air.

2. At least one strong bee colony should be kept per acre of land to keep sufficiently high bee population to insure that each flower is visited by maximum number of times.
3. In order to maintain the health and activity of the bee population, any pesticide should preferably be applied to the crop when bees are not present in the vegetable field, usually at dusk or after dark.
4. Selective use of pesticides since many of the traditional classes of insecticides will have a negative impact on the hive and systemic pesticides may damage the bee population.
5. In the green house cultivation of tomato, brinjal or sweet pepper, etc. the bee hives need supply of sugar water for the total life expectancy of the hive, since blossoms of these crops do not produce nectar.

5. Integrated pest management

Insect pests are one of the major factors undermining improved vegetable productivity particularly in intensive cropping system where the insect pests easily find their hosts through out the year and this situation of high biotic stresses by insect pests assumes serious proportion further with the threat of global warming. In general vegetables have low pest resistance and in general, the insect pests destroy more than one third of vegetable production with an average of about 40% yield loss which under severe infestation might cause total destruction. In the present era of chemical agriculture, different chemical pesticides are the basic armoury for insect pest control programme however, the hydrochlorin, organo phosphorous ,carbamate and other chemical compounds used as pesticides are highly toxic and pollute the environment and affect humans through their residue in food chain and in other various ways.

Majority of pesticide related health problems take place in the developing countries, because the users are not sufficiently aware of the dangers and the proper use of pesticides. Vegetables are more vulnerable to cause such health hazard because of the nature of maturity and their consumption. Many of the vegetables are harvested at very short interval and are consumed raw without much processing. So, the approach of controlling insect pests only by chemical pesticides is now posing social problems apart from being increasingly ineffective and at the same time expensive in the present complex cropping systems.

Integrated pest management

Different circumstantial evidences and backed up research findings have suggested that pesticides alone are inadequate for controlling insect pests. The use of pesticides to control pests involves significant degree of risk in polluting the environment, development of pest resistance to pesticides, pest resurgence, destruction of natural enemies and pesticide residue in the food chain. Integrated pest management (IPM) refers to the holistic approach for the management of pest population below economic threshold level (ETL) through efficient and judicious, utilization of the available resources with active participation of farmers.The output of IPM implementation would be that the farmers should become expert of IPM and should be able to make their own decision in dealing with pest problems. The concept of IPM stands basically on three strategies, Prevention, Observation and Intervention

Components of IPM

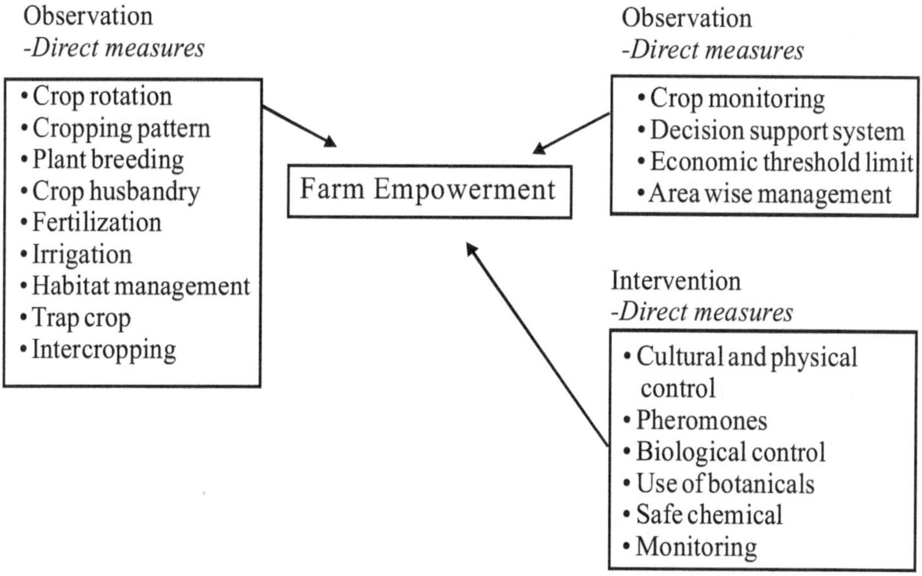

The fundamental approach for a sustainable IPM towards improvement of sustainable agriculture which includes (a) biological control using one organism to control another through manipulation of pest bioagent relationship (b) use of different botanicals that are derived from plants to control insect pests (c) behavioural management through manipulation of cultivated and natural environment to preserve the natural enemy balance of the insect pests and (d) development and utilization of insect pest tolerant varieties.

IPM is based on basic understanding and application of ecological principles integrating multi disciplinary methodologies to develop strategies for the management of agro-ecosystem. The objectives of IPM research in this context are (I) understanding of each component in the agro-ecosystem (2) to develop activity component of IPM (3) to develop basic understanding of a suitable IPM strategy against key pests of major vegetable crops and (4) to validate and improve the existing IPM technology.

During implementation of IPM approach, the farmers should be considered as a researcher in their own field becoming a part of research network supporting IPM. Therefore, the strategies to disseminate IPM research results to farm level include:

- Field training driven researches which at the same time impart training to the extension staff.
- On-farm research involving farmers' participation.
- Field demonstration through Krishi Vigyan Kendra/Government department /NGO for convincing the farmers regarding the usefulness of IPM in vegetable cultivation.
- Adequate Government interventions for making the avaiailable different biocontrol agents, botanicals, pheromone trap, etc to the farmers.

Biocontrol agents for pest management

Egg parasitoids

Trichogramma brasiliensis is a parasitoid specifically useful to control tomato fruit borer *(Helicoverpa armigera)* which attacks tomato, peas and beans, etc. With the onset of adult catch in pheromone trap or first flowering, release of the parasitoids as parasitized eggs @50,000/ha with repeated inoculative release at the same rate at weekly interval is very effective. Similarly, *T. chilonis* and *Trichogrammatodea bactrae* are effective to control okra shoot and fruit borer (*Earius* spp.) and diamondback moth (*Plutella xylostella*) on crucifers, respectively. Weekly release of *T. chilonis* @250,000 parasitized eggs /ha has positive effect in control of brinjal shoot and fruit borer. *Trathella flavoorbitalis* is a very effective larval parasitoid of brinjal shoot and fruit borer which can be preserved and restored under field condition by periodical collection of affected shoots along with the larvae in the bag made up of mosquito net. The larvae along with the twigs are kept in hanging condition in the shed. The adult parasitoids will emerged out with in a fortnight and dispersed in the open field for searching of their hosts.

Larval parasitoids

Cotesia plutellae and *Diadegma semiclausum* are dominant larval parasitoids of diamondback moth, among which the later one is susceptible to high temperature and is mostly active in areas having mild climate. Very high natural parasitization by these larval parasitoids indicates the importance of their role in pest suppression if conserved through proper management practices. Use of neem seed kernel extract (4%) helps to control the pest with least impact to these larval parasitoids. *Campoletis chlorideae* is the most common larval parasitoid of *Helicoverpa.* Usually second instar larvae are parasitized in the field. This parasitoid is mass multiplied on young larvae of *H. armigera* and *Spodoptera litura.*

Predators

Ladybird beetles *(Coccinella septempunctata* and *Chilomenus sexmaculatus)* are natural predators of aphids and other soft-bodied insects. These predators can easily be mass multiplied and used for suppression of aphids on different vegetable crops. Generally, adult beetles released @ 30 beetles/sq.m are sufficient to control the aphids. Green lace wings (*Chrysoperla carnea)* are commonly found to predate on soft-bodied insects like aphids, thrips, jassids, whiteflies and eggs of some lepidopteran pests. Mass release of early instars of these predators @ 50,000 eggs/ha is very effective in the management of sucking pests on vegetable crops.The predatory mite, *Amblyseius longispinosus* is very promising in suppressing the population of red spider mite in vegetable crops like pointed gourd, bhindi, brinjal etc. Mass release @ 20000 predatory mites/ha at weekly interval during February onwards, manage satisfactorily the population of red spider mite. Similarly, mass release of predatory mite species, *Agistemus flestchneri* @ 50,000/ha keep suppress the population of yellow mite, *Polyphagotarsonemus latus* in chilli, beans, cowpea etc. Besides, this predatory species voraciously predate upon scale insect, mealy bug and eggs and puparia of white fly.

Microbial biocontrol agents

Insect pathogenic bacteria, fungi and viruses are mostly used for the management of the pests of vegetable crops. Unlike natural enemies, these agents can be utilized as a prophylactic measure as done in insecticides and have an advantage of field applicability. The bacteria *Bacillus thuringiensis (Bt)* is most widely used microbial insecticide which is useful in the management of several leaf feeding lepidopteran pests (caterpillar) of vegetable crops. Several commercial formulations of *Bt* are now available in markets which are generally used as foliar spray @1.0 kg/ha. It is particularly effective against diamondback moth on cabbage however, it is also recommended for the control of shoot and fruit borer on okra. Nuclear Polyhedrosis Viruses (NPVs) are highly target specific entomopathogens with least impact on natural enemies. The NPV's have been widely used for the management of fruit borer *(Helicoverpa armigera)* on tomato and leaf eating caterpillar *(Spodoptera litura)* on cauliflower.It is recommended to spray Helicoverpa NPV and *Spodoptera* NPV @250-300 Larval Equivalent (LE) /ha mixed with 1% jaggery and 0.1% Teepol in 250 litres of water on tomato and cauliflower, respectively in the evening hours to get the optimum control of these insect pests.

Examples of IPM system for management of insect pests in different vegetable crops

Nursery bed management for different vegetable crops

- Soil solarization with transparent polyethylene sheet during summer is ideal to suppress the soil borne pathogens in nursery. Seed treatment with *Trichoderma* spp. @ 5-10 g/kg or soil treatment @ 10-25 g/ sq.m nursery area is effective against damping off diseases.

Brinjal

- Seed treatment with imidaclorprid @ 2.5g/kg reduces the little leaf and leafhopper incidence.
- Intercropping of brinjal (2 rows) with coriander (one row) or fennel (1 row) minimizes the incidence of brinjal shoot and fruit borer.
- *Trathella flavoorbitalis* is a very effective larval parasitoid of brinjal shoot and fruit borer which can be preserved and restored under field condition by periodical collection of affected shoots with the larvae in the bag made up of mosquito net. The larvae along with the twigs are kept in hanging condition in the shed. The adult parasitoids will emerged out with in a fortnight and dispersed in the open field for searching of their hosts.
- Integrated pest management of fruit and shoot borer includes raising the nursery under nylon net cover, removal and destruction of infested shoots along with larvae and periodical cleaning of dry leaves and old infested fruits at weekly interval along with the insect larvae starting from 25 days after transplanting and installation of sex pheromone baited plastic funnel traps at 10m distances just above canopy from 20 days after planting and need-based foliar spray of NSKE (4%) during late afternoon.

Tomato

- Seed treatment with imidacloprid @ 2.5g/kg, raising the nursery under nylon net cover, single spray of imidacloprid (Confidor) @ 0.3 ml/lit about 40 days before fruiting and subsequent spray 4% neeem seed kernel extract (NSKE) with 0.1% sticker minimizes whitefly population and reduce the incidence of tomato leaf curl virus (TLCV) and leafminor.
- Planting of two rows of marigold as trap crop with every 14 rows of tomato along with release of *Trichogramma brassiliensis* @50,000 parasitized egg/ha at flowering stage and spray of *Helicoverpa* NPV @ 350 LE/ha twice after flowering can manage the fruit borer *(Helicoverpa armigera)* satisfactorily.
- Use of synthetic pyrethroids and endosulfan alternatively with NSKE (4%) is also effective against fruit borer *(Helicoverpa armigera)*.

Okra

- Seed treatment with imidaclorprid @ 2.5 g/kg is effective against leafhopper in okra during early vegetative stage.

- Spraying of NSKE (4%) spray with sticker @ 0.1% minimizes the incidence of insect pests in the field.
- Planting of the stump of green manuring crop, *Glyricidia maculata* along the field bund is very effective for the management of many destructive pests of crops because the green shoots along with leaves having pest deterent properties add sufficient organic nutrients in the field.
- Spraying of endosulfan and neem formulation at their half doses at flowering stage can reduce borer (*Earius* spp.) infestation.
- Okra sown during 2nd week of June minimizes the incidence of both jassids (*Amrasca biguttula biguttula*) and shoot and fruit borer (*Earius* sp.).

Chilli

- Seed treatment with imidacloprid @ 2.5 g/kg is effective against thrips (*Scirtothrips dorsalis*) and mites (*Polyphagotarsonemus latus*) complex during initial stage.
- Alternate spray of dicofol @ 2.75 ml/l and wettable sulphur @ 2 g/l is highly effective against mite.

Cucurbits

- Effective fruit fly management include deep summer ploughing to expose the dormant pupae in the field, crop rotation incorporating non- cucurbitaceous crops, food bait consisting of rotten banana or smashed pumpkin pulp (1000g) + yeast (10 g) + malathion (10 ml) + citric acid (3 g) in flat earthen pot to attract and kill the adult flies and bait spray in one plant per 10 m^2 area with 20 ml malathion (50 EC) + 20 liters water +500 g molassess keeping rest of the plants remains free from insecticides to attract adult flies and control the population.
- Spraying of carbaryl 50 WP @ 2 g/l at cotyledon stage control red pumpkin beetles.

Cabbage

- Diamondback moth (DBM) population in cabbage can be managed by growing paired mustard rows with every 25 cabbage rows. The first rows of mustard should be sown 15 days before and second 25 days after cabbage planting. The trapped DBM larvae on mustard can be killed by spraying 0.1% dichlorovos or 5% NSKE. Such practices conserve the population of *Cotesia plutellae,* a dominant natural enemy of DBM.

6. Integrated Disease management

The five components of an IDM program are prevention, monitoring, correct disease and pest diagnosis, development and use of acceptable thresholds, and optimum selection of management tools. The management strategies available include genetic control, cultural control, biological control, and chemical control. What management strategy is most relevant depends in part on the particular pest. Diseases are controlled most by host-plant resistance, cultural practices, biological and chemical options. For disease management, it is important to understand the potential of a pathogen to infest and spread in the crop. The three parameters of disease progress are 1) the initial amount of inoculum, 2) the rate of disease increase, and 3) the time the crop is grown. These parameters interact in ways to produce a rapid pathogen population increase, manifested as exponential growth in many production systems.

Accurate Diagnosis of Diseases

Proper disease identification is critical for making correct disease management decisions. This will save time, money, and the environment. Effective use of different management options depends on correct identification of the problem.

Monitoring Pathogens

As mentioned, monitoring is a critical component of an effective IDM programme. Monitoring can be direct (looking for the pathogen or disease) or indirect (recording environmental conditions which affect disease development). Financial considerations weigh heavily in the choice of monitoring practice.

Prevention and Management Methods

Site selection and land preparation

Soilborne diseases remain a major limiting factor for the production of vegetables in most of the countries in the world. It is important to start with clean soil and proper sites for crops. Ploughing and disking will reduce pathogen carryover in old crop refuse. The longer the fallow period, the more pathogen populations are reduced. It is also essential to follow the latest recommendations for soil fumigation, cultural practices, and biological control options to eliminate or reduce initial inoculum of soilborne pathogens. It is important that soil compaction is avoided, since this interferes with root growth, encourages soil moisture retention and promotes root diseases. Preparation of raised beds generally allows for better drainage. Prior to planting, soil should be tested for nutrient levels and nematode populations. Histories of soilborne disease outbreaks are important in prediction of possible future problems. Planting times can be altered to avoid or reduce development of certain diseases.

Host resistance

It is very important to choose cultivars with multiple pathogen and nematode resistance whenever possible. In practice, control of many diseases of vegetables (Fusarium wilt, Verticillium wilt, and gray leaf spot for tomato) is achieved primarily by this method. Varieties and hybrids resistant to Tomato Leaf Curl Virus (ToLCV)

(TLBRH-6, BCTH-4, JKTH-3098), Bacterial wilt of tomato (Arka Abhijit, Arka Shrestha) and brinjal(Arka Nidhi, BB-54), pea powdery mildew (Arka Ajeet) and watermelon powdery mildew, downy mildew(Arka Manik) have been identified and commercially exploited in India.

Irrigation management

High soil moisture enhances the development of soilborne pathogens including *Phytophthora*, *Pythium* and the bacterial wilt pathogen. Excess water damages roots by depriving them of oxygen and creates conditions that favour infection by certain soilborne pathogens.

Irrigation management, based on plant needs will help to create an environment unfavorable for pathogen survival and disease development. Use of tensiometers or other devices for irrigation scheduling and avoidance of low areas can help in disease management.

Soil and fertilizer management

Plant nutrition and soil pH also influence some diseases. Fertilizers with a higher proportion of nitrate nitrogen (NO_3) than ammoniacal nitrogen (NH_4) will help to reduce the incidence of *Fusarium* wilt on tomato. Increasing soil pH by liming is a good management strategy to reduce Fusarium wilt incidence as well as severity of club root of cruciferous vegetables. Optimum calcium nutrition and higher soil pH may reduce the incidence of bacterial wilt in the field. Adequate calcium is necessary to minimize blossom end rot and to provide for overall healthy growth.

Cultural practices

Cultural practices serve an important role in prevention and management of plant diseases. The benefits of cultural control begin with the establishment of a growing environment that favours the crop over the pathogen. Reducing plant stress through environmental modification promotes good plant health and aids in reducing damage from some plant diseases.

Sanitation practices aimed at excluding, reducing, or eliminating pathogen populations are critical for management of infectious plant diseases. It is important to use only pathogen-free transplants, especially for late blight, bacterial spot, viral diseases, and early blight.

In order to reduce dispersal of soilborne pathogens between fields, stakes and farm equipment should be decontaminated before moving from one field to the next. Reduction of pathogen survival from one season to another may be achieved by destruction of volunteer plants and crop rotation. Removal of cull piles and prompt destruction of crops should be applied as a general practice.

Avoid movement of soil from one site to another to reduce the risk of moving pathogens. For example, sclerotia of *Sclerotinia sclerotiorum* and *Sclerotium rolfsii* are transported primarily through contaminated soil. Minimizing wounds during harvest and packing will reduce post-harvest disease problems. Depending on crops and other factors, sanitation of soil can be achieved to some degree by solarization.

Crop rotation is a very important practice, especially for soilborne disease control. For many soilborne diseases, at least a 3-year-rotation using a non-host crop will greatly reduce pathogen populations. This practice is beneficial for *Phytophthora* blight of pepper and *Fusarium* wilt of watermelon, but longer rotation periods (up to 5-7 years) may be needed. Land previously cropped to alternate and reservoir hosts should be avoided whenever possible. Vegetable fields should be located as far as away as possible from inoculum and insect vector sources.

Weed control is important for the management of viral diseases. Weeds may be alternate hosts for several important vegetable viruses and their vectors. Elimination of weeds might reduce primary inoculum. Cover crops help to reduce weed populations that may harbour pathogens between seasons. For this purpose, use cover crops that grow fast and provide maximum biomass. Non-host cover crops will help to reduce weed populations and primary inoculum for soilborne pathogens.

Excessive handling of plants such as in thinning, pruning and tying may be involved in spread of pathogens, particularly bacteria. It is advisable to handle plants in the field avoiding plant injury because some pathogens can only enter the host through wound. During pruning process and harvest, workers should periodically clean their hands and tools with a disinfectant preferably isopropyl alcohol.

If applicable, plants can be staked and tied for improved air movement in the foliar canopy. A more open canopy results in less wetness, discouraging growth of most pathogens.

Soil aeration and drying can be enhanced through incorporation of composted organic amendments in the soil. Build up of inoculum can be reduced by removing all plant materials (infected and apparently healthy) after harvest.

Polyethylene mulch can be used as a physical barrier between soil and above-ground parts of plants. This is an important practice for fruit rot control in the field. Highly UV-reflective (metalized) mulches repel some insects. It is beneficial to use metalized mulch during certain times of the year when insect vectors of some viral diseases are prevalent. Tomato spotted wilt virus (TSWV) incidence and associated vector thrips populations have been demonstrated to be effectively reduced by using metalized mulches on tomatoes.

7. Biological management of diseases

Many fungi, bacteria and actinomycetes have been found promising against different soil-borne pathogens causing root rot, damping off, collar rot and wilt in different vegetable crops. Important target pathogens are *Pythium* spp., *Sclerotium rolfsii, Macrophomina phaseolina, Rhizoctonia solani, Fusarium* spp., *Phytophthora* spp., *Phoma* sp., *Sclerotinia sclerotiorum, Aspergillus* spp., *Botrytis, Armillaria, Verticillium, Alternaria solani, Colletotrichum capsici and Pleospora beta.* Several species of *Trichoderma* have gained attention for their highly antagonistic property. Some of the antagonists are highly specific to a particular pathogen, but most of the fungal antagonists have broad-spectrum parasitic host range. Some antagonistic fungi also kill some plant parasitic nematodes infecting vegetables.

Potential biocontrol agents

Soil has maximum microbial biodiversity, which can be utilized for plant disease management. Several antagonists have been isolated and many of them are being utilized in management of soil-borne pathogen. The maximum biocontrol agents belong to fungal genera, viz. *Trichoderma* spp., *Aspergillus niger, A. flavus, Pythium nann, Trichothecium, Paecilomyces lilacinus, Penicillium funiculosus, Myriothecium, Corticium, Pythium cligandrum, Peniophora gigantea, Candida oleophila, Sporidesmium sclerotivorum, Coniothyrium minitans, Ampelomyces quisqualis, Chaetomium, Cladosporium, Fusarium semi tectum, Tuberculina, Phialophora, Catenaria* and *Verticillium.* Some bacterial antagonists like, *Pseudomonas* spp., *Agrobacterium radiobactor, Bacillus* spp. and Actinomycetes such as *Streptomyces griseus,* S. *rimosum* have also been utilized for the management of diseases. *Trichoderma* spp. has extensively been exploited for the management of soil borne diseases.

Crop	Pathogen	Antagonist
Tomato, brinjal, chilli, bean	*Rhizoctonia solani*	*Trichoderma harzianum, T. viridae*
Tomato, chilli, brinjal, bean, pea, cucumber	*Pythium* spp.	*T. viridae, T. harzianum, T. koningii*
Tomato, chilli, brinjal, bean, pea	*Sclerotium rolfsii*	*T. viridae, T. harzianum, T. koningii*
Pea, radish	*Pythium* spp., R. *solani*	*T. viridae, T. harzianum, T. hamatum*
Tomato, muskmelon	*F. oxysporum, P. aphanidermatum*	*T. viridae, T. harzianum*
Okra	*Macrophomina phaseolina*	*T. viridae, T. harzianum*
Cucumber	*Sclerotinia sclerotiorum*	*T. harzianum*

The best substrate is sorghum or mixture of sorghum with either rice bran or rice husk. Talcum powder is commonly used as carrier in commercial biopesticide product. The USA, Israel, Bulgaria, Australia, Sweden, Canada and India have commercialized the use of biopesticides in management of soil-borne diseases. Successful commercial use of *Trichoderma* spp. and *Pseudomonas fluorescence* has been demonstrated over past ten years in most part of the country. The biopesticides are either mixture of two bioagents or alone are available with different trade names, viz. *T. viridae* (Antagon TV, Trichoguard, Niprot, Bioderma, Tricontron) , *T. harzianum* (Trichodex, Trichodermin, F. stop, Bio-Coat), *T. virens* (Gliogard), *T. viridae* + *P. flourescens* (An tag or Combi), *T. harzianum* + *Bacillus subtilis* (Combat), *P. fluorescence* (Biomonas, Dagger G, Conquerer).

Mechanisms of biocontrol

Three antagonistic mechanisms are involved in biological control of soil-borne plant pathogens, viz. competition, antibiosis and mycoparasitism. Competition occurs when there is a demand of two or more microorganisms for the same nutrient resource in limited space and time. This is most potent mechanism to control *Fusarium oxysporum* f. sp. *melonis* by *Trichoderma harzianum.* Antibiosis occurs when production of toxic metabolites or antibiotics by the biocontrol agent has a direct effect on another organism and such toxic metabolites include gliotoxin produced by *T. virens,* viridin by *T. viride* and trichodermin by *T. harzianum.* In case of mycoparasitism or hyperparasitism, fungus is parasitized on another by chemotropic growth, recognition, attachment and degradation. Production of lytic enzymes, chitinases and â 1, 3 glucanases are most important for the lysis of pathogen cell wall. *Trichoderma* species compete aggressively, mycelium coils around pathogen hyphae and cause lysis of the cells of *Phytophthora* spp.

Flavobacterium balustinum and *Xanthomonas maltophila* are most effective in inducing suppression of damping off causing fungus *Rhizoctonia* sp. by ISR mechanism. Many of the antagonistic bacteria also act as plant growth promoting rhizobacteria (PGPR) and systemic acquired resistance (SAR) in plants against many disease causing pathogens of vegetable crops.

Application of biocontrol agents

Soil application and seed treatment are only feasible method of using bioagents under the field condition. Seed treatment is preferred in direct sown crops, whereas soil application and seed treatment are feasible in the crops which are transplanted in the main field. Best utilization of any biocontrol agents in soil is only possible when it is supplemented with well-decomposed organic amendments, congenial moisture and protected from extreme weather conditions till it establishes in soil.

Biological seed treatment can be done by priming on seed, soaking seeds, seedling dipping, and dry power treatments depending upon the nature of biocontrol agents.

Field studies clearly showed that *T. viride* is better than any other biocontrol agents in this respect. Seed treatments with *Trichoderma* @ 6-10g/kg of seed is effective provided spore concentration remain in between 106 and 109/ml. Dose for either seed or soil treatment is influenced by inoculum of pathogen present in soil and nature of seed. Soil application of *Trichoderma* @ 10-25 g along with 100 g neem cake/ sq.m is effective to manage nursery diseases of different vegetable crops.

Green manuring along with soil application of *Trichoderma* @5 kg/ha give good control of most of the soil borne diseases in *kharif* season grown vegetable crops. The *Trichoderma* must be applied soon after incorporating sunhemp preferably within three days depending upon moisture status and texture of soil.

Root dipping of vegetable seedlings by *Trichoderma* formulation @ 1% for 10-25 minutes is very effective in checking root rotting after transplanting in *kharif* season. Drenching of *Trichoderma* @ 1% near collar region should be carried out 20 days after transplanting to check root rot, collar rot and wilt incidence. Fast and vigorous seedling growth is recorded in biocontrol treated nursery beds where 20-25 days old seedlings are ready for the transplanting.

Foliar spray of *Trichoderma* on all ground portions in pointed gourd controls *Phytophthora* blight of both vine and fruit. One isolate of *Aspergillus niger* @ 1% along with sticker and molasses as foliar spray against *Alternaria* blight and anthracnose of chilli has given satisfactory control.

Sensitivity of the bioagents with pesticide

The fungicides, which are completely insensitive to the bioagents, but sensitive to the pathogen, can be used together to increase the efficiency of disease management strategies. Pencycuran and copper hydroxide up to 400 ppm concentration is absolutely insensitive for the bioagents. The bioagents show tolerable growth after 7 days of incubation with captan, chlorothalonil and propineb up to 200 ppm concentration.

Copper hydroxide shows synergistic effect at lower concentration, while fungistatic at higher concentration (400 ppm) for five days. Pencycuron is highly effective against *Rhizoctonia* spp. but insensitive to biocontrol agents and gives full protection when integrated both at a time. Different fungicides like, pencycuron, copper hydro oxide, captan, azoxystrobin, carboxin, sulfex and antibiotic like streptocyc1ine can safely be used in biocontrol strategies with *Trichoderma* for the management of soil-borne diseases in vegetable crops.

Imidacloprid can be integrated with *T. virens* and *T. viride* with slight adjustment in dose. Seed rate should be compensated according to soil type, insect endemic area and viral infection to get desired seedling stand of tomato. Soil application of carbofuran followed by seed treatment by antagonist can be integrated for insect pest and disease management without adverse effect on seed germination. Pesticides like cypermethrin, fenvalerate and herbicide like glyphosate are safe for integration with management by *Trichoderma*.

8. Weed management

Agricultural production depends on various factors like, genetic make up of the crop, soil and environmental factors and different production problems that interact either to enhance output or to limit production. Of the total losses due to different production problems caused by insect pests, disease causing pathogens, nematodes and weeds, one- third of the estimated crop loss is caused by weeds alone and much damage is caused during early stage of crop growth. So weed management is utmost important for successful production of vegetable crops. Intensity of weed population in different vegetable crops varies with climate, soil, management practices etc.

Weeds are no stranger to human beings and these have been recognized as a problem ever since cultivation of crops began. Jethro Tull first coined the term 'weed' in his writings "Horse Hoeing Husbandry" being the first book of agriculture in1731 in which 'weed' was termed for a plant where it is not desired. Since then various definitions of the weed has been suggested although, main feature of defining weed remains the same as the plant growing out of place and out of time and interfere with crop growth.

Characteristics of weed

- Weeds are competitive and aggressive in nature.
- Weeds can tolerate adverse climatic condition, e.g. *Calotropis procera, Saccharum spontaneum* tolerate high temperature and low moisture condition.
- Many weeds are morphologically similar to some crops they infest, e.g. *Phalaris minor* is similar to wheat and barley; *Asphodelus tenuifolius* is similar to onion and some weed seeds are morphologically similar to crop seeds, and thus cannot be separated after seed harvest.
- Weeds have high reproductive capacity and in adverse conditions, they even can reproduce asexually, e.g. *Chenopodium album* produces 10,000 to 15,000 seeds per plant while *Phalaris minor* produces 3000-4000 seeds per plant in one season.
- Weeds can resist their control or eradication due to possession of spines, thorns, etc. that hinder their removal from a particular site. e.g. *Zizyphus mummularia, Rumex spinosus* possess spines/thorns and *Cynodon dactylon, Sorghum halepense* can regenerate easily, hence continue infestation.
- Weeds are persistent in nature due to their high longevity, prolific and perennial nature, dormancy of seeds, resemblance with crop plants, ability of reproduction through both seeds and vegetative parts. Seeds of some weeds like, *Chenopodium album, Convolvulus arvensis* can germinate even after 20-25 years of burial in soil.
- Some weeds have the tendency of setting seeds early and prior to crop plants that create serious problem in following years.

- Repeated germination of weed seeds in different phases create problem for the crop throughout its life cycle. Weeds appear in different flushes in same crop till last harvest which helps in their persistence in the field throughout crop life.
- Some perennial weeds possess extensive and deep root system of 5-7 ft in *Cyperus rotundus* and even upto 20 ft in *Convolvlus arvensis* which makes it difficult for their uprooting.

Classification of weeds

Weeds can be classified in different manner as discussed below.

1. Classification based on morphology

- Broad leaf: *Amaranthus spinosus*, *Eclipta alba* , *Portulaca oleracea,* etc.
- Grasses : *Echinochola crusgalli*, *Cynondon dactylon*, etc.
- Sedges: *Cyperus rotundus*, *cyperus diformis,* etc.

2. Classification based on life cycle

- Annuals: Annual weeds which are usually small herbs with shallow roots and weak stem and propagated by seeds which mature in one season
- Rabi season annuals: *Avena fatua*, *Phalaris minor*, *Ageratum conyzoides*, *Chenopodium album*, *Argemone mexicana*, *Cirsium arvensis*, etc.
- Kharif season annuals: *Dactyloctenium aegypticum*, *Echinochola colonum*, *Amaranthus spinosus*, *Amaranthus viridis*, *Solanum nigrum*, etc.
- Biennials: Biennials require two year for completing their life cycles; in the first year they remain vegetative, and in the second year they produce flowers and set seeds. These types of weeds are mainly found in non-cropped areas. Common examples of biennials weeds are wild carrot, *Alternanthera echinata*, *Cirsium vulgare* , *Plantago* spp. , *Arctium* spp. etc.
- Perennials: Perennials weeds live more than two years and they reproduce both vegetatively and sexually forming seeds.
- Simple perennials: These perennials are propagated only by seeds e.g. *Sonchus arvensis*
- Bulbous perennials: These perennials possess a modified stem with scales and reproduce mainly by bulbs and seeds e.g. *Allium* sp.
- Corm perennials: These perennials possess a modified shoot and fleshy stem and reproduce mainly by corm and seeds e.g.Timothy
- Creeping Perennials: Creeping perennials are reproduced by seed, rhizome (*Sorghum halepense*), stolon (*Cynodon dactylon*), enlarged root system with numerous buds (*Convolvulus arvensis*) , tubers (*Cyperus* sp.), etc.

3. Classfication based on habitat

- Terrestial weeds: These weeds are mostly found in upland or dry land condition having enough moisture but the land is not water logged e.g. *Chenopodium album*, *Solanum nigrum* , *Celosia argentea* , etc.
- Aquatic weeds: These weeds grow in water and complete at least a part of their life cycle in water which are of several types.
- Free floating: *Eichhornia crassipes* , *Pistia stratiotes*,etc.
- Rooted floating: *Trapa bispinosa* . *Ludwigia adscendens* ,etc.
- Subemersed : *Ceratophyllum demersum* , *Hydrilla verticillata* ,etc.
- Immersed: *Jussieua repens* , *Nelumbium speciosum* ,etc.
- Amphibious : *Ranunculus aquatilis* , *Scirpus supinus* ,etc.

4. Classfication based on the origin of weeds

- Alien (Foreign origin): *Parthenium hysterophorus* , *Argemone mexicana*, etc.
- Apophytes (Indigenous): *Cynodon dactylon* , *Cyperus rotundus*, *Saccharum spontaneum* , *Acalypha indica* ,etc.
- Antrophytes (Introduced by man): *Phalaris minor*, *Corchorus acutangulua* .

Dispersal of weeds

Weeds develop various adaptive mechanisms for their continuous perpetuation and dispersal. Some seeds are smaller and lighter which are dispersed through blowing wind. Some heavier seeds are dispersed by animals and birds. Some weeds seeds possess specialized structure that extends buoyancy in the atmosphere e.g. parachute like pappus in weeds of Compositae family. Weeds of Graminae family possess fine long awns with arrow shaped pointed ends that get into clothes and cannot be removed easily and thereby dispersed to long distance. Presence of hooks and bars on surface of many fruits and seeds help in their attachment to animals and clothes of human beings. Sticky fruits are dispersed by animals and get attached to tillage equipment and thus are dispersed. Weed seeds are generally dispersed through the following modes.

- By wind
- By water current.
- By birds
- By animals
- By human beings and contact with different tillage equipment.
- Along with the farm produce.
- Along with silage, dung and farm yard manure.

Harmful effects of weeds

Successful production of vegetables is adversely affected by weeds in the different ways.

- Weeds are fast growing compared to the vegetable crops hence, compete with the crop for solar radiation, water, nutrients and space.
- Weeds reduce crop yield, quality and marketability of the produce.
- Weed infestation increases cost of cultivation because its control is basically a labour intensive proposition and require costly chemicals, equipment for chemical control measure.
- Some weed plants secrete harmful chemicals that may have adverse effects on crop plants, soil, human being and livestock. In India *Parthenium hysterophorus* cause contact dermatitis and asthma. *Rhus radicans* (Poison ivy), *Ambrosia sp.* (Ragweed) cause itching, hay fever and other debilitating allergies contributing to chronic human illness and suffering.
- The perennial rhizomatous grasses like, *Imperata cylindrica, Cynodon dactylon, Sorghum halepense, Saccharum spontaneum* cause considerable damage to the land due to lack of equipment strong enough to break sods and the land some times becomes unsuitable due to heavy infestation of these weeds.
- Presence of spiny and thorny weeds like, *Amaranthus spinosus, Argemone mexicana* etc. restrict movement of farm workers in carrying out farm practices.
- Some prominent aquatic weeds are *Eichhornia crassipes, Typha angustata, Hydrilla verticillata, Ipomoea aquatica,* etc. may block drainage and impede the flow of water in irrigation channels. They also result in seepage, breakage of canal banks, flooding, silting, and inadequate irrigation water delivery to farms located at a distance from main water source.
- Certain weeds provide hiding place for insect-pests and act as host plants for certain pathogens.

Examples of some weed host for different insect pests and pathogens

Weed	Insect pest	Crop host
Mustards	Cabbage root maggot	Cabbage, broccoli
Chick weed (*Stelaria media)*	Melon aphids	Most cucurbits
Wild morning glory (*Convolvulus arvensis)*	Aphids	Watermelon, pumpkin, some gourds (mostly ridge gourd)
Wild carrot (*Daucus carota)*	Carrot rust fly	Carrot, celery.
Night shade (*Solanum nigrum)*	Onion thrips	Onion
Lambsquarter (*Chenopodium album)*	Common stalk borer	Mostly tomato.
Pig weed (*Amaranthus viridis)*	Spinach mildew	Spinach, Swiss chard.
Night shade (*Solanum nigrum)*	Verticillium wilt, Tomato mosaic virus	Tomato
Night shade (*Solanum nigrum*	Potato mosaic	Potato
Mustards (*Brassica sp.)*	Clubroot	Cabbage, Cauliflower

Common weeds found in the field of vegetable crops

Some weeds are found most frequently in the field of different vegetable crops viz., wild oat (*Avena fatua*), gullidanda/Canery grass *(Phalaris minor*),chottis wank (*Echinochola colona*), cow foot grass(*Dactyloctenium aegyptium*), swank/ maddal (*Echinochola crusgalli*), doob grass /Bermuda grass (*Cynodon dactylon*), motha /nut grass (*Cyperus rotundus*), Kans (*Saccharum spontaneum*), goose grass (*Eleusine indica*), bathua (*Chenopodium album*), chaulai (*Amaranthus viridis*), tandla (*Digeria arvensis*), dodhak (*Euphorbia hirta*), fumatory / pitpapra (*Fumaria parviflora*), meadow pea/natri (*Lathyrus aphaea*), wild onion (*Asphodelus tenuifolius*), ivy leaved /bel (*Ipomoea pestigridis*), con spurry/jangli dhania (*Spergula arvensis*), dendelon/jangli gobhi (*Launea asplenifolia*), wild carrot (*Daucas carota*), milk weed (*Sonchus arvensis*), jangli palak (*Rumex dentatus*), thorny amaranthus (*Amaranthus spinosus*), night shade (*Solanum nigrum*), johnson grass(*Sorghum halepense*), orbanche (*Orbanchae aegyptica*), etc.

Methods of Weed Control

1. Preventive methods

- Care should be taken at the time of sowing of crops that weed seed free crop seed must be used.

- Movement of live stock should be checked from weed infested to clean area.
- Weed plants should be completely removed from the sides of irrigation and drainage channels.
- Use of clean farm machinery.
- Farm boundaries, roadsides, and other wastelands should be free from weeds.
- Planting of crop should only be done after cleaning of weed plants.

2. Curative or control methods

- *Mechanical methods:* Removing the weeds physically or with tools /implements is called mechanical method. Mechanical methods though traditional yet are very effective in controlling almost all the type of weeds. However, these methods are time consuming and labour intensive. Different mechanical methods adopted for controlling weed are as follows.
  a) Tillage: Tillage is done mainly with two objectives : to uproot the weeds growing in the field and to pulverize the soil.
  b) Hand pulling: Hand pulling is generally practiced to uprooting weed plants before flowering and seed setting along with their roots.
  c) Hand weeding: Small implements like khurpi , nirani , sickle, wheel hoe, etc can be used for uprooting plants.
  d) Burning: Burning is an economical and practical means of controlling weeds particularly in organic production system.
- *Cultural method:* In this method, selection of crop is one of the most important factors to check weeds up to the great extent. The crop should be quick growing with large leaf area and good branching habit. Most of the legume and leafy vegetables grow rapidly and develop into dense canopy compared to weeds which results in effective suppression of weeds. Proper crop rotation does not allow development of several weeds particularly associated weeds with crops, and hence do not pose a serious problem. Practice of raising green manure crop is helpful in reducing weed infestation as the weeds which grow along the green manure crop are buried and subsequently decomposed into the soil. Soil solarization is also an effective method of weed control in many vegetable crops.
- *Biological method:* Biological approaches involves various living agents such as fishes, snails, pathogens, insects, nematodes as well as botanicals i.e. bioherbicides to control weeds. Different bioherbicides like, Devine®, Biomal ®, etc are commercially available. Control of some perennial weeds by biological method has been done successfully viz., *Lantana camara* by lace bug , *Teleonemia scrupulosa;* cuscuta by *Melanogromyza cuscutae* , *Imperata cylindrica* by *Orseoliella javanica* ; *Cyperus rotundus* by a plant feeder *Bactra*

vermosana. However, biological method of weed control has been very little in practical use in vegetable crops.

- *Chemical methods:* Use of herbicides for the control of weeds offers the greatest feasibility for weed control. Chemical method of weed control is less costly than other methods besides its usefulness in minimizing tillage operations.

Herbicide

Versatility in vegetation management has been extended through the use of selective agricultural chemicals. The term "Herbicide" came from the latin words *herba* meaning plant and *caldere* meaning to kill .So, herbicide is the chemical that disrupt the physiology of a plant over a long enough period to kill it or severely reduce its growth. Herbicide changes the chemical environment of plants, which can be more readily manipulated than the climatic, edaphic or biotic environment. Herbicides reduce labour and machine requirements and modify crop production technology.

The first widely used herbicide was 2,4-dichlorophenoxyacetic acid (2,4-D) which was developed by a British team during World War II with first widespread production and use in the late 1940s. It still remains one of the most commonly used herbicides in the world. However, it exhibits relatively poor selectivity and is also less effective against some broadleaf weeds, including many vinous plants, and sedges. The 1970s saw the introduction of atrazine, which has the dubious distinction of being the herbicide of greatest concern for groundwater contamination because it does not break down readily (within a few weeks) after being applied. Instead, it is carried deep into the soil by rainfall causing the ground water contamination. Glyphosate was introduced in the late 1980s for non-selective weed control. It is now a major herbicide in selective weed control in growing crop plants due to the development of crop plants that are resistant to it. The combination of the herbicide with the resistant crop variety contributed to the consolidation of the seed and agro-chemical industry in the late 1990s.Many modern chemical herbicides for agriculture are specifically formulated to decompose within a short period after application which is desirable as it allows crops which may be affected by the herbicide to be grown on the land in future seasons. However, herbicides with low residual activity (i.e. which decompose quickly) often do not provide season-long weed control.

Classification of herbicides

Currently there are more than 250 herbicides in the world market. There are numerous classes of herbicides with different modes of action for killing weeds, as well as different potentials to have an adverse effect on health and the environment. Herbicides from different classes also differ in their environmental persistence and fate. Almost all herbicides can cause acute toxicity. Phenoxy herbicides are involved in acute symptomatic illnesses with relative frequency.

Different classes of herbicides

Class of Herbicide	Examples
Acetamides and analides	Alachlor, acetochlor, metolochlor, propachlor, propanil
Carbamates and thiocarbamates	Asulam, terbucarb, thiobencarb
Chlorophenoxy herbicides	2,4,-D, 2,4-DP, 2,4-DB, 2,4,5-T, MCPA, MCPB, MCPP, Dicamba
Dipyridyls	Paraquat, diquat
Heavy metals	Lead arsenate, arsenicals
Nitrophenolic and dinitrocresolic herbicides	Dinitrophenol, dinitrocresol, dinoseb, dinosulfon
Pentachlorophenol	Pentachlorophenol
Phosphonates	Glyphosate, glyfusinate, fosamine ammonium
Triazines	Atrazine, simazine, cyanazine, propazine
Urea derivatives	Diuron, flumeturon, linuron, rimsulfuron, tebuthiuron

Source: Farm Chemicals Handbook (1988)

The herbicides are basically classified on the basis of a) method of application, b) mode of action and c) mechanism of action

1. Method of application

- Soil-applied group: The herbicides in this group (Pre-emergent herbicide) are applied before planting and before emergences of weed or crop which are taken up by the roots of the target plant.
- Foliar-applied group: The herbicides of this group (Post-emergent herbicide) are sprayed on the standing plant part after the crop has emerged.

2. Mode of action

- Contact herbicides: These herbicides destroy only the plant tissue in contact with the chemical. Generally, these are the fastest acting herbicides but are less effective on perennial plants which are able to regenerate from roots
- Systemic herbicides: These herbicides are translocated through the plant, either from foliage down to the roots, or from soil application up to the leaves which can destroy a greater amount of plant tissue than contact herbicides.

3. Mechanism of action

This classification by mechanism of action indicates the first enzyme, protein, or biochemical step affected in the plant following application. Of this herbicide

- ACCase inhibitors: These herbicides generally kill grasses through inhibiting the enzyme Acetyl coenzyme A carboxylase (ACCase) which is involved in the first step of lipid biosynthesis. Thus, ACCase inhibitors affect cell membrane production in the meristems of the grass plant. The ACCases of grasses are sensitive to these herbicides, whereas the ACCases of dicot plants are not sensitive to these herbicides.
- ALS inhibitors: These herbicides kill the plant through inhibiting the enzyme acetolactate synthase (ALS) (also known as acetohydroxyacid synthase, or AHAS)which is involved in the first step in the synthesis of the branched-chain amino acids (valine, leucine, and isoleucine). These herbicides slowly starve affected plants of these amino acids which eventually lead to inhibition of DNA synthesis. They affect grasses and dicots in the same manner. The ALS inhibitor family includes sulfonylureas (SUs), imidazolinones (IMIs), triazolopyrimidines (TPs), pyrimidinyl oxybenzoates (POBs), and sulfonylamino carbonyl triazolinones (SCTs).
- EPSPS inhibitors: These herbicides kill the plant through inhibiting the enzyme enolpyruvylshikimate 3-phosphate synthase (EPSPS) which is involved in the synthesis of the amino acids tryptophan, phenylalanine and tyrosine. They affect grasses and dicots in the same manner. Glyphosate is a systemic EPSPS inhibitor but inactivated by soil contact.
- Synthetic auxin: These organic chemicals inaugurated the era of herbicides. Synthetic auxins mimic this plant hormone. They have several points of action on the cell membrane, and are effective in the control of dicot plants. 2,4-D, Clopyralid, Dicamba , Picloram, etc. are synthetic auxin herbicides.
- Photosystem II inhibitors: These herbicides kill the plant through reducing electron flow from water to NADPH2+ at the photochemical step in photosynthesis. They bind to the Qb site on the D2 protein, and prevent quinone from binding to this site. Therefore, this group of compounds causes electrons to accumulate on chlorophyll molecules. As a consequence, oxidative reactions occur in excess of those normally tolerated by the cell , and the plant dies. The triazine herbicides (including atrazine) are PSII inhibitors.

Organic herbicides

Organic herbicides are expensive and may not be affordable for general commercial production. They are also much less effective than synthetic herbicides but of course do not inject unnatural chemicals into the environment and can be used in a farming enterprise that has been classified as organic and organic herbicides include:

- Spice: Different spices are now effectively used in patented herbicides.
- Vinegar: The concentration of 5-20% solutions of acetic acid is effective which mainly destroys surface growth hence, respraying to treat the regrowth is needed.
- Steam: It is applied commercially in organic vegetable farms but now considered uneconomic and inadequate because it kills surface growth but not underground growth.
- Flame : It is considered more effective than steam.

Application of herbicides

Most herbicides are applied as water-based sprays using ground equipment. Ground equipment varies in design, but large areas can be sprayed using self-propelled sprayers equipped with a long boom, of 60 to 80 feet (20 to 25 m) with flat fan nozzles spaced about every 2.0 inch (500 mm). Towed, handheld, and even horse-drawn sprayers are also used. Inorganic herbicides are generally be applied in big farms of the European countries and The USA aerially using helicopters or airplanes, and can be applied through irrigation systems (chemigation).

Chemical weed management in different vegetable crops

Crop	Weed control measure	Time of application
Tomato	Metribuzin @ 0.25-0.75kg ai /ha butachlor or alachlor@ 2.0 kg ai/ha	Post-emergence; Pre-emergence, one week after transplanting
Brinjal	Pendimethalin 1.0kg ai/ha or fluchloralin 1.5kg ai/ha or butachlor 2.0kg ai/ha + one hand weeding 30 days after transplanting	Pre-emergence
Chilli	Pendimethalin 1.25 kg ai/ha + one hoeing 40days after transplanting	Pre-emergence
	Nitrofen @ 2.5 kg ai/ha or oxadiazon @ 1.0 kg ai/ha	Pre-emergence
Cabbage and cauliflower	Pendimethalin @ 0.56kg ai/ha or oxyfluorfen @ 0.25kg ai/ha or fluchoralin@ 1.2 kg ai/ha or nitrofen @ 2.0 kg ai/ha + one hand weeding 30 days after planting.	One day before transplanting seedlings
Knol khol	Butachlor @ 1.5kg ai/ha or fluchloralin @ 1.0kg ai/ha or Oxyfluorfen@ 0.25kg a.i./ha	Pre-planting
Onion	Pendimethalin @ 0.75 kg ai /ha or fluchloralin @ 1.5 kg ai/ha or oxadiazon @ 1.0 kg ai/ha or nitrofen @ 1.0 kg ai/ha + one hand weeding 45 days after transplanting	Pre-planting
Garlic	Pendimethalin @ 0.75 kg ai/ha or oxadiazon @ 1.0 kg ai/ha or fluchloralin @ 0.9 kg ai /ha + one hoeing 45 days after planting cloves	Pre-planting

Carrot and radish	Tenoran @ 2.5 kg ai/ha or alachlor @ 1.5 kg ai/ha Dalapon @ 1.0 kg ai/ha or fluchloralin @ 1.5 kg ai/ha or nitrofen 2.5 kg ai/ha	Pre-emergence, pre-planting One day after transplanting
Beet and turnip	Butachlor @ 1.5 kg ai/ha or fluchloralin @ 1.5 kg ai/ha	Pre-emergence
Pea	Alachlor @ 0.75 kg ai/ha or tribunil @ 1.5 kg ai/ha + one hand weeding 45 days after sowing	Pre-emergence
French bean	Alachlor @ 1.0 kg ai /ha	Pre-emergence
Cowpea	Alachlor @ 2.0 kg ai/ha or nitrofen @ 2.0 kg ai/ha	Pre-emergence
Okra	Pendimethalin @ 0.56-0.75 kg ai/ha or alachlor @ 2.0 kg ai/ha or fluchloralin @ 1.5 kg ai/ha + one hand weeding 45 days after sowing	Pre-emergence
Muskmelon	Alachlor @ 2.0 kg ai/ha or fluchloralin @ 1.2 kg ai/ha	Pre-sowing
Watermelon	Butachlor @ 2.0 kg ai/ha	Pre-sowing
Bottle gourd	Alachlor @ 2.5 kg ai/ha or fluchloralin @ 2.0 kg ai/ha or butachlor @ 2.5 kg ai/ha	Pre-sowing
Bitter gourd	Alachlor @ 2.5 kg ai/ha or fluchloralin @ 0.75 kg ai/ha or butachlor @ 2.5 kg ai/ha	Pre-sowing
Pumpkin and summer squash	Butachlor @ 2.0 kg ai/ha	Pre-sowing
Ridge gourd	Alachlor or butachlor @ 2.0 kg ai/ha	Pre-sowing
Fenugreek	Pendimethalin @ 0.75 kg ai/ha + hoeing	Pre-emergence

❑❑

3
Production Technology of Vegetable Crops

1. Potato

Scientific name: Solanum tuberosum L.
Family: Solanaceae
Origin: South America
Chromosome No.: 2n = 4x = 48

Introduction

Potato is one of the four major food crops of the world and is also a very popular vegetable crop. It is extensively cultivated in China, Russian Federation, Ukarine, Poland, Ireland, Great Britain, Germany, Netherlands, France, Spain, South America, India and USA. Potato was brought from Peru to Spain by the Spaniards in 1565 when it was widely grown in Spain and Italy. They were probably brought to England about 1586 by Sir Francis Drake. Potato was introduced in India in early 17th century probably by the Portuguese. Now it is one of the principal cash crops of India.

Importance and use

- Potato is the fourth major food crops of the world after rice, wheat and maize.
- Being a major source of carbohydrate it is often used as a substitute for cereals.
- It is one of the major vegetable crops of the world and is grown in almost every country.
- In India, it is considered as a staple vegetable and used either alone or mixed with other vegetables.
- Potato is used as vegetable in various forms such as, boiled, mashed, baked, fried and cooked in sambar and different curries of vegetable, fish and meat.
- Potato is processed into different commercially used dehydrated, canned and fried products like, shredded potato, sliced potato, chips, flakes, French fries, finger chips, granules, disc, cubes, flour, etc.

- It is used for several industrial purposes such as production of starch, alcohol, dextrin, glucose, etc.
- Potato starch is a large grained starch and contains 25 % amylase and 73 % amylopectin and high phosphate content.

Nutritional value

- Potato can be regarded as a wholesome food.
- Carbohydrates (20.6%) are the chief dietary constituent of potato.
- It also contains essential nutrients like, protein, minerals and vitamins.
- Major minerals present in potato are calcium, phosphorus and iron.
- It is also quite a rich source of vitamins especially vitamin B_1, B_2, B_6 and C.
- Potato also contains good amount of essential amino acids like, leucine, tryptophane and isoleucine.

Medicinal value

- Due to rich carbohydrate contents of potato, it is easy to digest and tasteful as well. This is the reason boiled potato is prescribed for babies, patients or for weak persons who require supply of natural energy from their light diets or for those who cannot take very heavy diet as meal.
- Potato contains natural fiber so it is helpful for appetite control and easy bowel movement.
- Potato is good for patients suffering from high blood pressure owing to presence of potassium and small amount of sodium.
- Potato consumption helps in supply of vitamin C and thus it works effectively against bleeding and spongy gums, frequent attack of cold, etc.
- Potato is an effective and instant antidote for inflammation. If the extract of potato is smeared on an external swelling it takes instant care for the pain, similarly consumption of potato helps in natural curing of digestive system and of intestinal problems. Consumption of potato for its neutral taste is a safe and comfortable food for the patients suffering from mouth ulcer.
- Potato pulps contribute miracle result in skin care routines. The raw pulps of crushed potato with honey make an excellent face and skin mask helping for natural removal of skin spots and pimple infections. Regular application of potato pulp makes the hard area of elbow joint and knees in soft and supple. Extract of potato is a good natural healer for minor burn injury.
- Potato contains high content of magnesium which creates natural prevention against depositions of calcium in kidney. Hence, consumption of potato helps for prevention from formation of kidney stone.

Adaptation in India

The most possible introduction of potato to India was with Portugese in Bengal. Firstly, it was introduced at the then Saptagram, Hooghly in West Bengal in 1535 A.D and later it was introduced to Burdwan, part of Hooghly, and Midnapur as commercial crop in 1570 A.D. Second evidence proposed that potato was introduced in India from Scotland by East India Company and distributed to Chakla Burdwan, Midnapore and Chitagaon (now in Banladesh). It was recorded to have grown in Surat and Karnataka around 1675 and later in the Nilgiri hills in 1822. It was also started to be cultivated in hills by the English during 1787. By the end of 19th century, the potato had spread throughout North India. Now potato is a ubiquitious crop in India.

Potato cultivation in India started with long duration European introductions of *Solanum tuberosum* background like, Magnum Bonum, Royal Kidney, Great Scot, etc. along with some variety like, Darjeeling Red Round of *Solanum andigena* background. These long duration varieties suitable for European conditions did not perform well under comparatively short growing seasons of the tropical and sub-tropical plains of India where high temperature prevails both at the time of planting and harvesting. Systematic breeding of potato suitable for Indian condition started with the establishment of Central Potato Research Institute at Shimla. All the present day potato varieties have been bred through hybridization involving different exotic and indigenously bred varieties along with different tuber bearing *Solanum* species. However, despite the adoption of hybridization between many tuber bearing *Solanum* species in breeding potato the present varieties are only being put under *Solanum tuberosum.*

Breeding behaviour

- Except for genetical and breeding purposes, the crop is generally propagated vegetatively by means of tubers.
- Vegetative propagation has distinct advantage to the breeders since superior genetic strains once identified can be increased and maintained by clonal propagation.
- Potato flowers sparingly, especially in certain favourable climates: cool climate, moderate humidity and long day conditions of more than 15 hours light period.
- Potato is basically self- pollinated crop, but some degree of cross-pollination occurs due to protruding stigma and self-incompatibility.
- Major pollinators are bumble bees (*Bombus Spp*) and honey bees (*Apis* species).
- Main obstacles to seed formation are high degree of male-sterility with no viable pollens, few sterile pollen and self-incompatibility, low pollen production and failure to produce viable pollens, dropping of buds and flowers either before or after fertilization.
- Anthesis generally starts from 5.00 am which continues up to 8.00 a.m.
- Anther dehiscence occurs through the development of pores at the tip of the anthers just after anthesis.

- The stigma receptivity is at its maximum during anthesis although it becomes effectively receptive a day prior to anthesis.
- The flowers emasculated one to two days earlier should have completely blossomed under the paper bag and are ready for cross-pollination in the morning hours and does not necessarily coincide with the anthesis.
- Pollens remains viable from the day of anthesis to the next morning.
- Dried pollens can be stored in sealed vial at 0°C for 1-2 months.
- It is possible to propagate potato through true seeds collected from ripe berries.
- True potato seeds (TPS) can be stored safely at ordinary room temperature and under low humidity for 2-3 years.

Origin and taxonomy

The potato is a native of tropical South America region where it grows wild in nature and presents the widest diversity of forms like, tuber shape, size, colour, taste, etc. The areas of origin and domestication of the cultivated diploid potatoes was in the high plateau of Peru-Bolivia in the region of the lake Titicaca. Cultivated tetraploids are also found in this region. Wild potato tubers are all bitter in taste and contain potentially toxic steroidal alkaloids. So, the first step in the evolution of the crop is the emergence of alkaloid free diploids which might have happened in the period 2000 to 5000 B.C. Among the cultivated potatoes, the two most important ones are the autotetraploides, *Solanum tuberosum* subsp. *tuberosum* and *Solanum tuberosum* subsp. *andigena* and two diploids, *Solanum phureja* and *Solanum stenotomum*.

There have been two views regarding the origin of cultivated potato i) it might have arisen from the diploid ancestral form *Solanum stenotomum* by simple chromosome doubling. *Solanum stenotomum* was also the first domesticated tuber bearing species originated from the wild *Solanum leptophyes*, ii) it might have arisen as a spontaneous amphidiploid hybrid from the ancestor, *Solanum stenotomum* and a diploid weed *Solanum sparsipilum*. In another school of thought, *Solanum vernei*, a wild diploid species occurring in the north-western Argentina, gave rise to the *andigena* forms of tuber bearing *Solanum* following autopolyploidy and further human selection led to the evolution of the species of commerce, *tuberosum*.

Potato belongs to the genus *Solanum*, series Tuberosa of the family Solanaceae. More than 200 tuber bearing *Solanum* species are grouped under section Petota, sub-section Potatoe and 19 series.Of the two most important cultivated autotetraploid species, *Solanum tuberosum* subsp. *tuberosum* is wide spread in Europe, USA, Canada and Asia while the other one, *Solanum tuberosum* subsp. *andigena* is grown only in the Andes of South America.

Solanum tuberosum subsp. andigena

Thin and long stem, small and narrow leaflets. Flowers are produced more profusely. They have long stolons and mostly coloured deep-eyed tubers. Tuberization takes place only under short-day conditions at high altitudes (above 2000m).

Solanum tuberosum subsp tuberosum

Shorter and thicker stem, larger and wider leaflets. Tubers are generally white and oval. Tuberization takes place both under long days in temperate climate as well as under short day conditions in the tropics.

The important diploid cultivated species used as source of resistance to some forms of virus and diseases are *S. demissum*, *S.stenotonum* and *S. phureja.* There are several wild species of the series *Tuberosa* such as, *S. sparsipilum, S. leptophyes, S. berthaultii, S. spegazzinii, S. vernei, etc.* However, all potato cultivars cultivated for commercial purposes outside South America belong to the species *Solanum tuberosum.* The ploidy levels of potato species varies from 2x to 6x with 73% of these being diploid, 4% triploids, 15 % tetraploid, 2 % pentaploid, and 6% hexaploid.

Improved varieties

- *Early maturing group:* (Requires 70-80 days after planting to give economic yield): Kufri Chandramukhi, Kufri Ashoka, Kufri Lauvkar, Kufri Kuber.
- *Medium maturing group:* (Requires 90-100 days after planting to give economic yield) : Kufri Jyoti, Kufri Bahar, Kufri Lalima, Kufri Sutlez, Kufri Pukhraj, Kufri Chipsona 1, Kufri Chipsona 2, Kufri Anand, Kufri Himalini, Kufri Sheetman, Kufri Moti, Kufri Kundan, Kufri Red, Kufri Alankar, Kufri Sherpa, Kufri Swarna, Kurphi Megha, Kufri Thenmalai, 92-PT-27, HPS-1/13.
- *Late maturing group:* (Requires 110-130 days after planting to give economic yield): Kufri Sindhuri, Kufri Badshah, Kufri Kanchan, Kufri Giriraj, Kufri Swarna, Kufri Jeevan, Kufri Dewa, Kufri Khasigaro, Kufri Naveen, Kufri Chamatkar, Kufri Kisan, Kufri Kumar, Kufri Safed, Kufri Neela, Kufri Neelamani.
- *Varieties moderately resistant to late blight disease:* Kufri Alankar, Kufri Jyoti, Kufri Swarna, Kufri Megha, Kufri Naveen, Kufri Neelamani, Kufri Muthu, Kufri Jawahar, Kufri Sutlej, Kufri Pukhraj, Kufri Giriraj.
- *Varieties moderately resistant* / resistant to early blight disease: Kufri Chamatkar, Kufri Sinduri, Kufri Jyoti, Kufri Badshah, Kufri Lalima, Kufri Jeevan, Kufri Khasigaro, Kufri Sherpa.
- *Varieties resistant to wart disease:* Kufri Jyoti, Kufri Sherpa, Kufri Naveen.
- *Varieties resistant to PVX virus:* Kufri Badshah, Kufri Sherpa.
- *Varieties moderately resistant to PVY virus:* Kufri Sherpa.
- *Varieties resistant to cyst nematode:* Kufri Swarna.
- *Varieties suitable for peocessing:* Kufri Chipsona-1, Kufri Chipsona-2, Kufri Chipsona-3, Atlantic, Diamant, Fritolay Hybrid 1333.
- *Red varieties:* Kufri Sindhuri, Kufri Lalima, Pimpernel, Kufri Kanchan.

Cultural requirements

- Potato is a cool season crop however, at low temperature vegetative growth is restricted and at temperatures nearing the freezing point there is permanent and often irrecoverable frost injury to the plant. For these reasons, potato is grown as a summer crop in the hills and as a winter crop in the tropical and subtropical regions.
- Potato is susceptible to frost.
- At higher temperature, the respiration rate increases and the carbohydrates produced by photosynthesis are consumed rather than stored in the tuber.
- In the vegetative growth period, the temperature may be higher but during tuberization phase the temperature should be lower.
- Long day coupled with high temperature conditions promote haulm growth without formation of tuber and short day with low temperature condition induce tuberization.
- The optimum temperature for good plant growth is 15°C to 20°C.
- Potato should be grown when maximum day temperatures are below 34°C and night temperatures are not above 20°C.
- Most potato genotypes do not tuberize when the prevailing night temperature is above 21°C.
- Well drained light soils like, loam and sandy-loam rich in organic matter are ideal for growing potato.
- Potato can be grown in all types of soil, alluvial, black, red and laterite having pH ranging from 6.5 to 7.5 and EC: < 0.5%.
- Gangetic plains with almost neutral soil are most suitable.
- Alkaline or saline soils are not suitable for potato cultivation.

Land preparation

- One deep ploughing followed by 2-3 cross and shallow ploughings are necessary for making the soil well pulverized, loose, friable, well levelled and good tilth.
- Add well rotten organic manure preferably farm yard manure or compost @ 30-35 t/ha during land preparation and level the land properly to provide good drainage..
- For amelioration of acid soil, apply 1.5 tonnes of dolomite or 1.0 tonne slacked lime/ha (depending upon the soil test).
- Apply phorate 10 G @ 15 kg/ha to control soil borne insects like cut worm, mole cricket, weevil etc.

- Pre planting irrigation is necessary if enough moisture is not available at the time of planting.

Planting time

1. Planting in the plains

   a. *Early crop*
      - Punjab and western Uttar Pradesh: 10-20 September
      - Central Uttar Pradesh : First week of October
      - Bihar: Second and third week of October
      - West Bengal: 2nd fortnight of October

   b. *Main crop*
      - Gangetic plains: November (north-western plains and West Bengal).
      - Plateau region of peninsular India: October – December

   c. *Late crop*
      - Gangetic plains: Last week November to middle of December.

   d. *Spring crop*
      - North-western plains: December- January

2. Planting in hills

   e. *North-eastern hills*
      - Valleys and high altitudes (3000-3500m) : May- June.
      - High hills (2500- 3000 m): April.
      - Mid hills (1000 – 1800 m): January – February (for spring crop) and August – September (for autumn crop)

   f. *North-eastern hills*
      - High hills: March.
      - Lower hills and valleys: January – February (for spring crop) and August – September (for autumn crop).

   g. *Nilgiri hills: April, August and January*

Planting of tuber

- Potato is traditionally propagated from tubers which have a dormancy of nearly 8-10 weeks.
- Multiple sprouting stage of the tuber is the best for planting because yield depends on the number of main stems.
- Collect Certified seed from Govt. approved organizations.

- Taking out of the whole seeds tubers from cold storage 10-15 days before planting and keep in shady place under diffused light condition in thin layer on the floor which helps in sprouting the tubers.
- Use at least 30 g sized (3-5 cm diameter), disease free healthy, whole or cut tubers having 2-3 sprouts.
- Cutting instruments must be cleaned and disinfected with 1% potassium permanganate solution each time after cutting any infected tubers.
- Potato is mainly infested with seed borne diseases. It is essential to treat the seed before planting. Treat the seed tuber with mancozeb @ 2.5 g / litre of water for 10 minutes or *Trichoderma viridiae* @ 4-5 g/ litre of water for 10 minutes or *Pseudomonas fluorescence* @ 2.0 g/ litre of water for 10 minutes.
- For treating one quintal of seed tubers, 50 litres solution is required and same solution can be used for 3-4 times after which fresh solution is to be prepared.
- Treated tubers should be dried under shade before planting.
- Average seed rate is about 25-30 quintals per hectare.
- Seed tubers may be planted either on ridges or on flat beds.
- Planting should be done at a depth of 3-5 cm keeping the sprouts upward.
- Planting in rows in flat beds is suitable in light soils. Tubers are planted and covered with soil to make ridge.
- During planting care should be taken so that the seed tubers do not come in direct contact of the fertilizers that have been applied as basal.

Planting of seedlings from TPS

- True potato seed (TPS) technology saves considerably the large quantity of seed tubers (25-30 quintals/ha) and their storage, space and transport.
- 100-150 g TPS is sufficient to produce seedlings for one hectare area.
- TPS can be stored safely at ordinary room temperature and under low humidity for 2-3 years.

Spacing

- 60 × 20 cm : Medium sized tubers (40 -50 cm); Plant population: 83,000/ha
- 60 × 25 cm: Larger sized tubers (50-55 cm); Plant population: 67,000/ha
- 60 × 15 cm: Small sized tubers (35-40 cm); Plant population: 1, 11, 000/ha

Nutrient management

- Potato responds well to nutrients.
- Nitrogen is the primary limiting nutrient in potato production.
- Potassium increases the size of tubers.

- Apply a general fertilizer dose of 180 kg N, 120 kg P_2O_5 and 120 kg K_2O per hectare. However, nutrient requirement depends on the type and nutrient status of the soil.
- Calcium ammonium nitrate (CAN) and ammonium sulphate are good sources of nitrogen in potato than urea. Similarly, di-ammonium phosphate (DAP) and super phosphate is effective source of phosphorous and the better source of potassium is the potassium sulphate than the muriate of Potash for potato.
- Application of half the dose of nitrogen and potash and full dose of phosphorus in furrows at the time of planting gives the best results.
- Rest half dose of nitrogen and potash should be top-dressed in two split doses at 25-30 days after planting and 40 days after planting followed by earthing-up.
- In case of micro-nutrient deficiency, apply zinc sulphate @ 25 kg/ha, sulphur @ 8.0 kg/ha and borax @ 10.0 kg/ha at the time of final land preparation.

Irrigation

- Adequate irrigation is essential for proper growth and yield of potato.
- Stolon formation, tuber initiation and tuber development stages are the critical stages for irrigation in potato.
- Give one light irrigation, if necessary, before planting to ensure uniform germination.
- Depending on the soil moisture, apply splash irrigation at 3-4 days interval up to 2-3 days before first earthing up.
- Irrigate the crop at an interval of 7-10 days after first earthing up.
- Level of irrigation water must not exceed $3/4^{th}$ part of the ridges.
- Total water requirement for the crop varies between 350 and 550 mm.
- Stop irrigation 10-15 days before harvesting the crop to allow the tuber skin to become firm.

Interculture and weed control

- Earthing up is necessary in potato because proper development of tubers depends upon aeration, moisture availability and proper soil temperature.
- Earthing up also prevents greening of tubers which occurs due to exposure of the tubers to sunlight.
- Mulching in the ridges with paddy straw, wheat husk, maize or jowar stalks is effective in inducing quick germination and conserving soil moisture.
- First Earthing up 3-4 weeks after planting
- Second earthing up10 days after 1st earthing up.

- Apply 2-3 hand weeding to make the plot weed free and for well pulverization of the soil.
- For chemical control of weeds, apply pendimethalin @ 1.0 kg a.i./ha 1-2 days after planting of tubers.
- Post-planting application of simazine @0.5 to 1.0 kg/ha over the ridges just after planting of potatoes is also effective.

Harvesting the crop

- Harvest potato before the temperature rise above 30°C to avoid rottage of tubers due to high temperature.
- Harvest the crop when haulms start yellowing and falling on the ground.
- It is better to cut the haulms 10-15 days before harvesting when irrigation has been stopped. Apply mancozeb @ 2.5 g/ litre of water on cut portion of haulms.
- Harvest the crop manually with the help of khurpi or spade or mechanically with the help of 1-4 row potato diggers.

Post harvest handling and storage

- Cure the tubers under shade for 10-12 days for hardening of the tubers. The height of the stack should be 1-1.5 m and width 3-5 m.
- Discard cut, rotten, insect damaged and rubbed tubers after harvesting.
- As per market demand, grade the tubers into small (<25 g), medium (25-50 g) large (50-75g) and extra large (> 75 g) on the basis of their size.
- Pack graded potatoes in gunny bags.
- Store the table varieties in cold storage at temperature of 2°C - 4°C and in 80-85% relative humidity.
- The low temperature checks sprouting and rot of tuber and high relative humidity reduces weight loss from the tubers.
- Store the processing varieties in cold storage at temperature of 10-12°C and 80-85% relative humidity.
- Potato stored at 10-12°C generally sprout and it is necessary to check sprout growth to reduce storage losses.
- To check sprout gowth, a sprout suppressant should be used. Most commonly used chemical is CIPC (chrolo isopropyl carbamate).
- For treating small quantity of potatoes, powder form of CIPC (25 mg dust/ tonne of tuber) is convenient.
- For a large scale treatment, liquid formulation of CIPC (Chloropropham 50% a.i marketed in the name of Oorja) of 30-35 ml is required for fogging one tonne of tuber. However, the dose can vary depending upon the loss of fog during treatment due to leakage, pore size of bags used, the quantity of potatoes stored, etc.

- After CIPC treatment, store should be kept air tight for 24-48 hours and maintain 10-12°C temperature. The treated potatoes are safe for consumption, 3-4 weeks after the treatment.

Plant protection measures

Physiological disorders

Uneven sprouting in the field: Adequate crop stand can not be maintained for this disorder which happens due to i) planting soon after their removal from cold storage, ii) soil moisture deficiency and iii)untreated cut tuber pieces if infected by the fungus.

Control measures

- Held tubers in storage till end of the dormancy period.
- Treat the cut tuber pieces with 0.25% Diathane M-45 for 10 minutes.
- Place tubers some what deep, particularly in light soil with the cut surface facing downwards.

Dormancy

Dormant or even semi-dormant tubers do not sprout readily after planting causing delayed and erratic crop stand. This physiological hindrance becomes a problem when two potatoes are taken in rotation and tubers produced in the hills are used as seed immediately soon after harvest.

Control measures

- In advanced stage of dormancy, cutting of tubers may terminate dormancy.
- Cut tuber pieces are immersed immediately in 1% thiourea solution for one hour and then plant them as soon as possible.
- Triple treatment: Spraying the whole tubers with 3% ethylene chlorohydrin @ 20 ml/kg of tuber in an air tight chamber for 4 hours; thereafter dip the cut pieces of tuber in 1% thiourea solution for one hour and finally dip the tubers in 1 ppm gibberellic acid solution for 10 minutes.Dry the treated tuber pieces in shade before planting.

Diseases

Late blight (*Phytophthora infestans*)

Late blight is the worst potato disease. It was first reported in the 1830s in Europe and in the US. It is famous for being the cause of the 1840s Irish Potato Famine, when a million people starved and a million and a half people emigrated. Late blight continued to be a devastating problem until the 1880s when the first fungicide was discovered. In recent years, it has reemerged as the most serious problem. This fungal disease is prevalent in almost all potato growing areas of the country. Seed tuber is the primary source of inoculums. Cool weather (10-20°C) and relative humidity of 80% and above induce the disease infestation. The dead areas appear at the tip or margin of the leaves and spread downward and inward.Decaying leaves emit offensive odour. In present day condition, stem infection is also very common, where dark brown lesions appear on the stem or on the branches and the whole plant die within few days.

The sign of tuber infection is brown to purple discolouration of the skin followed by a brown rot which extends to about half an inch below the surface of the tubers.

Control Measures

- Cultivation of tolenant varieties like Kufri Alankar, Kufri Jyoti, Kufri Swarna, Kufri Megha, Kufri Naveen, Kufri Neelamani, Kufri Muthu, Kufri Jawahar, Kufri Sutlej, Kufri Pukhraj, Kufri Giriraj.
- Use of disease-free, healthy and certified seed tubers.
- Restrict water supply particularly when weather condition is favourable for disease infection.
- Keep cull/compost piles away from potato growing areas.
- Destroy any volunteer potato plants.
- Keep tubers covered with soil throughout the season to prevent tuber infection.
- Remove infected tubers before storing to prevent the spread of disease in storage.
- Kill vines completely (dehaulming) before harvest to avoid inoculation of the tubers during harvest.
- If the late blight prevailing weather condition prevails, started spray with protectant fungicides Copper oxychloride (0.3%) / mancozeb (0.25%) / chlorothalonil (0.2%) / metiram (0.3%) at 10 days interval.
- If late blight present in field in pockets, started spraying with metalaxyl-mancozeb(0.25%) / cymoxanil-mancozeb(0.25%) / fenamidone –mancozeb (0.25%) /dimethomorph (1g) + mancozeb(2g)/ litre of water immediately, and alternate spray schedule with protectant and systemic fungicides at 7 – 10 days interval depending upon weather condition.
- Careful irrigation or avoid top dressing and irrigation under late blight prevailing condition.

Early blight (Alternaria solani)

This disease, also known as target spot, rarely affects young, vigorously growing plants. It is found on older leaves first. It is a common soil borne, fungal disease. Warm and moist weather favours the occurrence of the disease. Small, isolated, scattered, pale brown spots first appear on the leaflets. Later the spots grow in size and in the necrotic tissues, concentric rings develop.

Control measures

- Crop rotation and field sanitation
- Grow tolerant varieties like Kufri chamatkar, Kufri Sinduri, Kufri Jyoti, Kufri Badshah, Kufri Lalima, Kufri Jeevan, Kufri Khasigaro, Kufri Sherpa.
- Spray the crop with 0.3% mancozeb or 0.1% Bavistin or chlorothalonil 0.2% at the occurrence of intermittent rains in dry weather 3-4 times at 10 days interval.
- Two sprays of 0.2% chlorothalonil or difenconazole (0.5g /l) or metiram (0.3%) is very effective

Charcoal Rot (*Macrophomina phaseolina*)

It is fairly wide spread soil and tuber-borne fungal disease causes mostly tuber rot in storage. Dry and warm weather (28 -35°C) favours this disease infection. The disease may appear as stem blight, wet charcoal coloured tuber rot or as dry tuber rot. Small black spots appear at the site of lenticels and eyes of the infected tubers and foul smell emits from the rotten mass.

Control measurers

- Use of disease free tubers.
- Grow early variety so that the crop may be harvested earlier before the temperature reaches 28°C.
- Frequent irrigation to lower the soil temperature.

Black scurf (*Rhizoctonia solani*)

Characteristic symptoms of this disease are brownish-black sunken lesions on underground stems and stolons. The disease may cause non-uniform stands of weak, spindly-looking plants. Early-season infections often result in the pruning of young stolons where lesions girdle them completely. Dark stem lesions occurring below the soil line may girdle the main stem, resulting in yellowish or purplish leaves that curl upwards. On relatively healthy-looking plants, aerial tubers may form. During midseason, the fungus may develop a white powdery mold growth on the stems that extends just above the soil line. This often is associated with stem lesions below the ground. The fungus forms sclerotia (survival structures) on the tubers. The sclerotia vary from netted or scurfy residues to individual black masses on the tuber surface. Tubers may be misshapen, cracked, or may develop a russet like skin.

Control measurers

- Use of disease-free seed.
- Seed treatment with registered fungicides (pencyuron/validamycin) to reduce some infections, especially from an infected seed piece. Seed treatment with 3% boric acid is highly effective.
- Warm the seed prior to planting.

- Plant in warm soil (above 16°C).
- Any practice that promotes rapid emergence of sprouts will reduce the attack.
- Spraying the crop with validamycin (0.2%) is highly effective.
- Use proper crop rotation, preferably grasses or cereals

Soft rot (*Erwinia carotovora, E. carotovora* ssp. *atroseptica, E.carotovora* ssp. *carotovora)*

It is a common bacterial disease where infection in the field is often carried over to storage. In the field, striking brown-black or jet-black slimy areas on the stem just above the ground level appear which is called 'black leg'. In storage, the surface of the affected tuber becomes discoloured, sometimes wrinkled, depressed and the internal tissues turn soft and mushy resulting in disintegration of the whole tuber.

Control measures

- Dip the seed tubers in 0.01% streptocycline solution before planting.
- Drench soil with stable bleaching powder suspension @ 15 kg/ha before planting.
- Dip the tuber before storage in 0.3% Boric acid solution for 5 minutes.
- Store healthy tubers in dry cool (4°C temperature) and well-ventilated storage.
- Spraying with 0.01% Streptocycline + 0.3% Copper-oxychloride.

Common scab (*Streptomyces scabis*)

Two types of scab-like roughened lesions on tubers are caused by these bacteria viz., shallow scab and deep scab. Shallow scab consists of superficial rough areas, sometimes raised above and often slightly below the healthy skin. It consists of corky tissues which arise from abnormal proliferation of the cells of tuber periderm. Deep scab consists of lesions which are 1-3 mm deep and are darker than shallow scab and the tissue around turns corky.

Control measures

- Dip the seed tubers in 0.01% streptocycline solution for 30 minutes before planting.
- Keep the field in well-irrigated condition from the fifth week of planting up to the ninth week.
- Green manuring minimizes the incidence of the disease.

Leaf roll (*Potato leaf roll virus*)

This virus is transmitted by several aphids.Upward and inward rolling of leaflets and leaves become leathery and brittle. The infected plant produce short stolons and small tubers get clustered around the base of the plant.

Latent Mosaic (Potato virus X (PVX) and Potato virus S(PVS)

This sap transmitted virus cause mild mottling of leaves or light green discolouration

Severe infection leads to stunting of plants and cause degeneration in the performance of the varieties.

Control measures

- Use disease free certified seed stock.
- Uproot infected plants along with tubers showing the disease symptom.
- Avoids frequent movement in the field.
- Wash all tools and implements with 3% solution of trisodium phosphate before use them.
- Spraying the crop with systemic insecticides like, imidacloprid (1ml/3l) or fipronil (2ml/l) or monocrotophos (2ml/l) to control the aphids and leaf noppers.

Insect pests

Potato tuber moth (Phthorimaa operculella)

This serious pest both in storage and field was introduced in India through seed material imported from Italy. This pest completes 10-13 generations in a year under favourable environmental conditions. The moth lays eggs singly near the eyes of the exposed tubers and sometimes on the ventral surface of the leaves. The caterpillars after hatching bore into the tubers, petioles and terminal shoots. The caterpillars burrow into the tuber flesh, often tunneling it freely. Infected tubers become useless and ultimately rot by secondary infection.

Control measures

- Deep planting of healthy tubers and proper earthing up leaving no open cracks.
- Trap the adults by light trap and pheromone trap.
- Spray the crop with 0.2% carbaryl (Sevin WP 2 g/litre of water).
- Spray the crop with *Bacillus thuringiensis* (0.05%).
- Treating the gunny bags with 10% neem seed kernel extract.

Hadda beetle (Epilachna vigintioctopunctata, E. occellata)

Adults and grab of these polyphagous pests feed voraciously on the leaves by scraping in a characteristic manner leaving the veins intact.Characteristic skeletonised patches are developed on the leaves which later dry out.

Control measures

- Spray the crop with 2 g carbaryl or 1.5 ml endosulfan or 2.5 ml triazophos or 1.25 ml profenophos per litre of water.
- Collection and destruction of initial egg masses are useful to suppress the pest population.

Cut worms (*Agrotis ipsilon* and other *Agrotis* spp.)

The larvae after hatching feed gregariously on the foliage for few days and upon development, the caterpillars disperse and hide in the cracks and crevices of the soil. At night the caterpillars come out and cut the young potato plants near the ground level.

Control measures

- Clean cultivation, uproot stubbles and break big soil clots.
- Flood irrigation of infected plots expose the larvae for bird predation.
- Drench soil around the plants and the ridges with 2.5 ml chlorpyriphos or 2 g carbaryl per litre of water.

Mole cricket (*Grylotalpa africana*)

In some cases the mole cricket causes extensive damage by feeding on the developing tuber which reduces the market value of the crop.

Control measures

- Application of 2.5 ml chlorpyriphos / litre of water along with rice bran in the infected plots during evening hours reduce the pest population drastically.

Root-knot Nematode (*Meloidogyne incognita)*

This plant parasitic nematodes infest root and during feeding process, the cortical cells are stimulated into giant cells which cause root galls.The affected plants fail to draw nutrients and water from the soil which make them pale green and stunted.

Control measures

- Summer ploughing.
- Grow trap crop like marigold or wheat in rotation,
- Apply Furadon 3 G @ 20 kg/ha (1/2 during planting and rest part during earthing up).

Yield

Yield depends on the variety, climate and cultural practices.

- In plains, under well managed conditions: 25-35 t/ha.
- In the hills except valleys: 20 t/ha.

Solanaceous Fruit Vegetables

2. Tomato

Scientific name: Lycopersicon esculentum Mill.
(Solanum lycopersicum L.)
Family: Solanaceae
Origin: Mexico
Chromosome No.: 2n = 2x =24

Introduction

Tomatoes have been used as food by the inhabitants of Central and South America since pre-historic time. Long before the discovery of the Americas, the pre-Columbian and Aztecs developed larger fruited and new colours of tomato from the wild cherry tomato. Britishers believed to have introduced tomato in India in 1828 through the Royal Agri-Horticultural Society, Calcutta and afterwards it spread to other parts of the country.

Importance and use

- Tomato is one of the most versatile crops in the world because of its fast and wide climatic adaptation.
- Tomato is the world's most popular and widely cultivated vegetable crop and grown in all kinds of climate viz., temperate, subtropical and tropical.
- Tomato is basically warm season crop and has the potential to grow all the year round in mild climatic condition.
- It is universally treated as 'protective food' and provides almost all types of vitamins and minerals in quite fair amount.
- It is eaten directly as raw vegetable in sandwiches, *salad* etc. and used in a number of ways for culinary purposes.
- Its ripe fruits are utilized on a large scale in the preparation of a variety of processed products, such as puree, paste, powder, ketchup, sauce, soup, syrup, juice, drinks and canned whole peeled fruits.
- The mature green fruits are used for making pickles and preserves, and are also consumed after frying in oils.
- Fruit firmness, thick pericarp (0.5 cm and above), low locule number (2-3) and slow fruit ripening is preferred for good transportability and better shelf-life.
- High total soluble solids (^{o}Brix 5.5 and above), low pH (below 4.5), solid /acid ratio of 15 and sugar/ acid ratio of 8.5 and high viscosity of fruit pulp is ideal for processing purpose.

- Tomato fruits contain various types of flavoring compounds, which enrich the taste and color; thus, it is used in a number of ways to improve the flavor and characteristics of other foodstuffs.
- Red colour of tomato is due to the carotenoid pigment lycopene which is regarded as the powerful antioxidant.

Nutritional value

- It is rich source of minerals particularly potassium, vitamins and organic acids particularly citric and maleic acid.
- Ascorbic acid content ranges from 16-75 mg /100g of fruit weight.
- Total sugar content is 2.5 % in ripe fruit.
- Free sugars generally representing more than 60% of the total soluble solids are mainly glucose and fructose with traces of sucrose.
- The ripe tomatoes contain 20 amino acids, and the important ones are glutamic acid, aspartic acid, tryptophane and tyrosine.
- The orange-yellow color of tomato fruit is due to β-carotene and attractive red color due to lycopene, which are the potent non-nutrient bioactive substances and act as an anti-oxidant.
- It contains a steroidal glycoalkaloid, tomatine to the range of 130-150 mg per 100g fresh tomatoes.
- The strong odor of the leaves, stem and young fruits is attributable to this alkaloid and for a long time tomato was not consumed by human due to persistent superstitions of its poisonous nature.

Medicinal value

- Several epidemiological studies have shown that consumption of tomato protects against cancers of digestive tract, stomach, colon, rectum, gastrointestinal neoplasm and prostate, and this beneficial effect is due to presence of antioxidant compounds such as lycopene, Bcarotene and vitamin C.
- Among all the antioxidants lycopene- a major carotenoid without pro-vitamin A activity is considered responsible for their beneficial effects, and has been reported the most efficient quencher of singlet oxygen.
- Tomato is said to be useful in curing cankers and sores in the mouth.
- The secondary metabolites, i.e., colorless glycoalkaloids a-tomatine and dehydrotomatine significantly reduce in plasma cholesterol and feeding high tomatine green tomatoes induce a greater reduction in plasma LDL (bad) cholesterol and triglyceride levels than feeding low tomatine red tomatoes. Pickled green tomatoes are widely consumed in many countries have high tomatine content.

- The tomato pulp and juice are mild aperients (laxative), a promoter of gastric secretion and blood purifier and is considered to be the intestinal antiseptic.
- It is a very good appetizer and remedy for the patients suffering from constipation if used as soup.
- It stimulates torpid (sluggish, dull, inactivated) liver and is good in chronic dyspepsia.
- Tomatoes poultice is used in the treatment of edema during pregnancy.
- The people suffering from hyperacidity and from stones in gall bladder should avoid the excessive use of tomato in daily diet.

Adaptation in India

Britishers belived to have introduced tomato in India in 1828 through Royal Agri-Horticultural Socity, Calcutta. Tomato was introduced in India hardly 175 years back when tomato breeding in the world was mainly based on selection of chance variants resulting from mutation, natural out crossing and pre-existing recombination. Tomato spread rapidly through out India because of its wide adaptability in tropical environmental condition and during the course its spread, some selections from the introduced materials have also been made in different parts of the country. Few such old selections like, Annaji of Tamil Nadu, Meeruti of Uttar Pradesh, Patharkuchi of West Bengal are still being cultivated commercially.

Breeding behaviour

- Tomato is a mainly self-pollinated crop, but low percentage of cross-pollination also occurs.
- Anthesis starts from 6.00 am, continues up to 10.30 am and anther dehiscence from 7.00 to 10.00 am.
- Dehiscence of anther is longitudinal; it generally coincides with anthesis although dehiscence of some anthers in the same flower may occur even 1-2 days after anthesis.
- Generally flowers open after the self pollination has occurred (chasmogamy)
- A certain degree of cross-pollination is accomplished mainly by honey bees when the stigma protrudes beyond the anther cone.
- The optimum temperature for pollination is around 21°C.
- The stigma receptivity is maximum at anthesis although it become receptive 15-20 hours prior to anthesis, and the receptivity continues even upto five days after anthesis at an optimum temperature of 18 –25°C.
- Emasculated flowers are generally cross-pollinated in the morning hours next day or two days later depending on the bud maturity.

- Pollen remains viable for one day at moderate room temperature however, dried pollen can be stored in a sealed vial and kept at 0°C for a month.
- A gametophytic system of self-incompatibility is found in wild species, *L.peruvianum.*

Origin and taxonomy

The species belonging to *Lycopersicon* have been evolved *via* gene substitution as recorded in the biosystematic studies of *Lycopersicon* and closely related species of *Solanum*. The wild cherry tomato, *Lycopersicon esculentum* var *cerasiforme*, the ancestor of the modern cultivated tomato has originated in Peru-Equador-Bolivia region of the Andes. The Veracruz-Puebla region of Mexico is the centre of domestication of the cultivated tomato.

From Mexico, tomato was taken to Italy, Spain, Portugal and other European countries, Africa and Middle East by the explorers in the 16th century. It was taken to Philippines after its discovery by Magellan in 1521, and from there it was taken to other Asian countries. During the early years of its introduction tomato was considered to be poisonous in Britain, France, Spain and other European countries. Tomato moved to the USA from northern Europe around 1781 where it was first grown by Thomas Jefferson in Virginia.

The family Solanaceae having about 90 genera under which the genus *Lycopersicon* belongs is divided into two sub-families: Solanoideae and Cestroideae. The genus *Lycopersicon* belongs to the largest tribe Solaneae under the sub-family Solanoideae. The species of *Lycopersicon* is grouped into two sub-generic classifications i) *esculentum* complex: Refer to those species which can be crossed with the cultivated tomato eg., *L. esculentum*, *L. pimpinellifolium*, *L. cheesmanii* (all coloured fruited), *L. parviflorum*, *L. chemlewskii*, *L. hirsutum* (all green fruited) and ii) *peruvianum* complex: Refer to those species which are not crossable with the cultivated tomato eg., *L. chilense* and *L. peruvianum* (both green fruited).

Botanical classification of cultivated tomato according to fruit shape and plant spread:

L. esculentum var. *commune* (common round fruited tomato)

L. esculentum var. *grandifolium* (large potato leaved tomato)

L. esculentum var. *cerasiforme* (small fruited cherry tomato)

L. esculentum var. *validum* (upright tomato)

L. esculentum var. *pyriforme* (pear-shaped tomato)

Open pollinated improved varieties

- Determinate type: Pusa Early Dwarf, HS 101, HS 102, Hisar Arun, Punjab Chhuhara, Sonali, Roma, Pusa sheetal, Pusa Gaurav, CO-1, Punhab Keshri, La Bonita, Pusa Red Plum, Hisar Lalima, Narendra Tomato-2, Hisar Lalit, Arka Vikas, Arka Saurabh, Pusa 120, Arka Ahuti, Arka Abha, Arka Meghali, KS-2,Sakthi, TLB-111, Pusa Rohini
- Indeterminate type: Pusa Ruby, Sioux, Marglobe, Best of All, Krishnagar S-20, Pant Bahar, PKM-1, Utkal Kumari, Pusa Uphar, Bhagyashree, Dhanashree, NDT 5, Pant T 3, Hisar Lalit
- For fresh market: Pusa Early Dwarf, Pusa Ruby, PantT3, Arka Vikas, Arka Saurubh, CO.3, BT 12, Punjab Keshri, Pant Bahar , HS 110.
- For long distance market: Pusa Gaurav, Roma, Punjab Chhuhara, Pusa Uphar, Arka Saurabh, Pant T3, NDT 120. Pantharkuchi.
- Rootknot nematode resistant: Pusa 120, Hisar Lalit, Punjab NR-7
- Bacterial wilt resistant : BT-1, BT-10, BT-12, Arka Alok, Arka Abha, Sakthi
- Leaf curl virus resistant : Hisar Anmol, TLB-111
- Spotted wilt and leaf curl resistant: Dhanashree
- Abiotic stress resistant : Pusa Sheetal (low temp regime, upto 8°C night temperature), Pusa Sadabahar (both low and high temp regime, above 29°C average daily mean temperature)
- Round fruit: HS 102, Hisar Arun, CO-1, Hisar Lalima, NDT 120, HS 110, Hisar Lalit, Pant T3, Arka Saurabh, Marglobe, Best of All, Pant T1, Utkal Kumari, Pusa Uphar, Bhagyashree, Pusa Rohini
- Flattish -round fruit: Pusa Sheetal, Sweet 72, Azad T-2, Hisar Anmol (H-24), Pusa Ruby, Pant Bahar, PKM-1, Patharkutchi
- Pear shaped fruit: Punjab Chhuhara,
- Elliptical fruit: Roma, Pusa Gaurav, Punjab Keshri, Arka Ashish, Pusa Sadabahar, Sonali.

Public sector hybrids: Pusa Hybrid-1, Pusa Hybrid-2, Pusa Hybrid-4, Pusa Hybrid-8, Pusa Divya, Pant hybrid-1, Pant Hybrid-2, Arka Shreshta, Arka Vishal, Arka Vardan, Arka Abhijit, Rajashree, TH-2312, TH-802, COTH-1, BCTH-4, BCTH-62.

Cultural requirements

- Tomato is a highly adaptive and warm season crop and can be grown in the plains as well as in the hill condition.
- The optimum daily mean temperature is 21°C to 24°C.

- Average daily mean temperature below 16°C and above 28°C is harmful for the crop.
- Lycopene (pigment responsible for red colour) development in the fruits is highest at 21° to 24° C daily mean temperatures and drops off rapidly above 27°C.
- Night temperature is critical for fruit set, the optimum range being 15°C to 20°C and fruits generally do not set when night temperature is either below 13°C or above 20°C.
- There are some varieties and hybrids which can set fruits at low temperature (around 8°C night temperature) and high temperature (around 29-30°C average daily mean temperature).
- Well drained fertile, organic matter rich soils are ideal, however, sandy loam soil is best for early crop and clay loam soil for higher yield.
- The ideal soil pH is between 6.0 and 7.0 although it can tolerate a little soil acidity upto pH 5.5 but not below it.

Nursery management

- Seed beds are prepared finely, well drained, 15-20 cm raised, 1.0m wide and of convenient length.
- Fine and fully decomposed farmyard manure or compost @ 3-4 kg/m^2 should be well mixed to the beds.
- Drench the beds with formaldehyde (4.0%) or captan @ 2-3g/litre and cover with polythene sheet for 7 to 10 days to avoid damping off disease.
- 400 to 500g seeds of open pollinated varieties are required per hectare.
- 150 to 200g seeds of hybrids are required per hectare.
- Treat the seeds with captan or thiram @ 2-3 g per kg of seeds before sowing.
- Seeds are sown at shallow depth at 5.0 cm apart and covered with finely sieved leaf mould.
- After sowing cover the beds with straw or long dry grass till seed germination and sprinkle water regularly.
- Seedlings may be raised under poly tunnels having sides open to protect the young seedlings from heavy rain or scorching sunlight.
- Water the seedlings daily or in alternate day with rose can.
- Hardening of the seedlings by withholding water at least 4 to 6 days before transplanting.
- Stocky, 28-35 days old, healthy seedlings of 12-15 cm height with 3-4 leaves are ideal for transplanting.

Sowing time

Tomato is grown in three distinct seasons viz., early autumn, winter-spring and spring-summer in the plains of India although sowing and transplanting season vary with the region.In the hills tomato is sown in February-March or March-April.

- May-July (for early autumn crop in Maharashtra, Gujarat, Karnataka, Tamil Nadu, Andhra Pradesh and Rajasthan)
- September- October (for winter-spring crop in Maharashtra, Gujarat, Karnataka, Tamil Nadu and Andhra Pradesh, West Bengal, eastern Uttar Pradesh, Bihar, Orissa, Assam)
- February-March (for spring-summer crop in Punjab, Haryana, Delhi, Western Uttar Pradesh, Madhya Pradesh).
- December-February (for spring-summer crop in Maharashtra, Gujarat, Karnataka, Tamil Nadu , Andhra Pradesh and Rajasthan)

Land preparation

- Prepare the land in well advance with repeated ploughings (at least 4-5 ploughing) to a fine tilth.
- Remove all stubbles, weeds, etc. from the land.
- Add well rotten organic manure preferably farm yard manure or compost @ 20-25 t/ha during land preparation and level the land properly.

Transplanting

- 28-35 days old, healthy seedlings of 12-15 cm height are ready for planting.
- After hardening the seedlings for 4 – 6 days, irrigate the seed bed before lifting to facilitate easy pulling of young seedlings.
- Plant the seedlings in the field preferably during afternoon hours which give better field establishment.
- Irrigate the field just after transplanting.

Spacing

- Indeterminate hybrids: 90 cm × 90 cm; 90cm × 75cm
- Determinate hybrids: 75 cm × 75 cm
- Indeterminate varieties: 90 cm × 60 cm; 75 cm × 75 cm
- Determinate varieties: 60 cm × 60 cm; 60 cm × 45 cm

Nutrient management

- Adequate nutrition is necessary not only for obtaining higher yield but also for better fruit quality.
- Apply a general fertilizer dose of 100 kg N, 80 kg P_2O_5 and 80kg K_2O per hectare for open pollinated improved varieties.1/3 N along with other fertilizers should be given as basal and rest N should be top dressed in two split doses at 30 days interval after transplanting.
- Apply 200 kg N, 100 kg P_2O_5 and 100kg K_2O per hectare for the hybrids.1/4 N along with other fertilizers should be given as basal and rest N should be top dressed in three split doses at 30 days interval after transplanting.
- Two foliar sprays of 0.5% $ZnSO_4$ is beneficial for yield and quality.
- Two foliar sprays of 0.2% calcium chloride and two sprays of 0.3% borax check blossom end rot due to calcium deficiency and cracking of fruits due to boron deficiency, respectively.
- Calcium ammonium nitrate can be used as the source of nitrogen in calcium deficient soil.
- Apply vermicompost @ 4-6 t/ha as broadcasting during final land preparation.
- Apply Azotobactor @ 800 g/ha or Phosphate Solublizing Bacteria @800g/ha as seedling root dip.
- Prepare a suspension of bacterial inoculants in sufficient water.
- Dip the seedlings in the bacterial suspension in shade for 2-3 hours in the afternoon hours.
- Transplant the treated seedlings immediately.

Irrigation

- The plants require adequate moisture through out their growth period.
- During summer, apply irrigation at 4-5 days interval and during winter at 10-15 days interval.
- About 95 cm total water is needed for successful tomato crop and apply 2.5 cm irrigation water at every fortnight.
- Critical stages of irrigation are flower development, fruit set and after each harvest.
- Most widely adopted irrigation method is furrow method of irrigation.
- Nutrients can be applied through drip irrigation.
- Sprinkler irrigation, being popularized now a days, is more economical.
- Stop giving irrigation at fruit ripening stage during winter.

Interculture and weed control

- In indeterminate varieties and hybrids, fruit size can be increased and at the same time disease problem can be checked by training and pruning of the plants. Two or three strong basal stems are retained and all axillary branches on them are removed by disbudding, allowing only the main stems to grow and produce fruits.
- The plants are staked to avoid damage to fruits, the common practice being to stake individual plant or provide support to plants in a row.
- Practice shallow inter- cultivation during first 4 weeks to remove the weeds and conserve soil moisture.
- Give two hand hoeings at 1st and 3rd fortnight after transplanting and do earthing up at 2nd fortnight after transplanting.
- Stop intercultural operations when the plants have started flowering.
- Foliar application of 50 ppm PCPA increases fruit set at high and low temperatures.
- Regarding chemical weed control, a pre-planting treatment of pendimethalin1.0 kg a.i. /ha or fluchloralin 1.25 Kg ai/ha along with one hand hoeing 45 days after transplanting is effective.

Harvesting the crop

- Tomato fruits are harvested at different stages of maturity depending on the distance of the market and the purpose viz., mature green, pink and breaker, hard ripe and fully ripe.
- Breaker or pink stage is the ideal harvest stage which ensures good quality, better shelf-life and proper development of colour.
- The fully ripe fruits are harvested for processing purpose.
- Pick the fruits at an interval of 4-5 days in summer and 6-7 days in winter.
- Do not pull tomato fruits from the vine rather pick with a twisting motion of hand to separate the fruits from the stem.
- Pre-harvest sprays of 0.3% Difolatan at 10 days interval are effective in controlling post harvest diseases

Post harvest handling and storage

- Pre cooling of tomato after harvest at 12-13°C on the farm prolongs their storage-life.
- Application of 0.2% Benomyl is effective in preventing growth of most of the damaging fungi.
- Store mature green tomatoes in polythene bags of 100 gauge thickness at 12-14°C and 90-95% relative humidity for 18-21 days.
- Fully ripe tomatoes store well at 4-5°C and 85-90% relative humidity for 7 days in cold storage.

- Keeping pedicel intact with fruits increase shelf life by 3-4 days.
- Evaporative cool storage (Zero energy cool chamber) extends shelf life of tomato at breaker stage by 4-5 days.
- Tomato fruits treated with GA_3 at 10 ppm or BA at 25 ppm show minimum physiological weight loss and incidence of decay.
- Fruit dip in 0.5% calcium nitrate or calcium chloride extends the shelf life by checking CO_2 and ethylene production.
- Usually bamboo basket and wooden boxes of various size and shapes are used for packing of tomatoes. Wooden boxes are generally used for packing for long distance markets.

Plant protection measures

Physiological disorder

Blossom end rot

Brown, water-soaked discolouration appears at the blossom end of the fruit where senescent petals are attached while the fruit is still green. The spots enlarge and darken rapidly and the affected portion becomes sunken, leathery and dark coloured. This disorder is caused due to i) sudden change in the rate of transpiration specially in moisture stress condition, ii) continuously high evapo-transpiration regime and a large leaf area, iii)increasing level of nitrogen content in the fruits and iv) fall in the level of calcium content in the fruit.

Control measures

- Increase in the frequency of irrigation.
- Increase the level of phosphate which decreases the incidence of this disorder.
- Two foliar sprays of 0.2% calcium chloride at the time of fruit development.

Puffiness

Partially filled fruit become light in weight and lack in firmness. Cross-section of the affected fruits shows emptyness or pockets. This disorder is caused due to i) non fertilization of ovules, ii) embryo abortion after normal fertilization, iii) necrosis of vascular and placental tissue after normal development of the fruit, iv) high temperature and v) high soil moisture.

Control measures

- Avoid over irrigation.
- Three sprays of 0.3 –0.4% borax solution at initial plant growth, fruiting and fruit development stages.

Fruit cracking

In the severe form of cracking, surface of the mostly fully matured and ripe fruits cracks radially from the stem end. In another form of cracking, surface of the mature green fruits cracks concentrically around the shoulder of the fruit. Sometimes it is a genetically controlled character. However, it may be caused due to i)irrigation or rainfall after a long dry spell ii) exposure of the fruits to sun iii) boron deficiency

Control measures

- Irrigation at regular interval.
- Avoid pruning during summer season.
- Three sprays of 0.3-0.4% borax solution at initial plant growth, fruiting and fruit development stage.

Sunscald

Exposed portion of either green or nearly ripe fruits get blistered and water soaked. Rapid desiccation leads to sunken area, grey in green fruits and yellow in pink or red fruits. It is generally caused due to exposure of the fruits to extreme heat of scorching sunshine.

Control measures

- Avoid training and pruning in the summer months.
- Protect the plants from defoliation caused by diseases and insect pests.

Gold fleck

Tiny yellow spots appear as gold flecks around the calyx and fruit shoulder due to deposition of calcium oxalate. With high incidence fruits become unattractive. This disorder is caused due to i) higher supply of phosphatic fertilizers, ii) higher supply of calcium fertilizers and iii) increased magnesium concentration in the fruit.

Control measures

- Summer shading lowers the incidence of this disorder.

Diseases

Damping off (Pythium, Rhizoctonia, Phytophthora, etc.)

It is very destructive fungal disease in nursery beds, especially during early season. In pre- emergence damping off, the growing points are killed in the initial stages of seed germination before they come out through the soil. In post-emergence damping off, the seedlings topple over the ground due to collar rotting and rapid shrinking of the cortical tissue of the hypocotyls.

Control measures

- Hot water treatment of seeds (at 51.7° for 30 minutes)

- Treat the seeds with thiram/ captan / metalaxyl-mancozeb @ 2.5 - 3.0g /kg of seeds.
- Drenching the nursery beds 7 days before sowing with thiram / captan or any copper fungicide @ 3g / litre of water.
- Cover the beds with transparent polythene sheets before sowing and left to open sun for at least 20 days, known as soil solarization.
- Spraying the young seedlings with 0.2% Blitox or 0.25% metalaxyl-mancozeb or 0.25 % fosetyl-Al.
- Provision of proper drainage to the beds.
- Proper plant population should be maintainted.
- Remove the affected seedlings from beds as soon as the symptoms are visible.

Late blight (Phytophthora infestans (Mont) de Bary)

It is serious fungal disease when temperature lay between 10^0 to 20^0C with relative humidity of 85% and above for 18 hours continuously for at least two consecutive days. Brown to purple- black lesions appear on leaflets, petiole or stem, advanced rapidly to cause a severe blight in favorable condition white fructification of fungus appears on the lower side of the leaves. Grey water-soaked spots enlarged to indefinite size and shapes appear at stem end of the fruit.

Control measures

- If the late blight causing weather condition prevails, start spray with protectant fungicides copper oxychloride (0.3%) / mancozeb (0.25%) / chlorothalonil (0.2%) at 10 days interval.
- If late blight present in field in pockets, start spraying with metalaxyl-mancozeb(0.25%) / cymoxanil-mancozeb(0.25%) / dimethomorph (1g) + mancozeb(2g)/ 1 of water immediately, and alternate spray schedule with protectant and systemic fungicides at 7 – 10 days interval depending upon weather condition.
- Careful irrigation or avoid top dressing and irrigation under late blight prevailing condition.

Early blight (Alternaria solani (Ellis and Marv jones and Grout)

It is serious fungal disease when temperature lay between 18°C to 30°C and soil moisture is abundant. Circular to angular and dark brown to black spots appear on the leaflets. Usually, a narrow chlorotic halo is found around the necrotic spots. Fruits may be infected both at the green and ripe stage. At stem end, black or dark brown lesions usually develop with sunken spots which become leathery later.

Control measures

- Spray the crop with 0.3% mancozeb or 0.1% Bavistin or chlorothalonil 0.2% at 10 - 15 days interval starting form 45 days after transplanting as a prophylactic measure.
- Two sparys of 0.2% chlorothalonil or difenconazole (0.5g /l) is very effective.

Buckeye rot (*Phytopthora parasitica*)

This fungal disease becomes severe when soil moisture is plenty and soil temperature is around 18^{0}C. The disease symptoms appear on the lower fruits.A brownish spot appears often at the point of contact between the fruit and the soil. The spot enlarges and the surface of lesion assumes a pattern of concentric rings of narrow, dark brown and wide, light brown bands.

Control measure

- Spray the crop 3-4 times with 0.3% Blitox or mancozeb or metalaxyl-mancozeb (0.25%) / cymoxanil-mancozeb (0.25%) at 10 days interval starting from disease appearance.

Bacterial wilt (*Ralstonia solanacearum* E.E. smith)

It is a serious bacterial disease particularly in the sandy loom and lateritic soil. Drooping of the lower leaves followed by wilting of the entire plant. Leaf epinasty, marginal leaf necrosis and development of adventitious roots along the stem take place. A slimy grey material oozes out when the stem is cut.

Control measure

- Treat the seeds with hot water at 52°C for 20 minutes or with 0.01% streptomycin solution for 30 minutes.
- Proper crop rotation with non-solanaceous crops for at least 2 to 3 years reduces the infestation. Effective rotations are maize- okra-cabbage/cauliflower; maize-beans-maize, or okra-rice-mustard.
- Apply bleaching powder @ 15-20 kg /ha and 4.5q lime to the land and mix it properly to the soil atleast 3-4 week before planting.
- Proper drainage should be maintained.
- A combination of dhaincha as green manure crop and chemical soil amendment with lime and bleaching powder and application of bioagents as soil application is most promising in management of wilt of solanaceous vegetables.

Tomato leaf curl (*Tomato leaf curl virus*)

Tomato leaf curl virus is transmitted by white fly (*Bemisia tabaci*).Severe stunting of the plans with downward curling of leaves and excessive branching are the characteristic symptoms. Older leaves turn leathery and brittle and the newly emerging leaves look pale yellow and smaller in size. In severe infection, the plants remain unfruitful.

Control measure

- Destruction of all weeds around the field which serve as alternate host of the virus.
- Cover the nursery bed with 40 mesh fine nylon net and spray outside with imidacloprid (3.5 ml / 10 litre of water).
- Pre-sowing treatment of nursery beds with Furadon 3G @15g/100 sq.m.
- Apply Thimet 10G @ 10-15 kg /ha 10 days after transplanting followed by 2-3 foliar sprays with 0.05 % dimethoate (Rogor 1.5 ml /l) or imidachlorpid or acetameprid or thiomethoxam (3.5 ml / 10 litre) at 10 days interval and alternate the spray schedule with NSKE.
- Apply 2% power oil to the plants to prevent acquisition and inoculation of the virus by white flies.
- Sow 5-6 rows of border crops of maize, jowar and bajra all round the tomato plot at least 50-60 days before transplanting of tomato.
- Practice mixed cropping of tomato, French bean and brinjal.
- Use of Resistant cultivars/hybrids viz., H-24, BCTH-4, TLBRH-6, JKTH-3098.

Tomato mosaic (*Tomato mosaic virus*)

The tomato mosaic virus (TMV) is transmitted by contact with clothes, man, animal and machinery during cultivation or by contact of infected plants, plant debris and through infected seeds. Yellow-green mottling appears on leaves and occasionally on fruits. Infected leaflets are usually distorted, puckered and smaller than normal. Number of branches, inflorescence and flowers are substantially reduced.

Control measure

- Soak the seeds in trisodium phosphate solution (90g/l of water) for 15 minutes a day or two before sowing.
- Practice crop rotation excluding solanaceous crops.
- Wash all tools and implements with 3% solution of trisodium phosphate before using them again.

Insect pests

Tomato is infested by a numbers of insects and non insect pests of which some of them attained major pest status. The insect pests those are considered as major pests are discussed here.

Fruit borer (*Helicoverpa armigera*)

It is a widely distributed polyphagous insect pest causing devastating damage by boring the young developing fruits and feed inserting their head within it. Young larvae first feed on tender leaves and later infest the fruits by mining circular holes and feed therein.

Control measures

- Collection and destruction of initial invading larval population at early stage keeps the crop free from havoc damage.
- Plant 14 rows of tomato alternated with one row of African marigold as trap crop
- Inundative release of egg parasitoid *Trichogramma chilonis* and *Trichogramma brasiliensis* @ 50,000/ha /week suppress the pest population.
- Application of selective NPV is also very effective against this pest.
- Provision of birds perching site in tomato field suppress larval population as the predatory birds like drongo, starling, etc. successfully predate on them.
- Spray the crop with 0.05% chlorpyriphos (Dursban 2.5 ml /l) or 0.2% carbaryl (Sevin 2 g/l) at 7-10 days interval after fruit initiation.

White fly (*Bemisia tabaci*)

It is a dangerous pest of many important vegetable crops. Though it directly causes very little damage however, indirectly it causes profuse damage by transmitting viral diseases to the healthy crops. Apart from this, it also helps the yellow mite to disseminate by carrying them from infected crop to non-infected crops. The tiny nymphal stages of the whitefly directly suck sap from the ventral surface of leaves and the affected portion becomes yellowish, leaves wrinkle and curl downwards. Severe infestation causes stunted growth of the crops.

Control measures

- Encourage inundative release of nymphal-pupal parasitoids, *Encarsia formosa* to suppress population outbreak of the whitefly.
- The release of predatory mite, *Agistemus fleschneri* is very effective for managing destructive population of nymphal and pupal stage of whitefly.
- The crop should be sprayed with azadirectin @ 1.0 % (Neemazol 1ml/l).
- Spray the crop with diafenthiuron (Pegasus 0.5gm /l) or 0.03% phosphamidon (Demecron 1.5 ml/l), 3-4 round at regular 7 days interval targeting only young apical parts of plant canopy.

Tobacco caterpillar (*Spodoptera litura*)

The larvae feed gregariously during early instars and voraciously during fourth onward stages preferably at night on the foliages. Severe infestation of this pest in nursery bed causes entire seedlings to be defoliated overnight.

Control measures

- Collection and destruction of leaves along with the egg masses and early gregarious larval stages are one of the best ways to suppress the pest.
- Application of selective NPV is also very effective against this pest.
- Conserted programme should be undertaken periodically for inundative release of egg parasitoids, *Trichogramma chilonis, T. brasiliensis* and larval parasitoids, *Bracon hibator, Apantales* sp. to suppress the population outbreak of the pest.
- Spray the seedlings with 0.05% endosulfan (Thiodan 1.5 ml/l) or 0.2% carbaryl (Sevin 2 g/l) at 10 days interval.

Epilachna beetle: (*Epilachna vigintioctopunctata*)

The grubs and adults feed on the leaves along with the chlorophyll bearing tissues leaving the veins and cuticle intact resulting in the development of characteristic skeletonised patches on the leaves.

Control measures

- Collection and destruction of egg masses along with the grubs and adults in the initial invading population is very effective to check destructive form of this pest.
- Spray the crop with 0.2% carbaryl (Sevin 2g/l) or 0.05% endosulfan (Thiodan 1.5ml/l) at 10 days interval.

Roof knot mematode (*Meloidogyne incognita* and *M. javanica*)

The nematodes attack the roots and produce swellings or galls there. Infested plants become stunted and yellow in colour. Plants wilt in hot and dry weather and in severe cases, die.

Control measures

- Plough the field in summer months to expose the soil to high temperature to reduce the nematode population.
- Incorporate neem cake or neem leaf @ 500 g/m^2.
- Grow trap crops like marigold, wheat, etc. in rotation or intercropped with tomato.
- Apply Nemagon, Nematex or Vapam to the soil.
- Apply Thimet 10G (25 kg/ha) or Furadon 3 G (33.3 kg/ha) in the field at transplanting.

Yield

- 20-25 tonnes/ha (open pollinated determinate varieties).
- 25-35 tonnes/ha (open pollinated indeterminate varieties).
- 50-80 tonnes /ha (Hybrids)

3. Brinjal

Scientific Name: Solanum melongena L.
Family: Solanaceae
Origin: India
Chromosome No.: 2n = 2x =24

Introduction

Brinjal, also known as aubergine or eggplant is one of the important vegetables of the tropical countries particularly India, Bangladesh, Pakistan, China, Japan, Indonesia, Thiland, Malaysia and the Philippines. It is also commonly grown in France, Italy, Spain, Bulgaria, the U.S.A. and several countries of Africa. It was known in India from far ancient times. Besides the commonly grown species, *Solanum melongena* L as brinjal or eggplant, other species of *Solanum* are also grown for vegetable purpose in different parts of the world viz., *Solanum muricatum, Solanum quitoense* and *Solanum sessiliflorum* in central and South America, *Solanum macrocarpon* in south-east Asia, *Solanum aethiopicum*, *Solanum gilo* and *Solanum macrocarpon* in Africa.

Importance and use

- Brinjal is the most common, popular and principal vegetable crop grown widely in India for its various shape, size and colour of fruits.
- High productivity, wide adaptation and ease in availability make the crop to find its place as the poor man's crop.
- In India, it is cooked in vegetable and fish curry, sambar, mashed, fried, grilled and stuffed with spices and then fried (varta).
- Bitterness of fruits is due to presence of glycoalkaloids, 'solasodine' and purple skin colour of fruits is due to anthocyanin pigment.

Nutritional value

- Brinjal fruits are fairly good source of calcium, phosphorus, iron and vitamins particularly 'B' group.
- 100 g of edible part gives 1.4 g protein, 18 mg calcium, 47 mg phosphorus, 0.11 mg Riboflavin, 0.9 mg Niacin,124 IU Vitamin A, 12 mg Vitamin C.
- Pigmented purple coloured brinjal contains more vitamin C than green one.
- High amount of glyco alkaloids (20 mg/ 100g fresh weight) in the fruits produce a bitter taste and off-flavour.
- There is much variation in chemical constituents in the fruits of different cultivars.
- Cu content and polyphenol oxidase activity are highest in purple colour fruits and lowest in white coloured fruits. Similarly, Fe and catalase were highest in the green and lowest in the white cultivars.

- Green cultivars have the better processing qualities than the more popular purple cultivars

Medicinal value

- Fruits are excellent remedy for those suffering from liver troubles and high blood cholesterol problem.
- Fresh or dry product of fruit or leaf markedly drops in blood cholesterol level.
- The de-cholesterolizing action is attributed to the presence of higher amount of (65%) ploy-unsaturated fatty acids (linoleic and lenolenic acids) in the flesh and seeds of the fruit.
- Magnesium and potassium contents in the fruits further help in reduction of cholesterol.
- It is rich in bioflavonoids which can prevent stomach cancer.
- Piercing white brinjal fruits with a needle and fried in sesame oil is a remedy for toothache.
- White brinjals are said to be a remedy for diabetic patients.
- Aqueous extracts of fruit juice inhibits choline esterase activity of human plasma and dry fruit is reported to contain goitrogenic principles.

Adaptation in India

India is the primary centre of origin of brinjal and rich biodiversity exists in Eastern Ghats, north-eastern region, central India, eastern India particularly in Orissa and West Bengal. There are more than 100 local cultivars are grown in India in different names like, Muktakeshi, Bhangar, Patakata, Elokeski, Nababganj Local, Makra, Soyla, Kanta Makra, Nurki, etc. in West Bengal; Singhnath, Bholanath in Tripura; Ratan Mani, Burupada, Borigumma, Kutiguda, Dhenkanal, Batparsi, Banmalipur Local, Narayanpur Local, Nisan Local, Radha, Green Rocket, Chitrakonda, etc. in Orissa; Hydarpur, Improved Muktakeshi, Sharanpur Long, Kathvaia, etc. in Bihar; Muktakeshi, Ramnagar Giant, Batia, Sheonath, Bundelkhand, Dudhya baingan, Jafrabadi, etc. in Uttar Pradesh and north-central region; Manjari Gota, Bourad Local, Borgona Local, Khandesh, Khed Shivpur, Malkapuri, Shivpur Khandala, Chikhalgaon Local, Hingana Dorla, etc. in Maharashtra; Junagadh Bhattu, Surati Gota, Gulabi Dorla, etc. in Gujarat; Coorg Local, Irangeri, West Coast Green Round, Udupi Gulla, etc. in Karnataka; Nurki Round, Nurki Long, Banjari, etc. in Madhya Pradesh; Wynad Giant, Pattabiram, Annamalai, etc. in Tamil Nadu; Gola Baingan, Ludhiana Local, Batia, etc. in Punjab.

Breeding behaviour

- Brinjal is basically self- pollinated crop, but a high degree of cross pollination (up to 30%) which makes it some times as often cross pollinated crop. In fact, heterostyly present in brinjal cultivars has also helped in widening the genetic diversity through partial cross pollination occurring in nature, even though it is considered as autogamous.

- Major pollinators are bumble bees (*Bombus* Spp) and honey bees (*Apis* species).
- Anthesis generally starts from 7.30 am which continues up to 11.00 am with peak time between 8.30 to 10.30 am however, it depends on the prevailing temperature and humidity condition.
- Anther dehiscence occurs at least half an hour after anthesis. Anther dehiscence occurs through the development of pores on the anthers.
- The stigma receptivity is maximum at the time of opening of the flower although it become effectively receptive a day prior to anthesis, and the receptivity continues even upto four days after anthesis.
- The flowers emasculated one to two days earlier should have completely blossomed under the paper bag and are ready for cross pollination in the morning hours.
- Pollen viability is retained for 8-10 days at moderate temperature of 20-22°C with 50-55% relative humidity. Dried pollens can be stored in sealed vial at 0°C for 1-2 months.

Origin and Taxonomy

Brinjal is a native of India, believed to be Indo-Burma region and perhaps China is the secondary centre of origin of this crop. *Solanum incanum*, a wild species and having wide distribution in atleast 10 habitats in India is the progenitor of the cultivated species, *Solanum melongena*. The first record of brinjal in India was during 300 BC to 300 AD. Atleast 33 Sanskrit names for brinjal have been mentioned in the ancient literatures of India and the most common being Varttaka, Bhantaki and Vattingan. It was cultivated in Africa around 9th centuryA.D. It was known in Italy at the end of the 14th century and introduced into southern Europe in the 15th century and the name" eggplant" was probably derived from the white egg like fruits.

Brinjal, the non-tuberiferous *Solanum* (2n=2x=24) belongs to sub-genus *Leptostemonum,* section Melongena of the family Solanaceae, sub-family Solanoideae and Tribe Solaneae.

There are three botanical varieties of *Solanum melongena* characterized by fruit shape and size and growth habit of plant.

S melongena var esculentum – Fruit round or oval

S melongena var. serpentinum – Fruit long, slender.

S melongena var. depressum – Plant short and dwarf.

Improved open pollinated varieties

Long fruited variety: Pusa Purple Long, Mysore Long, Hisar Pragati, Pusa Bhairav, KS-331, JB-15, Green Long, Punjab Barsati, PH-4, Pant Samrat, Azad B-1, Arka Sheel, Arka Shirish, Punjab Sadabahar, NDB-25, Annamalai, Haritha.

Round or oval fruited variety: MDU-1, BR-112, Surya, CHBR-1, DBR-8, Azad B-2, Pusa Uttam, Pusa Upkar, Pusa Purple Round, Pant Rituraj, T-3, Jamuni Gole, Punjab Neelam, Hisar Shyamal, Azad Kranti.

Oblong fruited variety: CO-1, CO-2, Bhagyamati, Utkal Tarini, Hisar Jamuni.

Longish-oblong or cylindrical fruited variety: Utkal Madhuri, Utkal Keshari, Pusa Kranti, Pusa Anupam.

Small fruited variety: PKM-1, Shyamala, Aruna, Pusa Ankur, Pusa Bindu, Manjari Gota.

Cluster bearing variety: PLR-1, Arka Neelkanth, KKM-1, ARU-1, Arka Nidhi, Utkal Jyoti, Gulabi, Vaishali, Pragati, Pusa Purple Cluster, ARU-2C, Arka Kushumkar, Punjab Barsati.

Public sector hybrids: Arka Navneet, Pusa Hybrid-5, Pusa Hybrid-6, Pusa Hybrid-9, NDBH-1, NDBH-6, ABH-1, ABH-2, Krishna, COBH-1, Neelam.

Cultural requirements

- A long and warm growing season with average temperature range of 21°C – 27°C is most suitable.
- Extreme temperature (below 10°C and above 30°C) adversely affects the reproductive growth.
- Warm and humid climate favours luxurious growth, while poor growth is exhibited in cool climate.
- It grows on all types of soil ranging from light sand to heavy clay however, well-drained, fertile, silt or clay- loam soils are ideal.
- Light soils are good for early crop, while clay-loam soils are well-suited for high yield.
- It can tolerate slightly acidic soil athough, the ideal soil pH range of the soil is 5.5 to 6.6.
- Brinjal should follow a leguminous crop to enrich soil fertility.

Nursery management

- Seed beds are prepared finely, well drained, 15-20 cm raised, 1.0m wide and of convenient length.
- Fine and fully decomposed farmyard manure or compost @ 3-4 kg/m^2 should be well mixed to the beds.

- Drench the beds with formaldehyde (4.0%) or captan @ 2-3g/litre and cover with polythene sheet for 7 to 10 days to avoid damping off disease.
- 375 to 500g seeds of open pollinated varieties are required per hectare.
- 200 to 250g seeds of hybrids are required per hectare.
- Treat the seeds with captan or thiram @ 2-3 g per kg of seeds before sowing.
- Seeds are sown at shallow depth 5.0 cm apart in the row and covered with finely sieved leaf mould.
- After sowing cover the beds with straw or long dry grass till seed germination and sprinkle water regularly.
- Seedlings may be raised under poly tunnels having sides open to protect the young seedlings from heavy rain or scorching sunlight.
- Water the seedlings daily or in alternate day with rose can.
- Hardening of the seedlings by withholding water at least 4 to 6 days before transplanting.
- Stocky, 30-35 days old, healthy seedlings of 12-15 cm height with 3-4 leaves are ideal for transplanting.

Sowing time

1. May-July (for Autumn – Winter crop i.e., main season crop)
2. October-November (for Spring – Summer crop).
3. February-March (for Summer-Rainy crop)

The crop can be grown round the year in mild climate of Karnataka and West bengal. In the hills seeds are sown in April.

Land preparation

- Prepare the land in well advance with repeated ploughings (at least 4-5 ploughing) to a fine tilth.
- Remove all stubbles, weeds, etc. and level the land.
- Add well rotten organic manure preferably farm yard manure or compost @ 20-25 t/ha during land preparation and level the land properly.

Transplanting

- 30-35 days old, healthy seedlings of 12-15 cm height are ready for planting.
- After hardening the seedlings for 4 – 6 days, irrigate the seed bed before lifting to facilitate easy pulling of young seedlings.
- Plant the seedlings in the field preferably during afternoon hours which give better field establishment.
- Irrigate the field just after transplanting.

Spacing

- Round fruited improved varieties and hybrids: 90 cm × 90 cm.
- Vigorous growing cultivars: 75 cm × 60 cm.
- Semi-vigorous and midseason cultivars: 60 cm × 60 cm.
- Long fruited cultivars: 60 cm × 45 cm.
- Early and dwarf types: 45 cm × 45 cm.

Nutrient management

- Brinjal is long duration crop and is good feeder of macro nutrients.
- Apply a general fertilizer dose of 125 kg N, 80 kg P_2O_5 and 80kg K_2O per hectare for open pollinated improved varieties.1/3 N along with other fertilizers should be given as basal and rest N should be top dressed in two split doses at 30 days interval after transplanting.
- Apply 200 kg N, 100 kg P_2O_5 and 80kg K_2O per hectare for the hybrids.1/4 N along with other fertilizers should be given as basal and rest N should be top dressed in three split doses at 30 days interval after transplanting.
- Two foliar sprays of 0.5% $ZnSO_4$ and single spray of 0.15% $CuSO_4$ increase yield and quality of fruits.
- Apply vermicompost @ 4-6 t/ha as broadcasting during final land preparation.
- Apply *Azotobacter* @ 80 g/ha or Phosphate Solublizing Bacteria @800 g/ha as seedling root dip.
- Prepare a suspension of bacterial inoculants in sufficient water.
- Dip the seedlings in the bacterial suspension in shade for 2-3 hours in the afternoon hours.
- Transplant the treated seedlings immediately.

Irrigation

- Brinjal being a shallow rooted crop needs irrigation at frequent interval.
- During summer, apply irrigation at 4-5 days interval and during winter at 10-15 days interval.
- About 100 to 110 cm total water is needed for successful brinjal crop.
- Most adopted method of irrigation is furrow method. Drip irrigation is beneficial for reducing water requirement and weed control.
- Nutrients can be applied through drip irrigation.

Interculture and weed control

- Practice shallow inter- cultivation particularly a few days after every irrigation to remove the weeds and conserve soil moisture.
- 3-4 hoeings are normally needed to check the weeds.

- Earthing up after second top dressing of nitrogen fertilizer is helpful.
- Regarding chemical weed control, a pre-planting treatment of Pendimethalin 1.0 kg a.i. /ha is effective.
- Use of 2.5ppm 2,4-D or 20 ppm 4CPA or 60 ppm NAA as whole plant spray at flowering promote fruit set and increase yield.
- Spraying of the plant protection chemicals should be avoided in the morning hours to facilitate free movement of the pollinating insects like, bees.

Harvesting the crop

- Harvest the fruits when they develop good colour, marketable size, still immature and tender.
- Generally 15-19 days fruits are optimally matured for harvest.
- The good marketable quality fruits are bright, glossy appearance having freshness and optimum size without any fading or change in fruit colour.
- Fruits are harvested with stalk at joint.

Post harvest handling and storage

- After harvesting, treat the fruits with the solution containing 200 ppm NAA in combination with 90 ppm prochloraz to retard fruit senescence and decaying.
- Under ordinary condition, the normal storage life of brinjal fruits is 1-2 days in the winter months.
- 2-3 weeks storage life can be obtained when fruits are stored in cold storage at 8° -10°C temperature and 85-90% relative humidity.
- Waxing the fruits in 3% concentration help to reduce weight loss and retard the loss in acidity, starch, sugar and ascorbic acid contents.
- For long distance transport, fruits should be harvested towards evening hours and thereafter cooled them with sprinkling of water.
- Store the fruits in perforated polythene bags or PVC stretch film.

Plant protection measures

Diseases

Damping off (Pythium spp, Phytophthora spp)

It is very destructive fungal disease in nursery beds, especially during rainy season. Affected seedlings turn pale green and brownish lesion is found at the basal portion of the stem. The lesion girdles the stem and finally the seedlings topple over.

Control measures

- Hot water treatment of seeds (at 51.7° for 30 minutes)
- Treat the seeds with thiram/ captan @ 2.5g /kg of seeds.

- Drenching the nursery beds 7 days before sowing with thiram/ captan or with any copper fungicide @ 3g / litre of water.
- Cover the beds with transparent polythene sheets before sowing and left to open sun for at least10 days, known as soil solarization.
- Spraying the young seedlings with 0.2% Blitox or 0.25% metalaxyl-mancozeb or 0.25 % fosetyl-Al.
- Provision of proper drainage to the beds.
- Remove the affected seedlings from beds as soon as the symptoms are visible.
- Avoid flooding the beds to check spread of the disease.

Phomopsis blight (*Phomopsis vexans*)

This is a serious fungal disease of brinjal. Infected leaves show small, circular spots which become gray to brown with irregular blackish margin. The affected leaves may turn yellow and die. These lesions may also develop on the petiole and stem. On the fruits, soft, water decayed portion are observed which later turn black. The whole or major portion of the fruit may rot. The disease is a major problem in seed crop.

Control measures

- Use seeds obtained from disease-free plants.
- Treat the seeds with thiram or captan @ 2.5 g / kg of seed.
- Keep the fields neat and clean all the time.
- Hot water treatment of seeds before sowing at 50^{0}C for 30 minutes.
- Spray the crop with Diathane M-45 @ 0.2.5% or Bavistin @0.1% at 7-10 days interval.
- Collect and burn the diseased plants and fruits.
- Grow the tolerant varieties like Pusa Bhairab, Pant Samrat, etc.
- A suitable crop rotation with leguminous or cucurbits may also help to reduce the disease infestation.

Bacterial wilt (*Ralstonia solancearum*)

This is a severe bacterial disease of brinjal. The characteristic symptoms of the disease are yellowing and dropping of the foliage, stunting and wilting of the plant gradually, followed by collapse of the entire plant.

Control measures

- Treat the seeds with hot water at 52°C for 20 minutes or with 0.01% streptomycin solution for 30 minutes
- Proper crop rotation with non-solanaceous crops for at least 2 to 3 years reduces the infestation. Effective rotations are maize- okra-cabbage/cauliflower; maize-beans-maize, or okra-rice-mustard.

- Apply bleaching powder @ 15-20 kg /ha to the land and mix it properly to the soil atleast 3-4 week before planting.
- Proper drainage should be maintained.
- A combination of dhaincha as green manure crop and chemical soil amendment with lime and bleaching powder and application of bioagents as soil application is most promising in management of wilt of solanaceous Vegetables.
- Grow the tolerant varieties like Arka Nidhi and BB-54.

Little leaf

This disease is caused by a phytoplasma. The affected plants are shorter in stature but produce larger number of branches, leaves and roots than the healthy one. The leaves are malformed to small and yellowish in colour. Internodes get shortened and give the plans a bushy appearance. Infected plants normally do not bear any fruits. The disease is transmitted by leafhopper.

Control measures

- Remove the infected plants from the field and burn them to check the spread of the disease.
- Spray Rogor @ 1.5 ml/litre or carbaryl @ 0.3% to control the vector population.
- Early sown crop (mid June) become less affected due to prevalence of low leaf hopper vector population.

Insect pests

Shoot and fruit borer (Leucinodes orbonalis)

This is the most dangerous pest of brinjal causing devastating damage of brinjal. The warm seasoned crops are the worst sufferer than the winter crops. The larva initially attacks the terminal shoots and bores inside, as a result of which the shoots wither drooping down and dry. In the bearing stage, it prefers to bore into the young fruits by making holes and feeds inside. The drooping shoots of the plant is an indication of its presence. Affected fruits get rot and fall down.

Control measures

1. Nip off the withering shoots and pick up the bored fruits daily and bury them deep into the soil is one of the best way to manage the pest. The affected plant parts and fruits along with the larvae may be collected into bag prepared by mosquito net to conserve the larval-pupal parasite like *Trathela flavoorbitalis* and *Bracon* sp.

2. Application of carbofuran 3G one and half teaspoon full per plant 2-3 weeks after transplanting.
3. Spray the crop with spinosad (0.15ml/l), 0.25% carbaryl or 0.05% endosulfan at the early flowering stage and after harvesting of fruits is very effective.
4. Spraying the crop with profenophos or triazophos 1.25ml/l of water.
5. Crop rotation excluding brinjal.
6. Use of pheromone trap to destroy the adult males.

Epilachna beetle (*Epilachna vigintioctopunctata*):

It infects most of the solanaceous and cucurbitaceous crops. The adults and grubs are very destructive, feeding voraciously on the leaves and tender plant parts. Characteristic skeletonized patches are developed due to their feeding which turn Ladys'lace like appearance on leaves and later dry away.

Control measures

- Spray 0.2% carbaryl (Sevin 2 g/l) or endosulfan 0.05% (Thiodan 1.5 ml/l) at the early stage of infestation.
- Mechanical control by hand picking and destruction of gregarious population along with egg masses and adult are very effective to keep suppress the destructive population of the pest.

Jassids (*Amrasca biguttula biguttula*)

The adult and nymphs are destructive which colonise on the under surface of the leaves and cause damage by sucking saps. The infested leaves curl upwords along the margins which may turn yellowish, and show burnt up patches. Jassids also transmit little leaf disease.

Control measures

- Spraying of azadirectin 1% periodically as prophylactic measure against jassid is very effective to prevent the pest outbreak.
- Application of imidacloprid (Confidor) @ 0.3ml/l is very effective against the pest. Aldicarb granules @ 1.0 kg a.i./ha at 5 days after transplanting, followed by spraying of carbaryl 0.2% or diafenthiouron (Pegasus 0.5g/l) after fruit-set provide satisfactory result.

Thrips (Scirtothrips dorsalis)

Both the adults and nymphs of this tiny insect cause damage by sucking sap from both the surfaces of the leaves. Continuous sucking by huge population along the veins makes silver shiny and pale appearance of leaves. The severely infested leaves gradually wither.

Control measures

- Application of starthane (Acephate) 1ml/l or acetamiprid @ 0.4ml/l is very effective against this pest.

Red spider mite (*Tetranychus* spp.)

Four spider mite species, *Tetranychus urticae* Koch, *T. macferlanei* Baker & Pritchard, *T. ludeni* Zacher and *T. neocaledonicus* Andre are commonly found to attack vegetable crops. They are tiny, microscopic and morphologically alike without visual difference among these four species. However, *T. urticae* is found everywhere to attack the crops. Brinjal is severely infested by these mite species and found to feed from the dorsal surface of the leaf. All the active stages of the mite suck sap resulting whitish appearance of the crop. Severely infested crops are covered by their heavy webbing that affect the photosynthetic activity of the plants, the leaves gradually dry up and fall off prematurely. Summer crops are worst affected.

Control measures

- Use of balanced fertilizer and proper irrigation keeps the plants healthy and reduce mite attack.
- Spray the crops with neem based acaricides during first occurrence of mite in the crop.
- Release of predatory mite, *Amblyseius longispinosus* is very effective against these mites.
- Heavy infestation requires acaricidal application (dicofol @ 2.5 ml/l; propergite @ 1.5 ml/l or diafenthuron @ 1ml/l) to save the crop.

Stem borer (*Euzophera perticella*)

The caterpillars attack stems and often kill the young plants. The growth of infested plant is stunted or withers.

Control measures

- Remove and destroy the affected plant parts
- Spray the crop with 0.03% diazinon or 0.2% carbaryl.

Root-knot nematodes (*Meloidogyne* spp)

The nematodes attack the roots and produce swellings or galls there. The affected plants become stunted in growth, and leaves show chlorotic symptoms and become rough in texture.

Control measures

- Apply Nemagon or Nematex to the soil.

- Plant 6 week old *Tagetes erecta* (Marigold) seedlings in the alternate rows of brinjal, which acts as trap crop for nematode.
- Deep ploughing 2-3 times in the summer months are beneficial in minimizing the nematode population.
- Apply Furadon 3 G (15 kg/ha) in the field at transplanting.

Yield

Yield of brinjal varies according to the region, cultivar and duration of the crop.

- Early crop: 20-30t/ha
- Long duration crop: 35-40 t/ha
- Hybrids: 40-80 t/ha.

4. Chilli

Scientific name: Capsicum species
Origin: Southern and Central America
Family: Solanaceae
Chromosome No.: 2n =2x =24

Introduction

Chilli or hot pepper is a tropical vegetable crop commonly used through out the world as spice for its pungency and red colour of ripe dried fruits and also for its green fruits for pungency and flavour. India is one of the leading chilli producing and exporting countries in the world. Other major chilli growing countries are Mexico, Japan, Ethiopia, Uganda, Nigeria, Thiland, Turkey, Indonesia, China and Pakistan. By the turn of the 15th century when the Spanish and Portuguese discovered South America, chilli peppers were widely cultivated for human consumption. The chilli pepper has been cultivated in South America for nearly 10,000 years, so it has had plenty of time to evolve. The Portuguese brought capsicum form Brazil to India during 1584. Chilli or hot pepper or chilli pepper and sweet pepper or bell pepper belongs to the same species, *Capsicum annuum* L. Besides this most commonly grown species, *Capsicum annuum* as chilli of commerce, other species of *Capsicum* are also grown as chilli in different South American countries viz., *Capsicum baccatum* var. *pendulum* in southern Columbia, Equador, Peru, Bolivia, southern Brazil, northern Chile and Argentina; *Capsicum chinense* in West Indies, low land south America, southern Bolivia and southern Brazil; *Capsicum frutescens* in Mexico, central America, lowland south America, southern Bolivia and southern Brazil.

Importance and use

- Chilli is grown throughout the country and is used in almost all dishes.
- West Bengal and Rajasthan account for more than 80% of the total area and production.
- Chilli of commerce, *Capsicum annuum* is classified into several groups of cultivars on the basis of fruit shape and the manner in which it is prepared and used.
- They can be eaten fresh as biting chilli or in curries, sambar, rasum and other meat, fish and vegetable dishes to impart pungency, colour and flavour to food items.
- Dried fruits are used to make universal curry powder and curry paste.
- Fresh green and ripe chillies are used to make all kinds of pickles, different sauces and paste.
- The red colour, Capsanthin is used in high quality cosmetic preparations like, lipstick.
- The essential oil, oleoresin is used in the food and beverage industries.
- Chilli extractions are used in the preparation of ginger beer and other beverages.
- The pungency principle of chilli, capsaicin (C_{18} H_{27} NO_3) a condensation product of 3 – hydroxy 4 methoxy benzylamine and decylenic acid produces a highly irritating vapour on heating.
- A non-conventional use of chilli as self-defense sprays is gaining popularity in different countries.

Nutritional value

- The nutritional value of chilli lies in their vitamin (A and C) and phosphorus contents otherwise, chillies add more than just flavour to the diet.
- Chilli is particularly rich in vitamin C (ascorbic acid content 120-130 mg/ 100g fresh).
- Chilli also contains appreciable amount of vitamin A (280-290 IU/100 g fresh).
- It has high crude fibre content (25.0-27.0 %).
- Chilli peppers lose their vitamin content when they are cooked or pickled or if they are allowed to ripen too much.

Medicinal value

- The stimulation of the hot chilli causes the heart to beat faster thus causing the body to sweat and act as body's natural air conditioner.

- The warmth of chillies not only stimulates the taste buds but also the appetite and digestion through mouth watering effects of increasing saliva flow.
- Chilli has the protective effect against peptic ulcer disease.
- Chilli peppers get rid the body of enough fats to lower the blood cholesterol level and reduce the chances of heart attack.
- Capsaicin, the pungent principle has substantial antigenotoxic and anticarcinogenic effects.
- The carotenoides, predominantly, capsanthin and beta carotene have anti oxidant property.
- The flavonoides present in red pepper particularly, quercetin, kaempferol, myricetin and luteolin are responsible for induction of enzymes that detoxify carcinogens.
- Capsaicin, once ingested, cause the brain to release endorphins into the blood stream which can induce a natural feeling of well being similar to that achieved by long distance runners.
- Capsaicin is also used in many pharmaceutical preparations like, pain balms, vapour linements, skin ointments for cold, sore throat, chest congestion, etc.
- Capsaicin is also used in the preparations of prickly heat powder.
- The oil content of the capsicum, if dissolved in ether and applied with cotton wool is considered to be very useful in relieving rheumatic pains

Adaptation in India

Mixed stock of *Capsicum* species was introduced in India by the Portuguese about 425 years ago into Kerala. Christian missionaries also introduced *Capsicum* species in the North-eastern states long ago separately. High adaptation in the tropical Indian condition, wide spread almost through out the country and farmers' selections particularly for pungent and red types have resulted in the evolution of huge genetic diversity of chilli in India. Huge number of local cultivars of chilli are grown in India in different names like, Suryamukhi, Bhangar Local, Suti, Sili, Beldanga Local, Jateswar Local, Bullet, etc. in West Bengal; Oosimilagi, Lavang Mirche, Golgonda Mirche, Sadasivpet, Shivali, Mundu, Nallapadu, Warangal, Samba etc. in Andhra Pradesh; Salem, Ramanand, Samba Kodakal, Sattur Samba, Kovilpatti, Nambiyur Local, Kandangadu Local, Ramanathapuram Gundu, etc. in Tamil Nadu; Javari, Javageal, Byadgi, Dubba Bayadgi, Kadda, Sankeshwar, Chincholi, Kollegal, Arsikera, Madhugiri, etc. in Karnataka; Kandhari, Achar, Vella Kandhari, etc. in Kerala; Muslawadi, Sankeshwar, Sindhur, Dondicha, Achalpuri, Malkapuri, etc. in Maharashtra; Gholar, Patta, etc. in Gujarat; Mathania, Mandoria, etc. in Rajasthan; Elichip, Kalipeeth, etc. in Madhya Pradesh; Sirhindi, Sanauri Red, Sanauri Yellow, Rajpura Long, Jullundhur, Shokati, Longi, etc.

in Punjab; Kandhari, Suryamukhi, etc. in Uttar Pradesh; Rahri, Patna Red, Sabour, Angon, Longi, etc. in Bihar; Puri Red, Puri Orange, etc. in Orissa; Latabh, Baghi, Gauhati Black, etc. in Assam.

Breeding behaviour

- Chilli is basically self- pollinated crop, but a high degree of cross pollination (even up to 68%) which makes it most of the times as facultative cross pollinated crop. Hence, rouging of off types is considered as the most important operation in seed production.
- To maintain the purity of the varieties, isolation distance of 400-500 m is safe for seed production.
- Major pollinators are honey bees (*Apis* species), ants and wind.
- Anthesis generally starts from 5.30 am and majority of the flowers open by 9.00 a.m. however, it depends on the prevailing temperature and humidity condition.
- Anther dehiscence occurs at least one hour after anthesis.
- The pollen grains in the anther become fertile a day before anthesis with maximum fertility on the day of anthesis.
- The stigma receptivity is the maximum at the time of opening of the flower although it become effectively receptive a day prior to anthesis, and the receptivity continues upto two days after anthesis.
- At high temperature condition, fair crossing success can be achieved if hybridization is taken up either early in the morning or late afternoon.
- Pollen viability is retained for a day in ambient condition. Pollens can be kept in refrigerator under airtight condition for 2-3 days. Dried pollens can be stored in sealed vial at -5.0°C for 10 days.

Origin and taxonomy

The primary centre of origin of chilli is in Southern and Central America and secondary centre in Guatemala. The chilli of commerce, *Capsicum annuum* var. *annuum* was domesticated in Mexico. The archaelogical excavations at Tehuacan in Mexico have indicated the existence of *C. annuum* around 7000B.C. On his return from Americas, Columbus brought chilli, *Capsicum annuum* to Spain in 1493. It spread to other parts of the Mediterranean region. Later it went to north in England in 1548 and moved to central Europe in the 16th century. It was introduced to India by Portuguese in 16th century.

Both chilli or hot pepper and capsicum or sweet pepper belongs to the family Solanaceae under the same species *Capsicum annuum*. There are 20 wild species of the genus *Capsicum* and the five cultivated species include *Capsicum annuum* var. *annuum*, the chilli of commerce, *Capsicum baccatum* var. *pendulum*, *Capsicum frutescens* , *Capsicum chinenese* and *Capsicum pubescens* . Hot chillies and sweet peppers, *Capsicum annuum* var. *annuum* have been originated from the wild and weedy *Capsicum annuum* var. *minimum* distributed from the southern United states to the northern South America. Another cultivated species, *Capsicum baccatum* var. *pendulum* was originated from its wild progenitor, *Capsicum baccatum* var. *baccatum* in South America. The species *Capsicum frutescens* is wide spread as a wild, weedy or semi-domesticated species in low land tropical America including Mexico, central South America and lowland South America. Perhaps, *Capsicum chinense* has been originated from the wild type of *Capsicum frutescens*. No ancestral type could be identified for the other cultivated species *Capsicum pubescens*.

The five cultivated species of *Capsicum* can be identified on the basis of corolla, anther, calyx, peduncle and seed characters.

Capsicum annuum var. *annuum* (corolla: milky white, rarely purple; anther: blue; calyx: inconspicuous calyx teeth; peduncle at a node:one, occasionally two; seed: straw coloured.), *Capsicum baccatum* var. *pendulum* (corolla: cream, yellow, tan or brown spots; anther:yellow,brown; calyx:prominent calyx teeth; peduncle at a node:one or two; seed; straw coloured), *Capsicum frutescens* (corolla: pale-greenish or yellowish-white; anther:blue; calyx: no calyx teeth; peduncle at a node:two or more; seed; straw coloured), *Capsicum chinenese* (corolla: white, milky-greenish or yellowish-white; anther:blue; calyx: a constriction on calyx ; peduncle at a node: two or more; seed; straw coloured), *Capsicum pubescens* (corolla: purple; anther: purple; calyx: conspicuous calyx teeth ; peduncle at a node: one; seed: black).In fact, morphologically *C. frutescens* and *C. chinense* are alike except that the latter has a constriction below the calyx.

Domestication , human selection and breeding have produced hundreds of varieties in the chilli growing areas of the of world from the great natural diversity of the *Capsicum* species differing greatly in fruit size and shape, pungency and ranging in colours from orange , red , yellow, violet to green. Different cultivar groups of hot chillies under *Capsicum* species can be classified on the basis of fruit shape and pungency.

Ancho	*C. annuum*, large, heart shaped ,thin-walled fruits, mildly pungent
Cayenne	*C. annuum*, fruits tapered, slender, wrinkled with thin wall, highly pungent.
Cherry	*C. annuum*, spherical, small, slightly flattened, thin-walled fruits, sweet and pungent.
Chiltepin	*C. annuum*, very small, egg-shaped fruits, thin walled, highly pungent.
Cuban	*C. annuum*, fruits of irregular blunt shape, mildly pungent, thin walled.
Jalapeno	*C. annuum*, fruits small, almost cylindrical with rounded ends, thick walled, skin russeted in some varieties, highly pungent.
New Mexican	*C. annuum*, long, slender with a tapered end, thin walled, moderately pungent, some varieties have sweet fruits.
Serrano	*C. annuum*, fruit short, elongated, thin walled with blunt end, very pungent
Squash	*C. annuum*, fruits flattened scallop- shaped, medium to thick walled, pungent
Tabasco	*C. frutescens*, fruits small, slender, erect, thin walled, extremely pungent.
Habanero	*C. chinense*, fruits small, thin, puffy-walled, extremely pungent

Improved open pollinated varieties

Andhra Jyoti (G-5), Bhagyalaxmi (G-4), K-2, K-1, , J-218, Muslawadi, LCA 206b, JCA 283, Phule Jyoti, Phule Suryamukhi, Bhaskar, CO-1, CO-2, MDU-1, NP 46-A, Punjab Lal, Pusa Jwala, Punjab Surkh, Punjab Guchhedar, Pant C-1, Pant C-2, JCA-154, Kalyanpur Selection, Pusa Sadabahar, X-235, Kalyanpur Red, Agnirekha, Arka Lohit, Sindhur, Aparna,

Mild to medium of pungent (capsaicin content 0.20 to 0.55 mg/g fruit)

Kovilpatti –1, Kovilpatti – 2, Bhagyalakshmi (G-4), Pusa Jwala, CO-3, Jwala Mukhi, Jwala Sakhi, Ujwala, RHRC Pendant Cluster, Andhra Jyothi (G-5), PMK –1, PLR- 1 Muslawadi, CA 960, NP 46 A, Sabour Arun, Agnirekha, Phule Jyoti, Phule Suryamukhi, Arka Abir.

Pungent to highly pungent (capsaicin content more than 0.55 mg/g fruit)

CO-1, Pant C-1, PKM-1, Arka lohit, Pusa Sadabahar, Jawahar Mirch 218 (J –218), Punjab Lal, Hisar Vijay (H-28) , Hisar Shakti (H-44) , Bhaskar, G-2 , MDU-1 Sindhur, X-197, Phule Sai, Jawahar Mirch 283.

Samba type (long and pendant)

Kovilpatti-2, Bhaghalakshmi (G-4) , Pusa Jwala, Co-3, Jwala Muhhi, Muslawadi, NP 46 A, Sabour Arun, CO-1, Hisar Vijay (H-28).

Gundu type (short, conical, somewhat round or oblong in shape)

PKM-1, PLR-1, CO-2, Andhra Jyothi (G5).

Variety suitable for colour extraction

KTPL-19, Arka Abir.

Variety high in oleoresin content

Hisar Shakti, Punjab Lal, Bhaskar (LCA-235), Ujwala, CO-3, RHRC Pendant Cluster.

Varieties for pickling

GCA 154, PLR-1.

Varieties tolerant to leaf curl virus

Pusa Jwala, Jwala Sakhi, Pusa Sadabahar, Jawahar Mirch 218 (J-218), Punjab Lal, Hisar Vijay (H-28), Hisar Shakti (H-44).

Varieties tolerant to thrips

Bhaskar, Kovilpatti-2, Kovilpatti-1, Andhra Jyothi, Phule Suryamukhi.

Varieties suitable for export

Bhaskar, Bhagyalakshmi, Andhra Jyoti, G-3, Pant C-1, Arka Lohit, X-235, Panvel, Ansari, Musalwadi.

Public sector hybrids

CH-1, CH-3.

Cultural requirements

- Chilli performs well in warm humid tropical and sub-tropical regions.
- A temperature range of 20-33^{0}C is optimum for chilli.
- Extreme temperatures, below 15°C and above 35°C significantly reduce reproductive growth and pollen viability.
- Humidity favours growth but frost is injurious.
- Low temperature at fruit ripening stage likely to delay colour development in the fruits.
- Chilli grows best in a well-drained loam soil rich in organic matter.
- Chilli can also be grown in heavy soil rich in organic matter particularly for dry chilli production.
- The pH should be around 6.5 – 7.5.

- It can tolerate salinity to a considerable extent.
- Chilli performs well with a rainfall of 600-1200 mm spread over 4 to 5 months.
- Enough water should be available, but the soil should also have good drainage facility.
- The crop prefers moderate sunshine, high relative humidity and drizzling rains during early establishment.

Nursery management

- Seed beds are prepared finely, well drained, 15-20 cm raised, 1.0m wide and of convenient length.
- Fine and fully decomposed farmyard manure or compost @ 3-4 kg/m^2 should be well mixed to the beds.
- Drench the beds with formaldehyde (4.0%) or captan @ 2-3g/litre and cover with polythene sheet for 7 to 10 days to avoid damping off disease.
- 700 – 800 g seeds of open pollinated varieties are required per hectare.
- 350 to 400 g seeds of hybrids are required per hectare.
- Treat the seeds with captan or thiram @ 2-3 g per kg of seeds before sowing.
- Seeds are sown at shallow depth 5.0 cm apart in the row and covered with finely sieved leaf mould.
- After sowing cover the beds with straw or long dry grass till seed germination and sprinkle water regularly.
- Seedlings may be raised under poly tunnels having sides open to protect the young seedlings from heavy rain or scorching sunlight.
- Seedlings should have plenty of sunlight but should be given shade from direct sun.
- Thinning of the seedlings should be done at least 20 days before transplanting for proper growth of the seedlings.
- Water the seedlings daily or in alternate day with rose can.
- Hardening of the seedlings by withholding water at least 4 to 6 days before transplanting.
- Stocky, 40-45 days old, healthy seedlings with 3-4 leaves are ideal for transplanting.
- Clipping of the seedlings 10 days prior to transplanting helps in better establishment of the seedlings and also accelerates the growth of the axilary buds.

Sowing time

Chilli can be grown in West Bengal, Maharashtra and South India almost throughout the year.However, chilli is grown in three distinct seasons viz., early autumn-winter, winter-spring and summer-rainy in the plains of India although sowing and transplanting season vary with the region. In the hills chilli is sown in February-March or March-April.

- May-July (for early autumn-winter crop)
- September- October (for winter-spring crop)
- January -February (for summer- rainy crop)

Land preparation

- Prepare the land in well advance with repeated ploughings (at least 4-5 ploughing) to a fine tilth.
- Remove all stubbles, weeds, etc and level the land.
- Add well rotten organic manure preferably farm yard manure or compost @ 20-25 t/ha during land preparation and level the land properly.
- Apply vermicompost @ 4-6t/ha as broadcasting during final land preparation.

Transplanting

- 40 -45 days old, healthy and clipped seedlings of 12-15 cm height are ready for planting.
- In open pollinated varieties, approximately 50 – 60 thousand seedlings are required for one hectare land.
- After hardening the seedlings for 4 – 6 days, irrigate the seed bed before lifting to facilitate easy pulling of young seedlings.
- Plant the seedlings in the field preferably during afternoon hours which give better field establishment.
- Irrigate the field just after transplanting.

Spacing

- Dwarf and medium tall varieties for green chilli: 45-60 cm × 30-45 cm
- Tall varieties or hybrids for green chilli : 75 cm × 60 cm
- Early winter-winter season for dry chilli: 75 cm × 75cm; 90 cm x 90 cm

Nutrient management

- Chilli is long duration crop and is good feeder of macro nutrients.
- Apply a general fertilizer dose of 100 kg N, 60 kg P_2O_5 and 60kg K_2O per hectare for open pollinated improved varieties. 1/3 N along with other fertilizers should be given as basal and rest N should be top dressed in two split doses at 21 and 42 days after transplanting.
- Apply 150 kg N, 80 kg P_2O_5 and 80kg K_2O per hectare for the hybrids. 1/4 N along with other fertilizers should be given as basal and rest N should be top dressed in three split doses at 30 days interval after transplanting.

- Apply vermicompost @ 4-6 t/ha as broadcasting during final land preparation.
- Apply *Azotobactor* @ 80 g/ha or Phosphate Solublizing Bacteria @ 800 g/ha as seedling root dip.
- Prepare a suspension of bacterial inoculants in sufficient water.
- Dip the seedlings in the bacterial suspension in shade for 2-3 hours in the afternoon hours.
- Transplant the treated seedlings immediately.

Irrigation

- Chilli needs judicious irrigation for proper growth and yield.
- Chilli can not withstand excess moisture.
- Frequent and heavy irrigations induce lanky growth and cause flower shedding.
- During summer, apply irrigation at 4-5 days interval and during winter at 10-12 days interval.
- About 18 acre-inch of total water is needed for successful chilli crop.
- Critical stages of irrigation are tenth leaf to flower, fruiting and after periodical harvests.
- Most adopted method of irrigation is furrow method. Drip irrigation is beneficial for reducing water requirement and weed control.
- Nutrients can be applied through drip irrigation.

Interculture and weed control

- Practice shallow inter-cultivation particularly a few days after every irrigation to remove the weeds and conserve soil moisture.
- 3-4 hoeings are normally needed to check the weeds.
- Earthing up the seedlings 21 days after transplanting after first topdressing with nitrogenous fertilizer.
- Regarding chemical weed control, a pre-planting treatment of fluchloralin @1.0-1.5 kg a.i. /ha followed by one hand weeding 30 days after transplanting is effective.

Harvesting the crop

- Harvesting of chillies depend on the type and purpose for which they are grown.
- Harvesting of green chillies may be started 60 to 70 days after transplantation.
- Ripe fruits are picked at an interval of 1-2 weeks and harvesting coutinues over a period of about 3 months.

- The number of picking varies from 6-10 depending upon the season, cultivar and cultural practices.
- The packaging used for transportation should have air holes for ventilation. Alternatively, jute bag can be used.

Post harvest handling and storage

- Green chillies can be stored in the cold storage up to 40 days at 0°C and 95 –98 % relative humidity.
- Green chillies treated with 200 ppm GA_3 significantly increase the shelf life.
- For dry chilli, harvest when the fruits are well ripened and partially withered in the plant itself.
- Keep the harvested fruits in heaps either indoor or in shade away from direct sunlight for 2-3 days to develop uniform red colour.
- Dry the red ripe fruits in the sun for 8-15 days depending upon the weather condition on clean dry polythene sheets, cemented/ concrete drying yards, etc.
- Artificially red ripe fruits are dried at about 54.4°C for 2-3 days.
- Spread the pods in thin layers for uniform drying with frequent turning to prevent mold growth and discoloration.
- Heap and cover the dried pods by clean gunny bags or polythene sheets.
- The moisture content of the dry pods is to be kept at 8-10 %.
- Keep the well-dried pods after removing the extraneous matters like plant parts, etc in clean, dry gunny bags and store ensuring protection from dampness.
- Stock the bags at 50 to 60 cm away from the wall of the cold storage.
- Storing dry chillies for longer period may lead to deterioration in colour.
- The product may be stored for 8-10 months in cold storage.
- Insects, rodents etc are to be effectively prevented from getting access to the premises where chilli is stored.

Plant protection measures

Physiological disorder

Flower and fruit drop

It is one of the major constraints in chilli cultivation. It may be caused due to i) low humidity and high temperature condition, ii)decreasing light intensity and iii)short day during early flowing stages.

Control measures

- Irrigation at flowering and fruit set stage reduces blossom and fruit drop.
- Foliar application of 50 ppm NAA at full bloom stage effectively controls the drop.

Disease

Damping off (*Pythium, Rhizoctonia, Phytophthora,* etc.)

It is very destructive fungal disease in nursery beds, especially during early season. The attack usually starts on the germinating seed, spreading to the hypocotyls, basal stem and developing taproot. In pre-emergence damping off, the growing points are killed in the initial stages of seed germination before they come out through the soil. In post-emergence damping off, the seedlings topple over the ground due to collar rotting and rapid shrinking of the cortical tissue of the hypocotyls.

Control measures

- Treat the seeds with thiram/ captan @ 3.0g /kg of seeds.
- Drenching the nursery beds 7 days before sowing with thiram/ captan or any copper fungicide @ 3g / litre of water.
- Cover the beds with transparent polythene sheets before sowing and left to open sun for at least 20 days, known as soil solarization.
- Application of bioagents like *Trichoderma* @ 5g /m^2 reduces the pathogen load in seedbed.
- Thin sowing and proper thinning of the seedlings at early stage.
- Spraying the young seedlings with 0.2% Blitox or 0.25% metalaxyl-mancozeb or 0.25 % fosetyl-Al.
- Provision of proper drainage to the beds.
- Remove the affected seedlings from beds as soon as the symptoms are visible.
- Avoid flooding the beds to check spread of the disease.

Die-back or Anthracnose (*Colletotrichum capsici*)

It is the most damaging, seed borne, fungal disease. Elongated, ash coloured, angular spots develop on veins and ventral surface of the leaf. Leaves fall down and the branches die back from the tip. Usually circular and sunken lesions with black margin appear on ripe fruits, which spread forming concentric markings with dark fructifications as the disease advances. The disease spread rapidly under hot and humid condition.

Control measure

- Seed treatment with 0.2% Bavistin or 0.3% thiram.
- Spraying the seedlings in the seed bed before transplanting with 0.3% captan or 0.1% Bavistin or chlorothalonil 0.2% or hexaconazole 0.1%.
- Spray the crop with hexaconazole 0.1% followed by 0.3% Blitox after removal of the infected twigs at 10 days interval.

Chilli leaf curl (Chilli leaf curl virus)

It is a destructive virus disease transmitted by white fly (*Bemisia tabaci*). Upward and downward curling of the leaves, yellowing of leaves accompanied by puckering of interveinal areas and thickening of veins are the characteristic symptoms. In advanced stage of the disease, axillary buds are stimulated producing clusters of small leaves, and the plant appears bushy with stunted growth.

Control measure

- Destruction of all weeds around the field which serve as alternate host of the virus.
- Cover the nursery bed with 50 mesh fine nylon net and spray outside with imidacloprid (3.5 ml / 10 l).
- Pre-sowing treatment of nursery beds with Furadon 3G @15g/100 sq.m.
- Apply Thimet 10G @ 10-15 kg /ha 10 days after transplanting followed by 2-3 foliar sparys with 0.05 % dimethoate (Rogor 1.5 ml /l) or imidachloprid or acetameprid or thiomethoxam (3.5 ml / 10lt) at 10 days interval and alternate the spray schedule with NSKE.
- Apply 2% power oil to the plants to prevent acquisition and inoculation of the virus by white flies.
- Sow 5-6 rows of border crops of maize, jowar and bajra all round the chilli plot at least 50-60 days before transplanting of chilli.

Insect pests

Thrips (*Scirtothips dorsalis*)

Both nymphs and adults lacerate the tender leaf tissues and feed on the oozing sap resulted in curling and crumpling of the leaves. Severe infestations cause shortening of internodes, thickening and reduce size of leaves with corrugated leaf lamina. Buds and flowers are also attacked. Infestation becomes severe during warm and dry weather conditions.

Control measure

- Seedling root dipping before transplanting in the aquous solution of fipronil (2ml/l) or imidacloprid (1ml/3l) or monocrotophos (2ml/l).
- Spraying the crop with acetamiprid (0.4ml/l) on fipronil (1.5ml/l) or imidacloprid (5ml/15 lit.) or thiomethoxam (1g/5 lit.) or profenophos (1ml/l) at fortnightly interval.

Aphid (*Aphis* spp.)

The aphid species *Myzus persicae* and *Aphis gossypii* attack chilli crop during early season of crop growth during the month of January-February. They multiply rapidly and apart from continuous sucking from the leaves and young apical twigs, they also secret honeydew on the plant canopy resulting formation of sooty mould. Severely infested plants may wither gradually.

Control measures

- Release of lady bird beetles and hover fly reduce the aphid population to some extent.
- Applications of imidacloprid @ 0.3 ml/l or chlorpyriphos @ 2.5 ml/l control the aphids.

Yellow mite (Polyphagotarsonemus latus)

Among the different mite species attacking vegetable crops, the yellow mite, *Polyphagotarsonemus latus* is one of the most destructive causing enormous yield losses of chilli. Though it has hundreds of host plants but chilli is one of the most preferred one which is attacked by the mite through out the warm and humid period of the year starting from February to November under Bengal basin. The mites multiply very fast complete their life cycle within three days. Their populations remain suppressed only during rainy days when it continues for days together and in winter months. The mite attack young apical leaves producing characteristic curling and crumpling symptoms which might be confused with that produced due to infestation of thrips and leaf curl virus. In case of mite attack, the young leaves usually curl downward in an inverted boat shaped manner, the texture of the leaves become leathery, lower surface turn shiny silvery- whitish appearance which gradually become brownish and curl completely. At the early stages of crop growth the infested leaves get thickened and later crack. The apical shoot gets crumpled, become rossetted and the growth of the plant is stunted. Infestation at the flowering stage causes bud and flower shedding resulting complete loss of yield.

Control measures

- Save the plant from the early infestation of whitefly which help to disseminate yellow mite by spraying with imidacloprid @ 0.3 ml/l.
- Prophylactic spray should be given with azadirectin 5% @ 1ml/l at 7day interval.
- Conserve and release predatory mite, *Agistemus fleschneri* for sustainable management of yellow mite in chilli ecosystem.
- Select mite tolerant chilli cultivars like PC-1, BC-CH-Sl-4 and plant during September- October to obtain maximum yield desynchronizing the occurrence of mite.
- Application of carbofuran 3G @ 1 kg a.i/ha 15 days after transplanting.
- Spray the crop with acaricides like dicofol @ 2.5 ml/l; propergite @1.5 ml/l or diafenthiuron @ 1 ml/l during early budding stage of the crop at 15 days interval.

Yield

- Green chillies of open pollinated variety: 7.0 – 10 t/ha
- Green chillies of hybrid :12-15 t/ha
- Dry chillies of open pollinated variety :1.5-2.5 t/ha
- Dry chillies of hybrid : 3.0 – 3.5 t/ha

5. Capsicum *or* Sweet pepper

Scientific name: Capsicum annuum
Origin: Southern and Central America
Family: Solanaceae
Chromosome No.: 2n =2x =24

Introduction

Sweet pepper or capsicum or bell pepper requires slightly cooler climate than hot chillies and is grown through out the world for its thick and fleshy fruits having delicate flavour and taste. The major countries producing sweet peppers are China, Indonesia, Sri Lanka, Pakistan, Turkey, Korea, Spain, Bulgaria, Hungary, Romania, Italy, Yugoslavia, Nigeria, Ghana, Tunisia, Mexico, USA, Argentina and Peru. In India it is generally grown in the cooler weather particularly in the periphery of cities. Mixed stock of *Capsicum annuum* including both hot and sweet peppers was introduced in India by the Portuguese from Brazil in 1584. However, chillies or hot pepper spread much more widely and rapidly than sweet pepper in India.

Importance and use

- Capsicum is also known as sweet pepper, bell pepper, pimento, Simla mirch or paprika.
- Sweet pepper is grown as autumn-winter crop in the plains and summer crop in the hill regions of India.
- Major capsicum growing areas lay around Bangalore, Belgaun and Mysore of Karnataka, Pune and Thane in Maharashtra, Nilgiri hill areas of Tamil Nadu, Ranchi of Jharkhand, Darjeeling hill and some part of 24 Parganas of West Bengal, hill regions of Uttar Pradesh and Uttaranchal, Himachal Pradesh and Jammu and Kashmir.
- Capsicum or sweet pepper is internationally classified into several groups of cultivars on the basis of fruit shape and pungency.
- Capsicum is mostly consumed both in green matured and ripe form raw, in salads, cooked, mixed and stuffed vegetable.
- Ripe forms of capsicums are mostly used in Europe and USA.
- It is widely used in the preparations of pickles, sauces and soups.

- In stews and other preparations sweet pepper imparts a novel flavor.
- The varieties or hybrids having bell shaped fruit, with thick flesh, predominant lobing, firm and shinning green colour are suitable for export.

Nutritional value

- Capsicum is particularly rich in Vitamin A and C otherwise, it add novel and delicate flavour to the diet.
- Capsicum contains very high amount of vitamin C (175 mg/100 g fresh).
- Capsicum is also particularly rich in vitamin A (683 mg/100g fresh).
- It has high crude fibre content.
- Sweet peppers lose their vitamin content when they are cooked or pickled or if they are allowed to ripen too much.

Medicinal value

- The delicate flovour of capsicum stimulates the taste buds but also the appetite.
- Appreciably high Vitamin A and C contents in capsicum act as potential antioxidant.
- The carotenoides, predominantly, capsanthin and beta carotene have anti-oxidant property.
- The flavonoides present in red pepper particularly, quercetin, kaempferol, myricetin and luteolin are responsible for induction of enzymes that detoxify carcinogens.
- Capsicum preparations are used to counter irritants, in lumbago, neuralgia and rheumatic disorders.
- It has tonic and carminative action especially in dyspepsia.

Adaptation in India

Mixed stock of *Capsicum* species comprising of both hot and sweet pepper was introduced in India by the Portuguese about 425 years ago into Kerala. Christian missionaries also introduced *Capsicum* species in the North-eastern states long ago separately. High adaptation in the tropical Indian condition, wide spread, extensive course of cultivation almost through out the country and farmers' selections particularly for pungent and red types have resulted in the development of huge numbers of local cultivars of chilli under different names in India. In contrast, very few local cultivars have been developed in sweet pepper suggesting that Indian farmers were in favour of pungent and red hot pepper. Hence, most of the non-pungent or mildly pungent types, if at all have been introduced in India as mixed stock, possibly got eliminated. Very few local cultivars of sweet pepper having pungent or non-pungent, small, roundish fruits are still being grown by the farmers in Karnataka (Srirangapatanam area), Goa and coastal areas of Kerala. However, leading present day varieties in India are important introductions although, some pure line selections have also been done from exotic materials.

Breeding behaviour

- Capsicum is basically self- pollinated crop, but it may be referred to an often cross pollinated crop due to appreciable degree of cross pollination (on an average 16%) and open nature of floral morphology.
- To maintain the purity of the varieties, isolation distance of 400-500 m is safe for seed production.
- Major pollinators are honey bees (*Apis* species), ants and wind.
- Anthesis generally starts from 5.30 am and majority of the flowers open by 9.00 a.m. however, it depends on the prevailing temperature and humidity condition.
- Anther dehiscence occurs after anthesis.
- The pollen grains in the anther become fertile a day before anthesis with maximum fertility on the day of anthesis.
- The stigma receptivity is the maximum at anthesis and the receptivity continues upto two days after.
- At comparatively high temperature condition, emasculation is advisable 2-3 days prior to anthesis and cross pollination is carried out on the day of anthesis.
- At high temperature condition, fair crossing success can be achieved if hybridization is taken up either early in the morning or late afternoon.
- Pollen viability is retained for a day in ambient condition. Pollens can be kept in refrigerator under airtight condition for 2-3 days. Dried pollens can be stored in sealed vial at –5.0°C for 10 days.

Origin and taxonomy

Capsicum annuum var. *annuum* under which both hot and sweet pepper belong was domesticated in Mexico. Mexico is probably the centre for morphological diversity for this important species on a world scale since it includes all the commercially important sweet peppers and many hot peppers.

Domestication , human selection and breeding have produced many varieties of sweet pepper from the great natural diversity of the *Capsicum annuum* var. *annuum* differing greatly in fruit size and shape, pungency and ranging in colours from orange , red , yellow, violet to green. Different cultivar groups of sweet peppers can be classified primarily on the basis of fruit shape and size.

Bell	Large, blocky fruits with a blunt end of 3 to 4 lobes, thick walled, sweet, fruit colours, green, red, yellow, orange, violet.
Cheese	Mostly globose shaped fruits, small to medium in size and of various shape, mild, yellow to green maturing to red.
Cuban	Fruits of irregular blunt shape, mildly pungent, thin walled.
Long Wax	Fruits long, tapered but in some cases blunt, mildly pungent
Pimento	Fruits large, cone or heart shaped, thick walled, not pungent.

Improved open pollinated varieties

California Wonder, Yolo Wonder, World Beater, Early Bountiful, Chinese Giant, Giant Bullnose, Arka Basant, Arka Gaurav, Arka Mohini, Nishat-1.

Public sector hybrids

Pusa Deepti, Solan Hybrid-1, Solan Hybrid-2.

Cultural requirements

- Capsicum performs well in mild climate hence, considered a cool season crop.
- A temperature ranging from 26-28°C during daytime and 16-18°C at night is ideal.
- It can not withstand extreme temperatures, below 15°C and above 30°C significantly reduce growth and fruit set.
- Capsicum is susceptible to frost.
- High relative humidity is good for growth and yield, fruit set and yield however, this condition is favourable for the development of foliar and fruit diseases.
- It is highly sensitive to environmental factors.Hence, its cultivation under poly-tunnels or poly-houses is preferred.
- Use of misters and foggers inside the polyhouse to maintain the slightly higher relative humidity and marginally reduced temperature to the extent of 3–4°C significantly increase fruit set and yield.
- Providing shade through poly or net house during summer and only shade net in open during other seasons is beneficial.
- Well-drained red-loam and alluvial soil with slightly acidic pH is considered ideal for sweet pepper cultivation.
- On sandy loam soil, it can successfully be grown with proper manuring and timely irrigation.
- It can stand acidity to a certain extent.
- The pH should be around 5.5-6.8.
- Enough water should be available, but the soil should also have good drainage facility.

- Capsicum has bi-directional root system and it is important that there is free unhampered root growth and root aeration for successful crop growth.

Nursery management

- Seed beds should be prepared finely, well drained, 15-20 cm raised, 1.0m wide and of convenient length.
- Mix well decomposed farmyard manure or compost @ 3-4 kg/m^2 and a fertilizer dose of 0.5 kg NPK of 15:15:15 per bed, 1.0 m wide and 7.5 m long 10 days before sowing the seeds.
- Drench the beds with formaldehyde (4.0%) or captan @ 2-3g/litre and cover with polythene sheet for 7 to 10 days to avoid damping off disease.
- The seed beds may be covered with transparent polyethylene sheet for 7-10 days (solarization) for soil disinfection.
- 300 g seeds of open pollinated varieties are required per hectare.
- 250 g seeds of hybrids are required per hectare.
- Treat the seeds with captan or thiram @ 2-3 g per kg of seeds before sowing.
- Sow the seeds in rows 8-10 cm apart and more than 1 cm deep and covered with finely sieved leaf mould.
- After sowing cover the beds with straw or long dry grass till seed germination and sprinkle water regularly.
- Seedlings may be raised under poly tunnels having sides open to protect the young seedlings from heavy rain or scorching sunlight.
- The nursery should be covered with 40 mesh nylon net to protect the seedlings from feeding by virus carrying vector, white fly.
- Seedlings should have plenty of sunlight but should be given shade from direct sun.
- Thinning of the seedlings should be done at least 20 days before transplanting for proper growth of the seedlings.
- Spray the seedlings with 0.25% Dithane M-45 or difolaton at weekly interval to reduce the chance of post-emergence damping off.
- Water the seedlings daily or in alternate day with rose can.
- Hardening of the seedlings by withholding water at least 4 to 6 days before transplanting.
- Stocky, 35-40 days old, healthy seedlings with 3-4 leaves are ideal for transplanting.
- Clipping of the seedlings 10 days prior to transplanting helps in better establishment of the seedlings and also accelerates the growth of the axilary buds.

Sowing time

- July- September (for autumn-winter crop)
- October- November (for winter-spring crop in poly-tunnel in north Indian condition)
- February- March (for summer crop in the hills)

Land preparation

- Prepare the land in well advance with repeated ploughings (at least 4-5 ploughing) to a fine tilth.
- Remove all stubbles, weeds, etc and level the land.
- Add well rotten organic manure preferably farm yard manure or compost @ 20-25 t/ha during land preparation and level the land properly.
- Apply vermicompost @ 4-6 t/ha as broadcasting during final land preparation.

Transplanting

- 35-40 days old, healthy and clipped seedlings of 12-15 cm height and with 4-5 leaves are ready for planting.
- It is better to transplant the seedlings in one side of the ridges.
- After hardening the seedlings for 4 – 6 days, irrigate the seed bed before lifting to facilitate easy pulling of young seedlings.
- Plant the seedlings in the field preferably during afternoon hours which give better field establishment.
- Irrigate the field just after transplanting.

Spacing

- Open pollinated varieties: 45 × 45 cm
- Open pollinated varieties: 60 cm × 30 cm with 55,000 plants/ha
- Hybrids: 75 -80 cm × 30-45 cm

Nutrient management

- Capsicum is a good feeder of macro nutrients.
- Dip the seedlings in 1.5 % superphosphate solution before transplanting gives better result because capsicum is a shy rooter.
- Apply a general fertilizer dose of 120 kg N, 80 kg P_2O_5 and 60kg K_2O per hectare for open pollinated improved varieties. 1/3 N along with other fertilizers should be given as basal and rest N should be top dressed in two split doses at 21 and 42 days after transplanting.

- Apply 200 kg N, 150 kg P_2O_5 and 100kg K_2O per hectare for the hybrids. 1/4 N along with other fertilizers should be given as basal and rest N should be top dressed in three split doses at 30 days interval after transplanting.
- Nitrate form of nitrogen is more preferable than ammoniacal form of nitrogen.
- Apply vermicompost @ 4-6 t/ha as broadcasting during final land preparation.
- Apply VAM fungus, *Glomus itraradices* in the seed bed.
- Apply Azotobactor @ 80 g/ha or Phosphate Solublizing Bacteria @800g/ha or Azospirillium @ 2 kg/ha as seedling root dip.
- Prepare a suspension of bacterial inoculants in 10 litres of water.
- Dip the seedlings in the bacterial suspension in shade for 30 minutes in the afternoon hours.
- Transplant the treated seedlings immediately.

Irrigation

- Capsicum needs judicious irrigation for proper growth and yield.
- Capsicum is sensitive to both soil moisture deficit and excess.
- Frequent and heavy irrigations induce lanky growth and cause flower shedding and reduce yield.
- In sandy-loam soil, it is best to provide irrigation at 60% soil moisture depletion level in the top 30 cm soil profile.
- Generally apply irrigation during winter at 12-15 days interval.
- Average consumptive use is 446 mm.
- Critical stages of irrigation are flowering, fruiting and after periodical harvests.
- Most adopted method of irrigation is furrow method.
- Drip irrigation is beneficial for reducing water requirement and weed control preferably polyhouse condition.
- Nutrients can be applied through drip irrigation.

Interculture and weed control

- Practice shallow inter-cultivation particularly a few days after every irrigation to remove the weeds and conserve soil moisture.
- 3-4 hoeings are normally needed to check the weeds.
- Earthing up the seedlings 25 days after transplanting after first topdressing with nitrogenous fertilizer.

- Regarding chemical weed control, a pre-planting treatment of fluchloralin @1.0-1.5 kg a.i. /ha followed by one hand weeding 30 days after transplanting is effective.
- Basalin @ 0.48 kg a.i. /ha applied 5 days after transplanting followed by one hoeing 35 days after transplanting effectively controls weeds.
- Spray the plants twice at 10 days interval with 25 ppm GA or 50 ppm NAA at flowering stage to increase fruit set.
- Judicious pinching of the flowers to produce 8-10 big fruits of 90 to 100 g weight per plant.
- Provide stake to the plants to produce blemish less fruits.

Harvesting the crop

- Pick the large sweet peppers when they are still green and full grown, firm and crisp.
- Red and yellow pepper varieties can be harvested at the onset of colour change.
- Pimento and paprika cultivars are picked at dark red ripe stage.
- Harvesting of sweet peppers starts 35-50 days after flowering depending on the variety.
- Harvesting can be done once in 10-12 days with 5-6 pickings in open pollinated varieties and 8-10 pickings in the hybrids.
- Pick the fruits with an upward twist with a piece of stem attached.
- The packaging used for transportation should have air holes for ventilation.

Post harvest handling and storage

- Harvested fruits are kept in shade to avoid sunscald.
- Immediate pre-cooling by immersing preferably in ice-cold water enhances the shelf-life.
- Bell pepper fruits are wiped, cleaned in fresh water, graded in size and packed in wooden boxes, paper cartons or plastic/iron crates before marketing.
- It can be stored up to 14 to 21 days at 7-10°C and high RH of 90-95% and for at least 40 days at 0°C and RH of 95-98 % with minimum shrinkage (only 4%).
- Dipping the harvested fruits for 2 minute in 1 or 2% potassium bicarbonate solution reduces decay development after storage and stimulates shelf life.
- The fruits should be transported in cold temperature condition of 7-10°C.

Plant protection measures

Physiological disorder

Flower and fruit drop

It is one of the major constraints in sweet pepper cultivation. It may be caused due to i) low humidity and high temperature condition, ii) decreasing light intensity and iii) short day during early flowering stages.

Control measures

- Irrigation at flowering and fruit set stage reduces blossom and fruit drop.
- Foliar application of 50 ppm NAA at full bloom stage effectively controls the drop.

Skin cracking

Cracking occurs around the shoulder of the fruits. Fluctuations in temperature and humidity are responsible for this disorder. High day temperature and average relative humidity increase the incidence of cracking.

Sunscald

Soft, light coloured area appears on the fruit, which becomes slightly wrinkled, slightly sunken and papery later. It occurs when the fruits are exposed to scorching sunlight.

Blossom end rot

Water soaked spots appear at the blossom end which become light brown and papery as the lesions dry out. Heavy irrigation after a period of low soil moisture condition and heavy application of nitrogenous fertilizers causes this disorder.

Control measures

- Judicious and timely irrigation.
- Application of judicious amount of P and K fertilizers along with N fertilizer.
- Two foliar sprays of 0.2% calcium chloride the time of fruit development.

Diseases

Damping off (*Pythium*, *Rhizoctonia* ,*Phytophthora*, etc.)

It is very destructive fungal disease in nursery beds, especially during early season. The attack usually starts on the germinating seed, spreading to the hypocotyls, basal stem and developing taproot. In pre emergence damping off, the growing points are killed in the initial stages of seed germination before they come out through the soil. In post-emergence damping off, the seedlings topple over the ground due to collar rotting and rapid shrinking of the cortical tissue of the hypocotyls.

Control measures

- Treat the seeds with thiram/ captan @ 3.0g /kg of seeds.
- Drenching the nursery beds 7 days before sowing with thiram/ captan or any copper fungicide @ 3g / litre of water.
- Cover the beds with transparent polythene sheets before sowing and left to open sun for at least 20 days, known as soil solarization.
- Application of bioagents like *Trichoderma* @ 5g /m^2 reduces the pathogen load in seedbed.

- Thin sowing and proper thinning of the seedlings at early stage.
- Spraying the young seedlings with 0.2% Blitox or 0.25% metalaxyl-mancozeb or 0.25 % fosetyl-Al.
- Provision of proper drainage to the beds.
- Remove the affected seedlings from beds as soon as the symptoms are visible.
- Avoid flooding the beds to check spread of the disease.

Die-back or Anthracnose (*Colletotrichum capsici*)

It is the most damaging, seed borne, fungal disease. Elongated ash coloured, angular spots develop on veins and ventral surface of the leaf. Leaves fall down and the branches die back from the tip. Usually circular and sunken lesions with black margin appear on ripe fruits, which spread forming concentric markings with dark fructifications as the disease advances. The disease spread rapidly under hot and humid condition.

Control measure

- Seed treatment with 0.2% Bavistin or 0.3% Thiram .
- Spraying the seedlings in the seed bed before transplanting with 0.3% captan or 0.1% Bavistin or chlorothalonil 0.2% or hexaconazole 0.1%.
- Spray the crop with hexaconazole 0.1% followed by 0.3% Blitox after removal of the infected twigs at 10 days interval.

Fruit rot or Phytophthora blight (*Phytopthora capcisi*)

This fungal disease is the most destructive disease in sweet pepper. The disease mostly appears in comparatively high temperature and high relative humidity condition and is spread through rain splash, wind and irrigation water. Initial symptoms in the field occur as crown rot from soil inoculum. In the early stage, wilting of the plants occur with purplish-black lesions at the collar region.Dark green water soaked spots later appear on the fruits which enlarge rapidly to cover the entire fruit causing rotting.

Control measures

- Crop rotation excluding capsicum and chilli.
- Avoid giving excessive irrigation.
- Provide proper drainage facility in the field.
- Spray the crop with 0.3% Blitox or 0.25% mancozeb or 0.25% metalaxyl-mancozeb or 0.25 % fosetyl-Al at 10 days interval.

Bacterial leaf spot (*Xanthomonas campestris pv. vesicatoria*)

These seed borne bacterial disease favours 21-31°C and 85 –90% relative humidity coupled with continuous drizzling and cloudiness for its development. Yellowish green spots appear on young leaves while dark coloured water soaked areas on older leaves. The spots later appear with straw-coloured centers and dark margins. Small blisters like watery spots appear on infested fruits.

Control measures

- Hot water treatment of seeds at 50^0 C for 20 minutes.
- Spray the crop with a combination or 100 mg Agrimycin 100 + 3.0 g Blitox per litre of water during favourable weather condition..

Soft rot (*Erwinia carotovra*)

It is a common post harvest bacterial disease causing fruit decay which usually starts at wounds in the carpel or peduncles.

Control measures

- Avoid injury to the fruits during harvesting.
- Storage of fruits in low temperature condition.

Chilli leaf curl (Chilli leaf curl virus)

It is a destructive virus disease of both chilli and sweet pepper, transmitted by white fly (*Bemisia tabaci*). Upward and downward curling of the leaves, yellowing of leaves accompanied by puckering of interveinal areas and thickening of veins are the characteristic symptoms. In advanced stage of the disease, axillary buds are stimulated producing clusters of small leaves, and the plant appears bushy with stunted growth.

Control measures

- Destruction of all weeds around the field which serve as alternate host of the virus.
- Cover the nursery bed with 50 mesh fine nylon net and spray outside with imidachloprid (3.5 ml / 10 l).
- Pre-sowing treatment of nursery beds with Furadan 3G @15g/100 sq.m.
- Apply Thimet 10G @ 10-15 kg /ha 10 days after transplanting followed by 2-3 foliar sparys with 0.05 % dimethoate (Rogor 1.5 ml /l) or imidachloprid or acetameprid or thiomethoxam (3.5 ml / 10l t) at 10 days interval and alternate the spray schedule with NSKE.
- Apply 2% power oil to the plants to prevent acquisition and inoculation of the virus by white flies.
- Sow 5-6 rows of border crops of maize, jowar and bajra all round the capsicum plot at least 50-60 days before transplanting of capsicum.

Insect pests

Thrips (*Scirtothips dorsalis*)

Both nymphs and adults lacerate the tender leaf tissues and feed on the oozing sap resulting in curling and crumpling of the leaves. Severe infestations cause shortening of internodes, thickening and reduction in size of leaves with corrugated leaf lamina. Buds and flowers are also attacked. Infestation becomes severe during warm dry weather conditions.

Control measures

- Seedling root dipping before transplanting in the aquous solution of fipronil (2ml/l) or imidacloprid (1ml/3l) or monocrotophos (2ml/l).
- Spraying the crop with acetamiprid (0.4ml/l) or fipronil (1.5ml/l) or imidacloprid (5ml/15 l) or thiomethoxam (1g/5 l) or profenophos (1ml/l) at fortnightly interval.

Aphid (*Aphis* spp.)

The aphid species *Myzus persicae* and *Aphis gossypii* attack chilli and sweet pepper crop during early season of crop growth during the month of January-February. They multiply rapidly and apart from continuous sucking from the leaves and young apical twigs they also secret honeydew on the plant canopy resulting formation of sooty mould. Severely infested plants may wither gradually.

Control measures

- Release of lady bird beetles and hover fly reduce the aphid population to some extent.
- Applications of imidacloprid @ 0.3 ml/l or chlorpyriphos @ 2.5 ml/l control the aphid population.

Yellow mite (*Polyphagotarsonemus latus*)

Among the different mite species attacking vegetable crops, the yellow mite, *Polyphagotarsonemus latus* is one of the most destructive causing enormous yield losses of both chilli and sweet pepper. The mites multiply very fast complete their life cycle within three days. Their populations remain suppressed only during rainy days when it continues for days together and in winter months. The mites attack young apical leaves producing characteristic curling and crumpling symptoms which might be confused with that produced due to thrips and leaf curl virus. In case of mites attack, the young leaves usually curl downward in an inverted boat shaped manner, the texture of the leaves become leathery, lower surface turn shiny silvery- whitish appearance which gradually become brownish and curl completely. At the early stages of crop growth the infested leaves get thickened and later crack. The apical shoot gets crumpled, become rossetted and the growth of the plant is stunted. Infestation at the flowering stage causes bud and flower shedding resulting complete loss of yield.

Control measures

- Save the plant from the early infestation of whitefly which help to disseminate yellow mite by spraying with imidacloprid @ 0.3 ml/l.
- Application of carbofuran @ 1 kg a.i/ha 15 days after transplanting.
- Spray the crop with acaricides like dicofol @ 2.5 ml/l; propergite @1.5 ml/l or diafenthiuron @ 1 ml/l during early budding stage of the crop at 15 days interval.

Pepper weevil (*Anthonomus eugenii*)

Eggs are laid beneath the fruit surface and the grubs after hatching feed within the fruits. The affected fruits become misshapen and discolored.

Control measures

- Spray the crop with 0.5% endosulfan or malathion.

Pepper maggot (*Spilographa electa*)

The maggots feed on the small fruits from inside and damaged fruits may by destroyed by the secondary infection of soft rot causing bacteria.

Control measures

- Spray the crop with 0.2% carbaryl or 0.05% endosulfan from bloom stage at weekly interval.

Yield

- Open pollinated variety in open field condition: 10 – 15 t/ha
- Open pollinated variety in poly house fitted in mister and fogger: 20 – 25 t/ha
- Hybrids in poly house fitted in mister and fogger: 35 – 45 t/ha

Cole Crops

6. Cauliflower

Scientific Name: Brassica oleracea L var. botrytis
Family: Cruciferae or Brassicaceae
Origin: Cyprus
Chromosome No. 2n = 2x =18

Introduction

Cauliflower is the most popular vegetable among the cole crops. It is grown in all continents of the world, of which Asia is the leading one followed by Europe and covers small acreages in North America, South America, Africa, Australia and New Zealand. India is the largest producer of cauliflower in the world and other major producers are China, France, Italy, UK, USA, Spain, Poland, Germany and Pakistan. It is believed that it has been originated in the island of Cyprus from where it moved to other areas like Syria, Turkey, Egypt, Italy, Spain and North Western Europe. In the middle of the 16th century, the first illustration and description of cauliflower was presented by the herbalist Dodoens (1544). The crop became more commonly cultivated, particularly in the beginning of the 18th century. Cauliflower was introduced in India from England in 1822 by Dr. Jemson, a botanist from Royal Kew garden who was the incharge of the Company Bagh, Saharanpur, Uttar Pradesh (erstwhile United Provinces).

The Royal Agri-Horticultural Society, Calcutta also introduced cauliflower in 1824 from South Africa. Now cauliflower is a major vegetable crop of India and grown in almost all the states particularly Bihar, Uttar Pradesh, Orissa, West Bengal, Assam, Haryana, Punjab and Maharashtra.

Importance and use

- Cauliflower is the most important crop among the Cole crops in India, China and many other countries.
- India is the largest producer of cauliflower..
- The name cauliflower has originated from the Latin words 'Caulis' meaning stem and 'Floris' meaning flower.
- The edible part of cauliflower is 'curd' which is botanically the pre-floral fleshy apical meristem.
- Morphologically, the curd is made up of numerous divided hypertrophic branches which terminate the main stem of the plant. It is highly suppressed and extremely ramified hypertrophied flower stalk.
- Development of tropical Indian cauliflowers and introduction of F_1 hybrids by Japanese seed companies in addition to the temperate or Snowball type, it has now become possible to grow cauliflower almost throughout the year particularly in the northern and central part of India.
- All the F_1 hybrids of tropical types have basically been developed from the Indian materials.
- In India there is hardly any house where it is not used regularly as vegetable.
- It is generally used as cooked vegetable either singly or mixed with potato, peas, capsicum or other vegetables as fried or in curry form or in sambar.
- Cauliflower is also used in fish curry as a delicacy.
- It is also used in the preparation of cauliflower pickle or mixed pickle with other vegetables.
- Boiled cauliflower is also consumed in continental form applying only salt, black pepper powder.
- Grated cauliflower is used for the preparation of stuffed food items.
- Cauliflower is generally processed in frozen form.

Nutritional value

- It is not a rich source of nutrients however, a substantial amount of protein, carbohydrate, phosphorus, calcium, iron and ascorbic acid is present.
- Protein content of cauliflower is much higher (2.4 g/100g fresh) than cabbage.
- It contains fairly good amount of Vitamin C (75 mg/100 g fresh).

- It is a good source of calcium (30 mg), iron (17 mg) and phosphorus (76 mg) per 100 g fresh.
- It also contains fair amount of vitamin A (70 IU/100 g fresh).

Medicinal value

- Cauliflower is fairly high source of glucosinolates (40-80 mg/100g fresh), predominantly sinigrin and glucobrassicin which have prominent anti-carcinogenic property.
- The products of hydrolysis of these glucosinolates formed by enzymatic or non-enzymatic action are biologically active compounds with diverse effect on human health.
- The product of glucosinolate breakdown, particularly the isothiocyanates act as anti-carcinogens.
- Cauliflower also contains S-methylcysteine sulfoxide (14 mg/100 g fresh) which also show anti-carcinogenic property.
- Beta carotene present in cauliflower act as potential anti-oxidant.

Adaptation in India

The cauliflower introduced in India more than 190 years ago was Cornish group of cauliflower of England. Later other European types like, Roscoffs, Italians and Northerns were also introduced. During the period of 100 years (1822 – 1929) these introduced cauliflower varieties underwent selection by the local growers when seed production was attempted by them in north Indian plains. Selections were exercised for early maturity and wider adaptability in hot and humid weather conditions. Natural crossings between Cornish and other European types followed by selection by the farmers and adaptation in high temperature condition resulted in the development of a separate group of cauliflower, the Indian or Tropical cauliflower. These tropical cauliflowers are distinctly different from the temperate European types and are adapted to early sowing and early harvest. Thus a new development in cauliflower took place in India in the form of tropical cauliflower resistant to hot weather and high rainfall. These tropical or Indian types are grown in the plains during June to December and normally followed by the temperate types known as snowball or late cauliflower.

In India two separate groups of cauliflower are grown namely, Indian or tropical cauliflower and temperate annual or Snowball type of cauliflower. The typical tropical cauliflowers form curd and mature at average daily temperature around 20°C and above. These are long stalked plant with broadly wavy and loosely arranged leaves producing highly flavoured, flat to semi-spherical , uneven and some what loose and yellow to cream colored curds. Cultivars of maturity group I and II (early and mid-early or Kanwari, Katki, Aghani group) belong to typical tropical cauliflower. Cultivars of maturity group

III (mid-late or Poosi or Pousi group) also show most of the characters of tropical cauliflower group. There are several local cultivars in varying maturities, commonly named after the season of curd maturity. Seeds of these local varieties are marketed by different private companies of Bihar (particularly Hazipur), Uttar Pradesh (particularly Faizabad), Punjab and Haryana.

The annual temperate or Snowball group of cauliflower is not adaptive to high temperature condition. They generally form curd within the temperature range of 10°C and 16°C. This cauliflower group shows similarities with Erfurt and Alfa types of cauliflower originated in Germany and the Netherlands. These are dwarf plant with dark green, wavy, short, erect (sometimes spreading) leaves and producing solid, very deep, knobby, bright white, less flavoured and high quality curds which are less sensitive to riceyness. Cultivars of maturity group IV (late or Maghi group) come under this type in strict sense.

Breeding behaviour

- The tropical or Indian cauliflower types behave like annual and bolt in response to cool temperature followed by long day condition. They produce flowers and set seeds freely in the plains of Punjab, Haryana, Uttar Pradesh and Bihar where temperature may drop down to even 1-2°C in some period of winter.
- Seeds of the earliest group of cauliflower (Kanwari group) can even be produced in the mild winter condition in the plains of West Bengal.
- The temperate or Snowball types (group IV) behave like biennial crop which require 6-8 weeks of vernalization at mean temperature range of 4° to 8°C after they have passed the initial curding phase for the production of seed stalks. Seeds of these types of cauliflower are only produced in the high hill situations of Himachal Pradesh and Kashmir.
- Cauliflower bear hermaphrodite flowers but varying levels of cross fertilization takes place even upto 70 percent in different open pollinated varieties due to manifestation of self-incompatibility of different intensities.
- Self-incompatibilty system which is sporophytic in nature and governed by multiallelic single gene (*S* gene) with generally dominant effect. Intensities of self-incompatibility depend on the presence of *S* allele in the genotype.
- In the early bud stage, 2-3 days before anthesis when the stigma has already become receptive (protogyny), self-fertilization can be done by self-pollinating with the pollens from other opened flowers of the same plant.
- The *S* alleles are active in the stigma in flowering stage but remain inactive in the bud stage, i.e., 2-4 days before anthesis.
- The glycoproteins present in the stigma hinder the pollen germination and penetration of the pollen tubes through the styler tissue.

- Heat tolerant tropical cauliflowers show significantly high level of self-incompatibility.
- The European annual or Snowball types are basically self-compatible in nature.
- In cauliflower cross pollination is generally done by bees, more particularly honey bees.
- Bees and other insect pollinators frequently visit the flowers of cauliflower because of the presence of nectar in the flowers.
- To maintain the purity of the varieties, isolation distance of at least 1200 m is safe for seed production.
- Genic male sterility controlled by single recessive gene and duplicate dominant genes with cumulative effect have been found in cauliflower however, such sterility is not generally utilized in hybrid seed production.
- The male sterility transgene *barnase* from *Bacillus amyloliquifaciens* which destroys the tapetum layer in the anther sac has been utilized to develop stable cytoplasmic male sterile line in cauliflower.
- Hybrids of cauliflower have been developed mainly in Japan, Korea, Taiwan and USA utilizing self-incompatibility and cytoplasmic male sterility mechanisms.
- Anthesis starts early in the morning and continues up to 11.30 am depending on the prevailing weather conditions.
- Dehiscence of anther is longitudinal and it generally occurs few hours after anthesis.
- The stigma becomes receptive at least two days before anthesis although receptivity is the maximum at anthesis and it continues at least one day after.

Origin and taxonomy

The cole crops have been originated mainly by mutation and introgression from wild species during evolution, human selection and adaptation. It has descended through mutation and selection from a common ancestor *Brassica oleracea* var *sylvestris*, the leafy kale like wild cabbage. Probably kale was the first cole crop to be selected and adapted by man. The cultivated species selected by man had biennial habit with yellow flowered racemose inflorescence unlike the ancestral kale or broccoli of the Mediterranean region which had cymose inflorescence with white flowers. Natural crossing between ancestral broccolis (*Brassica cretica* ssp. *nivea*), the wild annual with cymose white flowers and the biennial wild types have resulted into many types of *Brassica oleracea* including various forms of broccoli and cauliflower several thousand years ago. However, cultivation of broccoli was not documented till 1660.

Cauliflower is comparatively of a later origin, about 500 years ago, probably through introgression within broccoli genepool having a close genetic affinity. It was probably originated in the island of Cyprus from where it moved to other areas like, Syria, Turkey, Egypt, Italy, Spain and north-western Europe. It was first introduced in Italy by the Genoese from Levant or Cyprus around 1490. Italy is the centre of genetic diversity for both cauliflower and broccoli where several land races of these two crops are still available even today. At the beginning of 17th century cauliflower reached Germany, France and England although, the crop was cultivated more commonly in Europe from the beginning of 18th century. Many types of cauliflowers were developed in Europe in 16th to 17th century. All the cauliflowers grown in the Europe were either temperate annuals (eg., Cornish, Alpha, Erfurt, Danish, Lecerf, Mechelse, etc.) or biennial types with winter hardiness and vernalization requirement for flowering (Old English, Walcheran, Roscoff, Angers, etc.). The Cornish types of cauliflower were first introduced in India in 1822 from England and later other biennial European types like, Roscoffs, Italians and Northerns were also introduced. Natural crossing between different European annual and biennial types during attempts of seed production followed by selection by farmers and adaptation through the course of more than 100 years resulted in the development of Tropical or Indian cauliflower that are distinctly different from the temperate European types. The early and heat tolerant cultivars were introduced to different countries like, Japan, South Korea, Taiwan, Sri Lanka, Isreal, Hawaii, Brazil, etc. Many heat tolerant improved varieties and hybrids have been developed particularly in Japan, South Korea and Taiwan from these genetic base originally developed in India.

The *Brassica oleracea* cytodeme is a polymorphic aggregate species with chromosome number, 2n = 18. Cauliflower is a monogenomic species having genomic constitution C and chromosome number n = 9. It belongs to the Order Cruciferae, Tribe Brassicae, Sub-Tribe Brassica under the family Cruciferae or Brassicaceae.

Many types of cauliflower have been developed in Europe particularly in France, U.K., Sweden, Denmark and the Netherlands from 16th century onwards in addition to different land races developed in Italy being the centre of diversity. Some of the cauliflower groups of Europe are Originals or Italians in Italy (Jezi, Naples, Romanesco and Florablanca), Effurt, Alfa and Snowball in Germany and the Netherlands, Cornish and Northerns in England and Roscoff and Angers in France. The European bianneals include Old English, Walcheran, Roscoff, Angers and Sent Malo and the annuals include Alfa, Erfurt, Danish, Lecerf and Mechelse. The Tropical or Indian or Asian and Australian group of cauliflowers have been originated as a result of intercrossing among European annuals and biennial groups and simultaneous selections.

Open pollinated improved varieties

Cauliflower varieties grown in India can be grouped according to their temperature requirement and time of curd maturity with respect to the plains of northern part of India.

Maturity group		Temperature requirement for curd initiation and proper curd development
Early crop	Mid September to mid October	20-27°C
	Mid October to mid November	20-25°C
Mid -early (Mid November to mid December)		16-20°C
Mid-late (Mid December to mid January)		12-16°C
Late (Mid January to mid February)		10-16°C

Mid September to mid October: Early Kunwari, Pant Gobi-3, Pusa Early Synthetic, Synthetic 78-1.

Mid October to mid November: Pusa Deepali, Pusa Katki, RC-JOB-1, Pant Gobi-2.

Mid November to mid December: Hisar-1, Improved Japanese, Pusa Sharad, Pant Gobi-4, Punjab Giant-26, Punjab Giant-35, Pant Shubra.

Mid December to mid January: D-96, Pusa Synthetic, Pusa Himjyoti, Pusa Shubhra, Punjab Giant-35.

Mid January to mid February: Dania, Pusa Snowball-1, Pusa Snowball-2, Pusa Snowball K-1, Snowball-16, Ooty-1.

Public sector hybrid: Pusa Hybrid-2 (Mid November to mid December maturity)

Cultural requirements

- Cauliflower is a thermo-sensitive crop.
- Both tropical and temperate varieties are grown in India.
- The tropical heat tolerant Indian cauliflowers in strict sense can be classified into three maturity groups depending on the average temperature requirements for curd initiation and development under north Indian condition.
- The maturity groups of Indian cauliflowers are September (20-30°C), October (20-25°C) and November (16-20°C).
- Well drained, fertile, sandy- loam and clay-loam soils rich in organic matter are ideal.
- Light soils are good for early crop, while clay- loam soils are well suited for getting high yield.
- The ideal soil pH is 6 - 7.
- Cauliflower is sensitive to soil acidity, boron and molybdenum deficiency.

Nursery management

- Seed beds are prepared finely, well drained, 15-20 cm raised, 1.0m wide and of convenient length.
- Fine and fully decomposed farmyard manure or compost @ 3-4 kg/m^2 should be well mixed to the beds.
- 10 nursery beds of 3.0m length and 1.0 m wide are required to produce sufficient seedlings for one hectare.
- Drench the beds with formaldehyde (4.0%) or captan @ 2-3 g/litre and cover with polythene sheet for 7 to 10 days to avoid damping off disease.
- 500 to 750g seeds of open pollinated varieties of early and mid-early maturity group and 300-500 g seeds of mid-late and late maturity group are required per hectare.
- Generally one gram contains 300-350 seeds.
- Treat the seeds first by soaking for half an hour in Streptocycline solution (250mg/5 litre) and then after drying in shade with captan or thiram @ 2-3 g per kg of seeds before sowing.
- Seeds are sown at shallow depth at 5.0 cm apart, 1-1.5cm depth and covered with finely sieved leaf mould or vermicompost.
- After sowing cover the beds with straw or long dry grass till seed germination and sprinkle water regularly.
- Seedlings of the early and mid-early varieties in particular should be raised under poly tunnels having sides open to protect the young seedlings from heavy rain or scorching sunlight.
- Water the seedlings daily or in alternate day with rose can.
- Hardening of the seedlings by withholding water at least 4 to 6 days before transplanting.
- Stocky, healthy seedlings of 6-8cm height, 4-6 leaves of 35-40 days old are ready for transplanting for the early and mid-early varieties.
- Stocky and healthy seedlings of 28-30 days old are ready for transplanting for the mid-late and late varieties.

Sowing time

The time of sowing depends on variety, season and location

- September maturity group: May-June
- October maturity group: June – first fortnight of July
- November maturity group: July-August
- December maturity group: September
- January-February maturity group (late): October.

Land preparation

- Prepare the land in well advance with repeated ploughings (at least 4-5 ploughing) to a fine tilth.
- Remove all stubbles, weeds, etc. from the land.
- Add well rotten organic manure preferably farm yard manure or compost @ 25-35 t/ha during land preparation and level the land properly.
- The land should have adequate drainage facility particularly for the varieties of early and mid-early maturity groups.

Transplanting

- Healthy and seedlings of 35-40 days old for early and mid-early varieties and 28-30 days old for mid-late and late varieties are ready for planting.
- After hardening the seedlings for 4 – 6 days particularly for early and mid-early varieties, irrigate the seed bed before lifting to facilitate easy pulling of young seedlings.
- Plant the seedlings in the field preferably during afternoon hours which give better field establishment.
- 35-37 thousand seedlings can be accommodated in one hectare land.
- Irrigate the field just after transplanting.
- Planting in ridges is ideal particularly for early and mid-early varieties when rainfall is high.
- Spray the 15 days old seedlings with the starter solution of ammonium sulphate (50g/10litres of water) for better establishment.
- Spraying the seedlings with 0.1ppm IBA or 10ppm NAA gives better establishment, enhance vegetative growth and ensure more yield.

Spacing

- Early and mid-early varieties: 60 cm x 30 cm
- Mid-late and late varieties: 60 cm x 45 cm

Nutrient management

- Adequate nutrition is necessary for obtaining higher yield and better curd quality.
- Apply a general fertilizer dose of 125 kg N, 80 kg P_2O_5 and 60kg K_2O per hectare for open pollinated improved varieties. 1/2 N along with other fertilizers should be given as basal and rest N should be top dressed at 35 days after transplanting.
- Mid-late and late varieties require comparatively more nutrients than the early and mid-early varieties.

- Apply 200 kg N, 100 kg P_2O_5 and 100kg K_2O per hectare for the hybrids.1/2 N along with other fertilizers should be given as basal and rest N should be top dressed in two split doses at 20 days interval after transplanting.
- Judicious application of phosphatic fertilizer increase curd yield and dry matter content.
- Cauliflower is responsive to sulphur application hence, it is better to apply ammonium sulphate as the source of nitrogen.
- Weekly spray of 1-2% urea 20 days after transplanting gives better growth and increases yield.
- Three foliar sprays of 0.3% borax after 20, 35 and 50 days after transplanting is beneficial in increasing yield and checking browning disorder.

Irrigation

- Cauliflower is a shallow rooted vegetable crop and requires proper soil moisture throughout the cropping period.
- The early and mid-early crop requires only protective irrigation depending on the rain.
- Over-watering should be avoided.
- Irrigation is generally required at 10-12 days interval during cool season.
- Total water requirement is about 350mm.

Interculture and weed control

- Curd yield is reduced to the tune of 36-69% if weeds are not controlled in the early stages of growth.
- The crop needs at least 4 weeks weed- free period after transplanting to prevent yield loss.
- Pre-planting application of Basalin (alachlor) 2-2.5 litre/ha or Stomp (pendimethalin) 3.3 litre /ha in the finally prepared field followed by one hand weeding 40-45 days after is the best.
- Clear plastic mulching results in faster growth and advances the harvest upto 10 days and suppresses weed growth.
- Ten centimeter thick mulching with paddy straw is beneficial.
- Plants should be earthed up about 4-5 weeks after transplanting after topdressing with nitrogen fertilizer.
- Blanching i.e. covering the curd through tying the outer leaves up over the curd improve curd colour and quality.

Harvesting the crop

- Harvest the curds when they reach prime condition.
- It is better to harvest early than late.
- Late harvest causes ricy or fuzzy and loosening of curds.
- Harvest above the soil surface with a large and sharp knife or sickle.

Post harvest handling and storage

- Large leaves are trimmed away leaving only sufficient jacket leaves to protect the curd from bruising and other mechanical injury in transport.
- More number of jacket leaves are kept when cauliflowers are transported loose in gunny bags or basket.
- When they are transported in crates, the jacket leaves are trimmed leaving a fringe of leaves projecting 2-3 cm above the curds.
- Tight packing is essential to prevent shifting and ruishing.
- Cellophane and transparent films are also used to pack cauliflower which are hydro-cooled to remove field heat.
- Use of polythene packing provides modified atmosphere and consequently reduces decaying, softening, loss of total solids and weight.
- Consumers' package should be ventilated to prevent CO_2 injury.
- Storage at 0° to 1.7°Cwith 85-90% RH for 30 days.
- Post-harvest spray of 5-15ppm BA (N6-benzyladenine) checks yellowing of curd in storage.

Plant protection measures

Physiological disorder

Browning

It is a boron deficiency disorder where water–soaked lesions first appear in the stem, leaf and on the surface of the curd which later turn rusty brown in color. Browning of curd is some times associated with hollow stem symptom. The affected curd is bitter both in fresh and cooked condition. Cauliflower is most sensitive to this micro-nutrient deficiency disorder.

Control measures

- Correction of soil reaction and salinity.
- Soil application of borax @ 10-15kg/ha
- Three sprays of 0.25-0.50% borax solution (1-2 kg/ha) along with a sticker 20, 35 and 50 days after transplanting.

Whip tail

It is a molybdenum deficiency disorder particularly in the soil pH below 5.5 where young plants become chlorotic and may turn white, particularly along the leaf margins and finally becomes cupped and wither. Leaf blade of the older plants does not develop properly and may be strap like and severely savoyed. In extreme cases only mid rib develops. Growing point of the plant is usually deformed and no marketable curds are produced.

Control measures

- Raising of the soil pH upto 6.5 by proper liming.
- Soil application of 1.5kg/ha sodium or ammonium molybdate mixed with fertilizer or irrigation water when the plants are set in field.
- Three sprays of 0.1% ammonium molybdate solution along with a sticker 15, 30 and 45 days after transplanting.

Buttoning

This disorder is marked by the development of small curd or button which can not be developed to full size of the curd which is caused due to i)nitrogen deficiency, ii) planting with over aged seedlings, iii) planting early variety late in the season and iv) exposure of the seedlings to poor light conditions

Control measures

- Right selection of the varieties for different temperature regimes.
- Transplanting 35-40 days old seedlings of early and mid-early group.
- Supply of adequate nutrients.

Blindness

In this disorder the plants are lacking in terminal buds and curds and only large, dark green, thick and leathery leaves develop in these plants which is caused due to i) mechanical or insect injury of the terminal bud, ii) planting late variety early in the season and iii) exposure of the seedlings to very low temperature.

Control measures

- Careful handling of the plants and good protection against insects
- Avoid very low temperature exposure

Riceyness

This disorder is marked by velvety or granular appearance on the curd surface and premature initiation of the flower bud which is caused due to i) exposure to the temperature higher or lower than the optimum required for particular variety in curd development stage, ii) temperature fluctuation during curd development and iii) poor seed stock.

Control measures

- Selection of proper variety for the particular growing season.
- Good seed stock and favorable weather condition.

Diseases

Damping off (*Pythium*, *Rhizoctonia* ,*Phytophthora*, *Fusarium*, *Sclerotium*, *Botrytis*, *Sclerotinia*)

It is very destructive fungal disease in nursery beds, especially during early season. In pre- emergence damping off, the growing points are killed in the initial stages of seed germination before they come out through the soil. In post-emergence damping off, the seedlings topple over the ground due to collar rotting and rapid shrinking of the cortical tissue of the hypocotyls.

Control measures

- Hot water treatment of seeds at 50°C for 30 minutes.
- Treat the seeds with thiram / captan @ 3.0g /kg of seeds.
- Drenching the nursery beds 7 days before sowing with thiram / captan or any copper fungicide @ 3g / litre of water.
- Cover the beds with transparent polythene sheets before sowing and left to open sun for at least 20 days, known as soil solarization.
- Spraying the young seedlings with 0.2% Blitox or carbendazim(1g) + mancozeb(2g)
- Provision of proper drainage to the beds.
- Remove the affected seedlings from the beds as soon as the symptoms are visible.
- Avoid flooding the beds to check spread of the disease.

Downy mildew (*Peronospora parasitica*)

It is fungal disease where discoloration occurs in the young seedlings and in severe cases the whole plants die. In mature plants, purplish or yellow-brown spots appear on the upper surface of the leaf and white fluffy downy growth occurs on the lower leaf surface.

Control measures

- Hot water treatment of seeds at 50°C for 30 minutes.
- Crop rotation excluding cruciferous crops.
- Spraying the crop with 0.25% mancozeb, metalaxyl-mancozeb(0.25%)and cymoxanil-mancozeb(0.25%) with a sticker at weekly intervals.

Yellowing (*Fusarium oxysporum* f. sp. *conglutinans*)

It is a serious fungal disease where the fungi attack the seedlings through the rootlets. Yellowing starts from the lower leaves and progress towards the upper leaves very fast, leaves drop prematurely and the plants remain stunted.

Control measures

- Application of adequate organic matter in the soil.
- Application of lime to raise the soil pH to 7.2.
- Seed treatment with *Trichoderma viride* @ 5g/ kg of seeds
- Treat the seeds with thiram / captan 3g /kg of seeds.
- Spraying the crop with 0.1% Bavistin at 10 days interval.

Curd blight (*Alternaria brassicae*)

It is primarily a seed borne fungal disease where small, brown to black circular to slightly elongated spots appear on the leaves, petioles and stem and later blight patches occur on the curd.

Control measures

- Treat the seeds with thiram/ captan @ 3.0g /kg of seeds.
- Spraying the crop with copper-oxychloride (0.3%) or mancozeb(0.25 %)/ chlorothalonil (0.2%) or difenconazole (0.5g/lt) at 10 days interval.

Black rot (*Xanthomonous campestris pv campestris*)

This seed borne bacteria infects the plants in the nursery as well as in the main field. The infected plants become stunted, cotyledonary leaves turn yellow to black and drop prematurely which results in killing of the seedlings. In the main field, first sign of the disease appears near the leaf margin, characterized by chlorosis which progress towards the centre of leaf blades in V – shaped lesions. In the affected portion, veins and veinlets turn brown and finally black. In severe cases leaves wither, curds become discolored and subsequently soft rot may set in.

Control measures

- Grow resistant varieties like, Pusa Subhra, Pusa Snowball K–1.
- Seed treatment with hot water at 50-52°C for 30 minutes followed by a dip in 0.01% streptocycline solution for 30 minutes.
- Spraying the crop with 0.01% streptocycline or Agrimycin-100 at transplanting and curd initiation stage.
- Crop rotation excluding cruciferous crops.
- Spraying of copper-oxychloride (0.3%) + streptomycin sulphate (100 ppm).

Soft rot (*Erwinia carotovora*)

This bacterial rot occurs after the occurrence of black rot or after mechanical injury to the matured curd. Water soaked patches appear on the upper surface of the curd which expands and cover the whole curd. The water soaked portion turns to light brown and ultimately dark brown within a week. Sticker and spreader need to be added with spray mixture for spraying on cabbage and cauliflower

Control measures

- Spraying of copper oxychloride (0.3%) + streptomycin sulphate (100 ppm).

Insect pests

Cabbage borer (*Hellula undalis*)

The caterpillars after hatching mine into the leaves along the side veins and make it a white papery structure filled with their excreta and later they bore into the curd. The crop loss potentiality is very high in heavily infested crops.

Control measures

- Collection and destruction of larvae at early invading population by wire devices is one of the best ways to reduce the destructive form at later stage of crop.
- Application of DDVP @ 0.75 ml/l may control the pest.

Diamond back moth (*Plutella xylostella*)

The caterpillars feed on lower side of the leaves producing typical whitish patches and later they bore small holes in the leaves giving a shot hole appearance all over. This pest has appeared as a major and destructive pest in South Indian conditions.

Control measures

- Picking and destruction of the larvae at the early stages of the crop.
- Growing mustard as a trap crop in the field.
- Spraying the crop with 1.25 ml profenophos or 1.0 ml cypermethrin per litre of water.

Aphid (*Lipaphis erysimi*)

The aphid species attack all the cruciferous crops during seedling as well as later stage of crop growth. Severe infestation reduces the vigour of the crop due to their continuous sucking.

Control measures

- Encourage population growth of lady bird beetles and hover fly in the field.
- Spray with 3-4% neem pesticides along with sticker to control the pests.

Yield

- 10-15 tonnes/ha (open pollinated early and mid-early varieties).
- 15-20 tonnes/ha (mid-early hybrids)
- 18-20 tonnes/ha (open pollinated mid-season varieties).
- 20-25 tonnes/ha (mid-season hybrids).
- 25-30 tonnes /ha (open pollinated late varieties)

7. Cabbage

Scientific Name: Brassica oleracea L. var. capitata
Family: Cruciferae or Brassicaceae
Origin: Eastern Mediterranean region and Asia Minor
Chromosome No. 2n = 2x =18

Introduction

Cabbage, also known as white cabbage is one of the most popular vegetables in the world. It is grown widely in Europe, Russia, Asia, USA, Canada, South America, Japan and East Africa. The hard headed white cabbage evolved in Germany by human selection descended from the wild non-heading cabbage, originated in eastern Mediterranean region and Asia Minor which was probably brought into Western Europe by the Celts. The present day white headed cabbage was probably selected in Germany around 1150 A.D. and it moved to England in the 14th century. Cabbage was introduced in India in the Mughal period by the Portuguese however, it became popular during the British period. Traditionally cabbage is grown to a larger extent in the southern, eastern and coastal areas of India and cauliflower occupies more area in northern India. However, after the development of tropical hybrids it now produced almost round the year widely throughout the country. In India the major cabbage producing states are Uttar Pradesh, Bihar, Orissa, West Bengal, Assam, Maharashtra and Karnataka and other important growing states are Haryana, Rajasthan, Gujarat, Uttaranchal, Himachal Pradesh and Nilgiri hills of Tamil Nadu.

Importance and use

- Cabbage was one of the oldest cultivated vegetables, used by the ancient Greeks and Romans.
- Cabbage is the most important vegetables in the world.
- It is also called white cabbage or headed cabbage.
- India is the second largest producer of cabbage in the world.
- The edible part of cauliflower is 'head' which is formed by thickening of edible bud with thick tightly packed overlapping leaves.
- The present day heading cabbage descended from leafy, non-heading cabbage perennial in nature.
- It is biennial in nature and requires vernalization for seed production.
- Unlike tropical or Indian cauliflowers no such tropical cabbage types have been developed in India.
- All the F_1 hybrids of tropical cabbage hybrids have basically been developed in Japan.

- It is generally used as cooked vegetable either singly or mixed with potato, peas, or other vegetables as fried or in curry form or in sambar.
- It is used as stewed or boiled items in continental form.
- It is also eaten raw as salad and used for the preparation of soup.
- It is preserved as a well known fermented product called “sauerkraut”.
- Shredded cabbage after blanching in steam or boiled water is used for the preparation of dehydrated product by freeze drying or hot air drying.
- Blanched cabbage is canned in 1% salt solution as covering liquid.
- Cabbage after shredding in small pieces or grated is mixed with different vegetable oils or vinegar, spices and creams, called cole slaw and is taken as salad however, this preparation can not be stored for long period.

Nutritional value

- Cabbage is rich source of Ca, P, Na, K, S and Vitamin A, Vitamin C and dietary fibre.
- It contains good amount of Vitamin C (100 mg/100 g fresh).
- It is a good source of calcium (46 mg) and phosphorus (38 mg) per 100 g fresh.
- Good amount of phosphorus content is helpful in utilization of calcium and assimilation of carbohydrates and fats in human body.
- It also contains fair amount of vitamin A (80 IU/100 g fresh).

Medicinal value

- Cabbage is fairly high source of glucosinolates (40-90 mg/100g fresh), predominantly sinigrin and glucobrassicin which have prominent anti-carcinogenic property.
- The products of hydrolysis of these glucosinolates formed by enzymatic or non-enzymatic action are biologically active compounds with diverse effect on human health.
- The product of glucosinolate breakdown, particularly the isothiocyanates act as anti-carcinogens.
- Cabbage also contains S-methyl cysteine sulfoxide (18 mg/100 g fresh) which also show anti-carcinogenic property.
- Cabbage protects against bowel cancer due to presence of indole-3 carbinol.
- Beta carotene present in cabbage act as potential anti-oxidant.
- Rich in vitamin C and fairly high fibre and calcium contents reduce the risk of colon cancer.

- Cabbage used against ailments like gout, diarrhoea, stomach and coeliac troubles.
- Cabbage juice is remedy against poisonous mushrooms and as a gargle against hoarseness.
- The leaves are traditionally used to cover ulcers and wounds.
- Fermented product of cabbage-shredded leaves (sauerkraut) has curative effect on scurvy disease.

Adaptation in India

The exact date of introduction of cabbage into India is not certain but it is known to be cultivated during 16th century in the Mughal period as has been mentioned in the Ain-i-Akbari. However, it is not known whether cabbage seed production had been attempted during the Mughal period. The only recorded evidence of organized seed production of temperate vegetables including cabbage is in the early forties of twentieth century, when experiments were conducted at Quetta in Baluchistan and later in Kashmir valley, which later extended to Himachal Pradesh where it fulfills its chilling requirement to enter into the reproductive phase after the vegetative phase, being biennial crop in nature. This was necessitated due to stoppage of seeds import from Europe during World War II. Whether Moghul rulers regularly imported cabbage seeds and seeds of other temperate vegetables like carrot, turnip, beet root etc. or these were being presented by the European traders during their visits to the Moghul courts, is neither known nor could be established.

Although cabbage was introduced in India much earlier than cauliflower yet no tropical types were developed in India like that happened in cauliflower mainly because of its biennial nature for seed production. No doubt Japanese seed trade were the first to develop tropical cabbage varieties and the classical "KK" and "KY" crosses of 1950's spread the cabbage cultivation to the tropical regions of South-east Asia. The present day hot weather hybrids form compact heads under tropical conditions of day temperatures of 30-35^0 C and have been accepted widely by the farmers for growing in warmer conditions of east, south and western India. However it should not be overlooked that seed production of hot weather cabbage requires cool temperature of North Indian hills for brief periods to ensure good seed quality. Although we depend fully on the Japanese and Korean seed trade for these tropical cabbage hybrids.

Breeding behaviour

- Cabbage is a biennial crop and requires 6-8 weeks of vernalization at mean temperature range of 4° to 8°C after they have passed the juvenile phase (at least 10 weeks) for the production of seed stalks. Hence, seeds of cabbage can only be produced in the high hill situations of Himachal Pradesh and Kashmir.

- Cabbage bear hermaphrodite flowers but varying levels of cross fertilization takes place as high as 75 percent in different open pollinated varieties due to manifestation of self-incompatibility of different intensities.
- Self-incompatibilty system which is sporophytic in nature and governed by multiallelic single gene (*S* gene) with generally dominant effect. Intensities of self-incompatibility depend on the presence of *S* allele in the genotype.
- The *S* alleles are active in the stigma in flowering stage but remain inactive in the bud stage, i.e., 2-4 days before anthesis.
- The glycoproteins present in the stigma hinder the pollen germination and penetration of the pollen tubes through the styler tissue.
- The stigma becomes receptive five days before anthesis due to protogyny and remains so four days after anthesis.
- Due to protogynous nature, self-fertilization can be done by self-pollinating with the pollens from other opened flowers of the same plant.
- In cabbage cross pollination is generally done by bees, more particularly honey bees and flies.
- Bees and other insect pollinators frequently visit the flowers of cabbage because of the presence of nectar in the flowers.
- To maintain the purity of the varieties, isolation distance of at least 1200 m is safe for seed production.
- Genic male sterility controlled by single recessive gene and even single dominant gene have been found in cabbage however, this sterility is not utilized in hybrid seed production.
- Cytoplasmic male sterility system in which male sterility is manifested due to interaction of cytoplasm and nuclear gene is generally utilized in hybrid seed production.
- The cytoplasm of the Ogura male sterile radish has been successfully utilized in developing cytoplasmic male sterility in cabbage.
- Anthesis starts early in the morning between 7.30 and 8.00 am and continues up to 11.30 am depending on the prevailing weather conditions.
- Dehiscence of anther is longitudinal and it generally occurs few hours after anthesis.

Origin and taxonomy

All cole crops have been originated mainly by mutation and introgression from wild species during evolution, human selection and adaptation. It has descended through mutation and selection from a common ancestor *Brassica oleracea* var *sylvestris*, the leafy kale like wild cabbage. Probably kale was the first cole crop to be selected and adapted by man. The wild diploid species *Brassica oleracea* var. *sylvestris*, leafy kale-like plant which occurred in the northern shore of the Mediterranean sea and on the lime cliffs in the north-eastern England and along the North sea is the progenitor of the cultivated kale and non-heading cabbage. Such cabbage was probably in general use, around 2000-2500 BC by the Greeks and Romans. It is said that cabbage held in high esteem by the ancient Greeks and even worshipped by the Egyptians.

The hard-headed cabbage developed through selection in the North Europe descended from this wild leafy non-heading cabbage, originating in the eastern Mediterranean region and Asia Minor which was probably brought into Western Europe by the Celts. The wild perennial forms occur in south-western Britain, western France, northern Spain and the Mediterranean region. The present day white headed cabbage was developed in Germany through selection around 1150 AD and by 14th century it moved to England. Cabbage was introduced in India by the Portuguese during the Mughal period in the 16th century however, it became popular during the British period.

Savoy cabbage was originated earlier in Italy, probably from Portuguese Kale (*Brassica oleracea* var. *costata*).

The *Brassica oleracea* cytodeme is a polymorphic aggregate species with chromosome number, 2n= 18 having cabbage, cauliflower, Brussels sprouts, broccoli, knolkhol and kale. Cabbage is a monogenomic species having genomic constitution C and chromosome number n = 9. It belongs to the Order Cruciferae, Tribe Brassicae, Sub-Tribe Brassica under the family Cruciferae or Brassicaceae.

Genetic diversity in cabbage is mainly found in north-west Europe. The cultivated types of cabbage show great variation with respect to size, shape and colour of leaves and size, shape, colour and texture of the head. Three cultivated forms of cabbage have been recognized.

- *Brassica oleracea* var. *capitata* f. *alba* (white cabbage)
- *Brassica oleracea* var. *capitata* f. *rubra* (red cabbage)
- *Brassica oleracea* var. *sabauda* (savoy cabbage)

White cabbage

Cultivars of this group are the only commercial cabbage grown both in India and the world. Leaves are generally smooth and light green in colour. Inside colour of the head is creamy white. Varieties of this cabbage group are available in three different head shape namely, round, flat or drum head and pointed.

Savoy cabbage

Leaves of this cabbage group are blistered and puckered in texture and light green or green in colour. Varieties of this cabbage group are available in three different head shape namely, round, flat or drum head and pointed.

Red cabbage

Leaves of this cabbage group are smooth, red and having distinct coating of wax. Varieties of this cabbage group are available in round head shape.

Open pollinated improved varieties

White cabbage varieties grown in India can be grouped on the basis of days required for maturity of heads after transplanting.

White cabbage

Early (60-75 days)

Golden Acre, Pride of India, Copenhagen Market, Pusa Mukta, Pusa Agethi, Pusa Synthetic, Express, Drumhead Early.

Mid season (80-95 days)

Glory of Enkhuizen, Kalimpong English Ball

Late season (90-130 days)

September, Pusa Drumhead, Late Large Drumhead, Kalimpong Eclipse Drumhead

Pointed headed white cabbage

Jersey Wakefield, Charleston Wakefield

Red cabbage

Mammoth Rock Red, Red Acre

Savoy cabbage

Chieftain

Tropical hybrids

Most of the high temperature tolerant tropical hybrids are early maturing and are produced by different private seed companies of Japan, Korea, Taiwan, Vietnam, etc.

Cultural requirements

- Cabbage is basically a cool season crop and thrives best in cool and moist conditions.
- Cabbage is hardier than cauliflower and can withstand frost and extreme cold weather.
- The minimum temperature for growth of cabbage is just above 0°C.
- 15-20°C temperature is optimum for growth and head formation and growth is checked when temperature exceeds 25°C.
- The tropical heat tolerant hybrids form tight heads at 25° to 35°C temperature regimes.
- Cabbage looses its flavour when grown under warm and dry conditions.
- Cabbage is grown on a wide range of soil.
- Well drained, fertile, sandy-loam and clay-loam soils rich in organic matter are ideal.
- Light soils are good for early crop, while clay- loam soils are well suited for getting high yield.

- Due to slow growth on heavy soils, keeping quality of heads is improved because of compactness.
- The ideal soil pH is 6.0 – 6.5 since phosphorus availability is more in this pH range.
- Cabbage does not grow well in highly acidic soils and pH should not be below 5.5.
- Most of cabbage types are moderately tolerant to salt.
- Plants grown in saline soils are more susceptible to diseases like black leg.

Nursery management

- Seed beds are prepared finely, well drained, 15-20 cm raised, 1.0m wide and of convenient length.
- Fine and fully decomposed farmyard manure or compost @ 3-4 kg/m^2 should be well mixed to the beds.
- 10 nursery beds of 3.0m length and 1.0 m wide are required to produce sufficient seedlings for one hectare.
- Drench the beds with formaldehyde (4.0%) or captan @ 2-3 g/litre and cover with polythene sheet for 7 to 10 days to avoid damping off disease.
- 500 g seeds of open pollinated early varieties and 375 g seeds of late varieties are required per hectare.
- Generally one gram contains 300-350 seeds.
- Treat the seeds first by soaking for half an hour in Streptocycline solution (250mg/5l) and then after drying in shade with captan or thiram @ 2-3 g per kg of seeds before sowing.
- Soaking the seeds in 0.1% boric acid before sowing enhance yield.
- Seeds are sown at shallow depth at 5.0 cm apart, 1-1.5cm depth and covered with finely sieved leaf mould or vermicompost.
- After sowing cover the beds with straw or long dry grass till seed germination and sprinkle water regularly.
- Seedlings of the early and mid-early varieties in particular should be raised under poly tunnels having sides open to protect the young seedlings from heavy rain or scorching sunlight.
- Water the seedlings daily or in alternate day with rose can.
- Hardening of the seedlings by withholding water at least 4 to 6 days before transplanting.
- Stocky, healthy seedlings of 6-8cm height, 4-6 leaves of 30-40 days old are ready for transplanting.

Sowing time

The time of sowing depends on variety/ hybrid, season and location. With the selection of proper variety and high temperature hybrid, cabbage can be grown almost round the year in mild climate of Karnataka, West Bengal and Maharashtra.

- High temperature hybrid in eastern India: June - August
- Early variety in eastern India: Mid to late September
- Early variety in northern India: August - September
- Late variety in northern India: September – October
- High hill condition: March – June.
- Mid-hill condition: Almost throughout the year.

Land preparation

- Prepare the land in well advance with repeated ploughings (at least 4-5 ploughing) to a fine tilth.
- Remove all stubbles, weeds, etc. from the land.
- Add well rotten organic manure preferably farm yard manure or compost @ 25-35 t/ha during land preparation and level the land properly.
- The land should have adequate drainage facility particularly for the high temperature hybrids and early varieties.

Transplanting

- Healthy and seedlings of 30-35 days old for early and mid-early varieties and 28-30 days old for mid-late and late varieties are ready for planting.
- After hardening the seedlings for 4 – 6 days particularly for early and mid-early varieties, irrigate the seed bed before lifting to facilitate easy pulling of young seedlings.
- Plant the seedlings in the field preferably during afternoon hours which give better field establishment.
- 40-45 thousand seedlings can be accommodated in one hectare land.
- Irrigate the field just after transplanting.
- Planting in ridges is ideal particularly for high temperature hybrids and early varieties when rainfall is high.
- Spray the 15 days old seedlings particularly of the high temperature hybrids with the starter solution of ammonium sulphate (50g/10litres of water) for better establishment.

Spacing

- High temperature hybrids and early varieties: 60 cm × 30 cm; 45 cm × 45 cm.
- Mid season varieties: 60 × 45 cm
- Late varieties: 60 cm × 45 cm; 60 cm × 60 cm; 75 cm × 30 cm.

Nutrient management

- Cabbage is shallow rooted heavy feeder and requires adequate nutrient supply for profitable yield.

- Apply a general fertilizer dose of 150 kg N, 60 kg P_2O_5 and 80kg K_2O per hectare for open pollinated improved varieties. 1/2 N along with other fertilizers should be given as basal and rest N should be top dressed at 30-35 days after transplanting.
- Apply 200 kg N, 125 kg P_2O_5 and 150kg K_2O per hectare for the hybrids. 1/2 N along with other fertilizers should be given as basal and rest N should be top dressed in two split doses at 20 days interval after transplanting.
- Weekly spray of 1-2% urea 20 days after transplanting gives better growth and increases yield.
- Two sprays of sodium or ammonium molybdate 20mg/litre of water at 25 and 40 days after transplanting increase the volume and weight of head and enhance the maturity.
- Application of non-symbiotic nitrogen fixing bacteria *Azotobacter* is helpful in increasing the vegetable and seed yield.
- Combination of 60 kg nitrogen / ha as inorganic source and *Azotobacter* as biofertilizer helps in increasing both vegetable and seed yield.

Irrigation

- Cabbage is a shallow rooted vegetable crop and requires proper soil moisture throughout the cropping period.
- The optimum growth occurs at moisture content of the soil varying from 60-100% field capacity with an average of 80 per cent.
- Irrigation should be given when the moisture content of soil had dropped below 50% of field capacity.
- Irrigation should be given immediately after transplanting and thereafter is generally required at 10-12 days interval during cool season.
- Critical stages of irrigation are at head formation and enlargement stage.
- Total water requirement is about 350 to 400mm.
- Excessive and frequent irrigations have an adverse effect on growth and yield.

Interculture and weed control

- Yield could be reduced significantly if weeds are not controlled in the early stages of growth.
- Critical crop weed competition period 35-40 days after planting.
- Two weedings at 30-40 days and 60 days after transplanting is optimum.
- Due to shallow root system, deep hoeing should always be avoided as it does harm to the feeder roots.

- Pre-planting application of Basalin (alachlor) 2-2.5 litre/ha or Stomp (pendimethalin) 3.3 litre /ha in the finally prepared field followed by one hand weeding 40-45 days after is the best.
- Black polythene mulch is effective in controlling weeds, conserving moisture and enhance growth and higher yield of cabbage.
- Ten centimeter thick mulching with paddy straw is also beneficial.
- Plants should be earthed up about 4-5 weeks after transplanting after topdressing with nitrogen fertilizer.

Harvesting the crop

- Harvest when the heads are suitable size, firm and tender and before they start cracking.
- Delay in harvesting results in cracking and rotting of heads.
- High temperature hybrids stay long in the field without the problem of head cracking.
- Heads are cut with a sharp knife attached with some wrapper leaves.
- Heads should not be directly kept on the soil, which gives bad look to the heads.
- Large non-wrapper leaves are then trimmed away leaving only sufficient jacket leaves to protect the head from mechanical injury during transport.
- Proper handlings of cabbage at the time of harvesting prevent huge post-harvest loss by mechanical, physio-biochemical and microbial loss.

Post harvest handling and storage

- Best storage at 0° to 1.7°C temperature and 90-95% relative humidity for 4-6 weeks.
- Late varieties with firm heads can be stored upto 12 weeks at 0° to 1.7°C temperature and 90-95% relative humidity.
- Under control atmosphere storage (3-4%CO_2 + 2-3%O_2 and 93-95%RH) for 265 days with a 17% total mass loss.
- Cabbage can also be stored for some period in zero energy cool chambers applying the principle of evaporative cooling.
- Tight packing in sacks is essential to prevent bruising particularly for the early varieties having loose heads.
- Transportation in crates is the best.

Plant protection measures

Physiological disorder

Browning

It is a boron deficiency disorder where water–soaked lesions first appear in the stem and leaf. Browning of the stem also occurs along with thickening and brittleness leaves and top portion of the seedlings. However, cabbage is less sensitive to this micro-nutrient deficiency disorder.

Control measures

- Correction of soil reaction and salinity.
- Soil application of borax @ 10-15kg/ha.
- Three sprays of 0.25-0.50% borax solution (1-2 kg/ha) along with a sticker 20, 35 and 50 days after transplanting.

Whip tail

It is a molybdenum deficiency disorder particularly in the soil pH below 5.5 where growing point gets distorted and leaf area is reduced.

Control measures

- Raising of the soil pH upto 6.5 by proper liming
- Soil application of 1.5kg/ha Sodium or ammonium molybdate mixed with fertilizer or irrigation water when the plants are set in field.
- Three sprays of 0.1% ammonium molybdate solution along with a sticker 15, 30 and 45 days after transplanting.

Diseases

Damping off (*Pythium*, *Rhizoctonia* ,*Phytophthora*, *Fusarium*, *Sclerotium*, *Botrytis*, *Sclerotinia*)

It is very destructive fungal disease in nursery beds, especially during early season. In pre- emergence damping off, the growing points are killed in the initial stages of seed germination before they come out through the soil. In post-emergence damping off, the seedlings topple over the ground due to collar rotting and rapid shrinking of the cortical tissue of the hypocotyls.

Control measures

Management schedules same as mentioned in cauliflower.

Downy mildew (*Peronospora parasitica*)

It is a fungal disease where discoloration occurs in the young seedlings and in severe cases the whole plants die. In mature plants, purplish or yellow-brown spots appear on the upper surface of the leaf and white fluffy downey growth occurs on the lower leaf surface.

Control measures

Management schedules same as mentioned in cauliflower.

Yellowing (*Fusarium oxysporum* f. sp. *conglutinans*)

It is a serious fungal disease where the fungi attack the seedlings through the rootlets. Yellowing starts from the lower leaves and progress towards the upper leaves very fast, leaves drop prematurely and the plants remain stunted.

Control measures

Management schedules same as mentioned in cauliflower.

Black leg (*Phoma lingam*)

It is a seed borne fungal disease which attacks mostly the seedlings. Seedlings topple over due to distortion of the vascular bundle and root system. Stem of the affected plants when splits open vertically, shows severe black discoloration of the vascular bundles.

Control measures

- Hot water treatment of seeds at 50ºC for 30 minutes.
- Seed treatment with thiram or captan @ 3g/kg of seeds.
- Spraying the plants with 0.2% captan or chlorothalonil (0.2%) or thiophanate-methyl (0.1%).

Club root (*Plasmodiophora brassicae*)

It is a soil borne fungus causes swelling of roots and rootlets like the shape of a club and xylem vessels are plugged. It declines the vigour and imparts pale green coloration to the plants. The plants tend to wilt in hot sunny condition.

Control measures

- Solarization of the field during summer by ploughing and covering with transparent polythene film.
- Liming the soil to raise the soil pH (7.2) as the disease appears more in acidic soil condition.
- Treat the seed bed with 1% formalehyde.
- Seed treatment with thiram @ 3g/kg of seeds.
- Seedling root treatment with 4% Calomel (mercurous chloride) before transplanting.

Black rot (*Xanthomonous campestris pv campestris*)

This seed borne bacteria infects the plants in the nursery as well as in the main field. The infected plants become stunted, cotyledonary leaves turn yellow to black and drop prematurely which results in killing of the seedlings. In the main field, first sign of the disease appears near the leaf margin, characterized by chlorosis which progress towards the centre of leaf blades in V – shaped lesions. In the affected portion, veins and veinlets turn brown and finally black. In severe cases leaves wither, curds become discolored and subsequently soft rot may set in.

Control measures

Management schedules same as mentioned in cauliflower.

Soft rot (*Erwinia Carotovora*)

This bacterial rot occurs after the occurrence of black rot or after mechanical injury to the matured curd. Water soaked patches appear on the upper surface of the curd which expands and cover the whole curd. The water soaked portion turns to light brown and ultimately dark brown within a week.

Control measures

Management schedules same as mentioned in cauliflower.

Insect pests

Cabbage borer (*Hellula undalis*)

The caterpillars after hatching mine into the leaves along the side veins and make it a white papery structure filled with their excreta and later they bore into the curd. Severely infested crops are subjected to havoc economic yield loss.

Control measures

- Collection and destruction of larvae mechanically using wire device.
- Spraying with DDVP @ 0.75 ml/l control the pest.

Diamond back moth (*Plutella xylostella*)

The caterpillars feed on lower side of the leaves producing typical whitish patches and later they bore small holes in the leaves giving a shot hole appearance all over. It causes heavy damage to cabbage.

Control measures

- Picking and destruction of the larvae at the early stages of the crop.
- Growing mustard as a trap crop in the field.
- Spraying the crop with 1.25 ml profenophos or 1.0 ml cypermethrin per litre of water with sticker.

Yield

Cabbage yield ranges widely depending on the variety or hybrid and climate

- 25-35 tonnes/ha (open pollinated varieties).
- 30-45 tonnes/ha (open pollinated late varieties).
- 50-80 tonnes /ha (hybrids)

8. Broccoli *or* Sprouting broccoli

Scientific Name: Brassica oleracea L. var. italica
Family: Cruciferae or Brassicaceae
Chromosome No.: 2n = 2x =18

Introduction

Broccoli (from the Italian *piccoli bracci*, meaning "little arms") is a plant of the family *Brassicaceae*. Broccoli possesses abundant arboreal, luscious, fleshy, flower heads, usually green in color, arranged in a tree-like fashion on branches sprouting from a thick, edible, sturdy, meaty stalk. The mass of flower heads is surrounded by lavish leaves. Broccoli is one of a vast array of plants that provide value to human beings. Broccoli is a good source of vitamin C, iron, fiber, potassium, vitamin A, calcium, zinc, magnesium, carotene, and vitamin B.

United States of America is the largest producer of broccoli in the world and other major producers are Italy, northern Europe and cooler regions of Far East. Broccoli cultivation in India has gained momentum in the recent times especially around big cities. Initially cultivation of broccoli started in India mainly to cater the demands of the big hotels for the delicacy of the foreigners. Now it has become quite popular particularly among the city dewellers of India mainly because of its high nutritive values and cancer prevention properties. In India it is mostly cultivated in the hilly areas of Himachal Pradesh, Uttar Pradesh, Jammu & Kashmir, Nilgiri hills and northern plains of India.

Importance and use

- Broccoli is well known not only for its delicate taste but also for its high nutritive, antioxidant and cancer prevention properties.
- Morphologically, broccoli resembles cauliflower, but the plant forms head rather than curd, consisting of green or purple or white buds and thick fleshy floral stalk.
- Besides the main head, long slender smaller heads which are called spears are developed in the axils of the leaves.
- Both heads and fleshy stems and spears are eaten either as cooked vegetables single or mixed with other vegetables particularly potato or as salad.

Nutritional value

- Among the cole crops, broccoli is the richest source of nutrients.
- It has about 50 times more vitamin A contents than cauliflower and cabbage.
- It is a rich source of protein, minerals and vitamins.
- It is a rich source of vitamin A (3500 IU/ 100g fresh).

- Protein content of broccoli is much higher (3.3 g/100g fresh) than cauliflower and cabbage.
- It contains much higher amount of Vitamin C (137 mg/100 g fresh) than cauliflower and cabbage.
- It is a good source of calcium (80 mg), iron (17 mg) and phosphorus (79 mg) per 100 g fresh.
- It also contains fair amount of riboflavin (0.12 mg/100 g fresh).

Medicinal value

- It is thought to be important in the prevention of certain types of cancer, along with diabetes, heart disease, osteoporosis, and high blood pressure.
- Broccoli has high antioxidant activity mainly because of the presence of considerable amount of flavonoids like kempherol, β- carotene and ascorbic acid.
- Broccoli is rich source of glucosinolates (40-80 mg/100g fresh), predominantly sinigrin and glucobrassicin which have prominent anti-carcinogenic property.
- The products of hydrolysis of these glucosinolates formed by enzymatic or non-enzymatic action are biologically active compounds with diverse effect on human health.
- The product of glucosinolate breakdown, particularly the isothiocyanates act as anti-carcinogens.
- Broccoli is also a rich source of sulphoraphane, a compound associated with reducing the risk of cancer.

Adaptation in India

The greatest genetic diversity of cole crops including broccoli is the Mediterranean gene centre including Greece, Syria, Cyprus, Sicily, Italy, Spain and Portugal. The annual types of broccoli and cauliflower occur in Italy and surrounding areas. In India broccoli is a recent introduction particularly from USA and Japan for commercial cultivation by the private seed companies around 1990. All the three varieties developed in India are selections from the exotic materials.

Breeding behaviour

- Broccoli is a biennial crop and requires 6-8 weeks of vernalization at mean temperature range of 4° to 8° C after they have passed the juvenile phase (at least 10-12 weeks) for the production of seed stalks. Hence, seeds of broccoli can only be produced in the high hill situations of Himachal Pradesh and Kashmir.

- Broccoli bear hermaphrodite flowers but varying levels of cross fertilization takes place as high as 90 percent in different open pollinated varieties due to manifestation of strong self-incompatibility.
- Self-incompatibilty system which is sporophytic in nature and governed by multiallelic single gene (*S* gene) with generally dominant effect. Intensities of self-incompatibility depend on the presence of *S* allele in the genotype.
- The *S* alleles are active in the stigma in flowering stage but remain inactive in the bud stage, i.e., 2-4 days before anthesis.
- The glycoproteins present in the stigma hinder the pollen germination and penetration of the pollen tubes through the styler tissue.
- The stigma becomes receptive five days before anthesis due to protogyny and remains so four days after anthesis.
- Due to protogynous nature, self-fertilization can be done by self-pollinating with the pollens from other opened flowers of the same plant.
- In broccoli cross pollination is generally done by bees, more particularly honey bees and flies.
- Bees and other insect pollinators frequently visit the flowers of broccoli because of the presence of nectar in the flowers.
- To maintain the purity of the varieties, isolation distance of at least 1200 m is safe for seed production.
- Cytoplasmic male sterility system in which male sterility is manifested due to interaction of cytoplasm and nuclear gene is generally utilized in hybrid seed production.
- The cytoplasm of the Ogura male sterile radish has been successfully utilized in developing cytoplasmic male sterility in broccoli.
- Anthesis starts early in the morning between 7.30 to 8.00 am and continues up to 11.30 am depending on the prevailing weather conditions.
- Dehiscence of anther is longitudinal and it generally occurs few hours after anthesis.

Origin and taxonomy

The cole crops have been originated mainly by mutation and introgression from wild species during evolution, human selection and adaptation. It has descended through mutation and selection from a common ancestor *Brassica oleracea* var *sylvestris*, the leafy kale like wild cabbage. Probably kale was the first cole crop to be selected and adapted by man. The cultivated species selected by man had biennial habit with yellow flowered racemose inflorescence unlike the ancestral kale or broccoli of the Mediterranean region which had cymose inflorescence with white flowers. In was thought that many types of *Brassica oleracea* including various forms of broccoli and cauliflower might have originated by natural crossing between ancestral broccolis (*Brassica cretica* ssp. *nivea*), the wild annual with cymose white flowers and the biennial wild types, followed by human selection several thousand years ago. The broccoli developed in eastern Mediterranean and Italy is the main centre of its diversification. As all the flower buds open in broccoli compared to prefloral fleshy apical meristem of cauliflower, it is thought that broccoli must have preceded cauliflower in the evolution of species. However, cultivation of broccoli was not documented till 1660.

Improved varieties

The green sprouting broccolis are classified in accordance to their maturity, e.g. "early", "medium" and "late" cultivars. There is more demand for green sprouting broccoli having green, firm and compact crown heads. The side shoots or heads are less preferred in the Indian market. They are grouped into early, mid and late types. Important Indian varieties are Pusa Broccoli, Palam Samridhi, Palam Haritika, Palam Vichitra and Punjab Broccoli 1.

- Early season type (maturity 65 – 70 days after transplanting): Early Danish Giant, Decicco, Green–Bud, Spartan Early, Coastal and Atlantic, Punjab Broccoli 1.
- Mid season type (maturity 85 – 90 days after transplanting): Green Sprouting Medium, Pusa Broccoli, Palam Samridhi.
- Late season type (maturity 115 - 130 days after transplanting): Waltham 29, Green Mountain, Coastal and Atlantic, Palam Haritika, Palam Vichitra.

Cultural requirements

- It is a cool season crop resistant to mild frost in most parts of India.
- Some cultivars of sprouting broccoli exhibit resistance to frost to a certain extent and that too dependent on the size of the plant.
- In the hills it is grown as spring-summer crop.
- The temperature of 20°–25°C is optimum for its proper growth, while 15°–20°C for heading stage.
- The heads become loose with rise in temperature.
- High temperature also gives rise bracts during developmental stage of head.
- These are generally biennial, while the annual cultivars are somewhat sensitive to frost. In such cultivars even a light frost causes considerable damage to the buds.

- Early and mid season types are annual in nature while the late types requires vernalization for the development of head initiation of flower stalk.
- Broccoli can be cultivated in a wide range of soils.
- The soils should be well-drained and sufficiently fertile.
- The crop is best suited in deep loamy soil.
- It requires moist soil for fast and appropriate growth.
- The shoots become more fibrous under dry soil.
- It is slightly tolerant to acidic soil but the optimum pH of soil for its cultivation is 5.0-6.5.

Nursery management

- Seed beds are prepared finely, well drained, 15-20 cm raised, 1.0m wide and of convenient length.
- Fine and fully decomposed farmyard manure or compost @ 3-4 kg/m^2 should be well mixed to the beds.
- 10 nursery beds of 3.0m length and 1.0 m wide are required to produce sufficient seedlings for one hectare.
- Drench the beds with formaldehyde (4.0%) or captan @ 2-3g/litre and cover with polythene sheet for 7 to 10 days to avoid damping off disease.
- For cultivation of broccoli 400-500 gm seed would be needed for one hectare, although for the early crops, the requirement may be more, to the tune of 1 kg.
- Treat the seeds first by soaking for half an hour in Streptocycline solution (250mg/5 litre) and then after drying in shade with captan or thiram @ 2-3 g per kg of seeds before sowing.
- Seeds are sown at shallow depth at 5.0 cm apart, 1-1.5cm depth and covered with finely sieved leaf mould or vermicompost.
- After sowing cover the beds with straw or long dry grass till seed germination and sprinkle water regularly.
- Seedlings of the early and mid-early varieties in particular should be raised under poly tunnels having sides open to protect the young seedlings from heavy rain or scorching sunlight.
- Water the seedlings daily or in alternate day with rose can.
- Hardening of the seedlings by withholding water at least 4 to 6 days before transplanting.
- Stocky, healthy seedlings of 6-8cm height, 4-6 leaves of 4-6 weeks old are ready for transplanting.

Sowing time

The time of sowing depends on variety, season and location

Early: From mid July to first week of august. It is best suited to Sikkim conditions (Matures by October).

Mid: From first week of September to last week of October (Matures by December – January).

Late: Up to first week of November (Suitable for the plain of northern and eastern India (Matures by February- March).

High hill situation: March – April (Mature by July-August).

Land preparation

- Prepare the land in well advance with repeated ploughings (at least 4-5 ploughing) to a fine tilth.
- Remove all stubbles, weeds, etc. from the land.
- It is advisable to apply slaked lime in every 3 years according to the soil test result. Lime should be applied at least 30 days before planting.
- Add well rotten organic manure preferably farm yard manure or compost @ 25-35 t/ha during land preparation and level the land properly.
- Since drainage is a problem for early crops in North India and even for mid season crops in longer monsoon season areas, bed should be prepared after the last cultivation and followed by manure applications.
- Beds should be prepared in such a way that excessive water drains out rapidly but does not cause erosion.
- Planting is done on ridges to avoid water logging problem.

Transplanting

- Healthy seedlings are ready for planting within 4-6 weeks.
- Irrigate the seed bed before lifting to facilitate easy pulling of young seedlings.
- Plant the seedlings in the field preferably during afternoon hours which give better field establishment.
- Irrigate the field just after transplanting.
- Planting in ridges is ideal particularly for early and mid-early varieties when rainfall is high.
- Spray the 15 days old seedlings with the starter solution of ammonium sulphate (50g/10litres of water) for better establishment.
- Spraying the seedlings with 0.1ppm IBA or 10ppm NAA gives better establishment, enhance vegetative growth and ensure more yield.

Spacing

- The optimum spacing for broccoli in the main field is 45 cm x 45 cm.
- In very rich soils, spacing can be reduced to 45 cm x 30 cm to avoid stem hollowness due to rapid plant growth.
- At a wider spacing, plants produce more laterals.
- The closer spacing is preferred for mechanical harvesting of the central head however, closer spacing delays maturity.

Nutrient management

- Use of optimum doses of fertilizers is important for its proper growth since both rapid and slow growth is undesirable.
- The bud clusters become loose and hollow-stem results from rapid growth, however, slow growth affect yield adversely.
- Generally, application of 15–20 tonnes of farmyard manure, 125 kg N, 80 P_2O_5 and 60 kg K_2O per hectare are recommended. The doses differ from place to place depending upon the fertility status of the soil.
- The full dose of P, K and half of N are applied at the time of preparation of land. The remaining dose of N should be top dressed in 2 equal split doses. The first is applied 4-5 weeks after transplanting, whereas second before head formation.
- A high yield of side shoots can be obtained by liberal use of N after harvesting central bud cluster.
- Micronutrient requirement of broccoli is fairly high. Molybdenum and Boron may be supplied by soil application or foliar sprays.
- Soils with pH above 7.0 affect the availability of boron.
- In soil application 500 g ammonium molybdate and 10-15 kg borax helps in controlling the deficiency.
- Weekly spray of 1-2% urea 20 days after transplanting gives better growth and increases yield.
- Three foliar sprays of 0.3% borax after 20, 35 and 50 days after transplanting is beneficial in increasing yield and checking browning disorder.

Irrigation

- Broccoli requires sufficient moisture in the soil for uniform and continuous growth.
- Frequent irrigation at 10-15 days interval is generally given depending on weather conditions.
- The dry conditions adversely affect the quality and yield of shoots by being more fibrous.
- Water logging conditions should always be avoided as it depresses plant growth.
- It is better to follow furrow system of irrigation.

Interculture and weed control

- The crop should be kept weed free in early stage.
- Hoeing is necessary in order to provide good aeration as well as to make the plot weed–free before irrigation.
- Since it is a shallow rooted crop, hoeing should not be done beyond the depth of 5 – 6 cm close to the plant to avoid injuries to the roots.
- A light earthing up at final hoeing is beneficial.

Harvesting the crop

- When the head has attained proper size, it should be harvested immediately.
- Usually, only the central head along with 15 cm long fleshy stem is harvested.
- It is very important that broccoli is harvested at the correct time, i.e. it is picked up before the buds open and when the bud clusters are compact.
- Generally harvesting continues for 4-6 weeks.
- After cutting, part of the foliage is removed from the harvested shoots.
- The head may be 15 – 20 cm in diameter and weighs 250 – 600 gm.

Post harvest handling and storage

- After harvesting, its heads should be immediately sorted, graded, packed in baskets and sent to markets.
- A high rate of respiration results in deterioration of its quality.
- They should be cooled at 4°C and then packed with ice in crates and stored in refrigeration.
- They can be stored well for 7 – 10 days at 4°C.
- Broccoli can also be preserved in glass jars after lactic acid fermentation.
- The side sprouts or spears are tied together after harvesting to give them a shape of single head before marketing.

Plant protection measures

Physiological disorder

Browning

It is a boron deficiency disorder where water–soaked lesions first appear in the stem, leaf and on the surface of the curd which later turn rusty brown in color. Browning of curd is some times associated with hollow stem symptom. The affected curd is bitter both in fresh and cooked condition. Both broccoli and cauliflower are most sensitive to this micro-nutrient deficiency disorder.

Control measures

- Correction of soil reaction and salinity.
- Soil application of borax @ 10-15kg/ha.
- Three sprays of 0.25-0.50% borax solution (1-2kg/ha) along with a sticker 20, 35 and 50 days after transplanting.

Whip tail

It is a molybdenum deficiency disorder particularly in the soil pH below 5.5 where young plants become chlorotic and may turn white, particularly along the leaf margins and finally becomes cupped and wither. Leaf blade of the older plants does not develop properly and may be strap like and severely savoyed. In extreme cases only mid rib develops. Growing point of the plant is usually deformed and no marketable heads are produced.

Control measures

- Raising of the soil pH upto 6.5 by proper liming.
- Soil application of 1.5kg/ha sodium or ammonium molybdate mixed with fertilizer or irrigation water when the plants are set in field.
- Three sprays of 0.1% ammonium molybdate solution along with a sticker 15, 30 and 45 days after transplanting.

Diseases

Damping off (Pythium, *Rhizoctonia* , *Phytophthora*, *Fusarium*, *Sclerotium*, *Botrytis*, *Sclerotinia*)

It is very destructive fungal disease in nursery beds, especially during early season. In pre- emergence damping off, the growing points are killed in the initial stages of seed germination before they come out through the soil. In post-emergence damping off, the seedlings topple over the ground due to collar rotting and rapid shrinking of the cortical tissue of the hypocotyls.

Control measures

- Hot water treatment of seeds at 50°C for 30 minutes.
- Treat the seeds with thiram/ captan @ 3.0g /kg of seeds.
- Drenching the nursery beds 7 days before sowing with thiram/ captan or any copper fungicide @ 3g / litre of water.
- Cover the beds with transparent polythene sheets before sowing and left to open sun for at least 20 days, known as soil solarization.
- Spraying the young seedlings with 0.2% Blitox or carbendazim (1g) + mancozeb (2g) or validamycin (0.2%)
- Provision of proper drainage to the beds.
- Remove the affected seedlings from beds as soon as the symptoms are visible.
- Avoid flooding the beds to check spread of the disease.

Downy mildew (*Peronospora parasitica*)

It is fungal disease where discoloration occurs in the young seedlings and in severe cases the whole plants die. In mature plants, purplish or yellow-brown spots appear on the upper surface of the leaf and white fluffy downy growth occurs on the lower leaf surface.

Control measures

- Hot water treatment of seeds at 50°C for 30 minutes.
- Crop rotation excluding cruciferous crops.
- Spraying the crop with 0.25% mancozeb, metalaxyl-mancozeb (0.25%) and cymoxanil-mancozeb (0.25%) with a sticker at weekly intervals.

Black rot (*Xanthomonous campestris pv campestris*)

This seed borne bacteria infects the plants in the nursery as well as in the main field. The infected plants become stunted, cotyledonary leaves turn yellow to black and drop prematurely which results in killing of the seedlings. In the main field, first sign of the disease appears near the leaf margin, characterized by chlorosis which progress towards the centre of leaf blades in V – shaped lesions. In the affected portion, veins and veinlets turn brown and finally black. In severe cases leaves wither, heads become discolored and subsequently soft rot may set in.

Control measures

- Seed treatment with hot water at 50-52°C for 30 minutes followed by a dip in 0.01% streptocycline solution for 30 minutes.
- Spraying the crop with 0.01% streptocycline or Agrimycin-100 at transplanting and head initiation stage.
- Crop rotation excluding cruciferous crops.
- Spraying of copper oxychloride (0.3%) + Streptomycin sulphate (100 ppm).
- Micronutrient (B+Mo) spray is beneficial after application of antibiotics.

Soft rot (*Erwinia carotovora*)

This bacterial rot occurs after the occurrence of black rot or after mechanical injury to the matured head. Water soaked patches appear on the upper surface of the head which expands and cover the whole head. The water soaked portion turns to light brown and ultimately dark brown within a week.

Control measures

- Spraying of Copper oxychloride (0.3%) + Streptomycin sulphate (100 ppm).

Insect pests

Cabbage borer (*Hellula undalis*)

The caterpillars after hatching mine into the leaves along the side veins and make it a white papery structure filled with their excreta and later they bore into the head. Severely infested crops are subjected to havoc economic yield loss.

Control measures

- Collection and destruction of larvae mechanically using wire device.
- Spraying with DDVP @ 0.75 ml/l control the pest.

Aphid (Lipaphis erysimi)

The aphid species attack all the cruciferous crops during seedling as well as later stage of crop growth. Severe infestation reduces the vigour of the crop due to their continuous sucking.

Control measures

- Encourage population growth of lady bird beetles and hover fly in the field.
- Spray with 3-4% neem based pesticides along with sticker to control the pests.

Yield

Yield vary greatly depending on the cultivars, maturity period and method of harvest.

- Open pollinated varieties: 10-15 t/ha.
- Hybrid: 15-20 t/ha.

Cucurbits

9. Watermelon

Scientific Name: Citrullus lanatus
Family : Cucurbitaceae
Origin : Tropical Africa
Chromosome number: 2n = 22

Introduction

Watermelon is the most widely cultivated cucurbit in the world. It is popularly known as tarbuj, tarmuj, kalinda and kalindi in different parts of India. Though it can be grown in garden land, it is a major river-bed crop of Uttar Pradesh, Rajasthan, Gujarat, Maharashtra and Andhra Pradesh. As a common summer season crop, it is grown from the lower Himalayan region to southern parts of India. Punjab, Haryana, Karnataka, Assam, West Bengal, Orissa, Himachal Pradesh, Uttar Pradesh, Tamil Nadu and Rajasthan being major watermelon growing states. An excellent dessert fruit, it is relished by rich as well as poor. The fruit juice makes an excellent refreshing and cooling beverage after adding a pinch of salt and black pepper.

Importance and use

- It is an important cucurbitaceous vegetable extensively grown in different parts of the country.
- It is also cultivated in different river beds of the country.

- The fruits may be long, oblong or round depending on the variety.
- Cross-cultural uses of watermelon are quite varied.
- They are primarily consumed fresh for their sweet and juicy fruits and are often eaten as desserts.
- The fruit can be prepared and eaten in many ways, drinks, appetizers, dessert, and the rind can be used as aphrodisiac.
- Watermelon seeds are also used as a source of oilseed, eaten as snack or can be added to other dishes.
- In some African cuisines, however, they are served as a cooked vegetable.
- The rind may be consumed in pickled or candied form.
- In parts of the former Soviet Union and elsewhere watermelon juice is fermented into an alcoholic beverage.
- Roasted seeds of this crop are eaten throughout Asia and the Middle East, and watermelon seeds are ground into flour and baked as bread in some parts of India.
- Watermelons are also sometimes used as feed for livestock.

Nutritional value

- The crop is comparatively low in nutritive value.
- The fruits contain 92% water, 0.2% protein, 0.3% minerals and 7.0% carbohydrate in 100 g edible flesh.
- The fruit is a rich source of lycopene and beta-carotene, the powerful antioxidant, which is the scavenger of free radicals.

Medicinal value

- Fruit is used in cooling, strengthening, aphrodisiac, astringent to the bowels, indigestible, expectorant, diuretic, stomachic, purifies the blood, cures biliousness, good for sore eyes, scabies and itches.
- The seeds are tonic to the brain.
- Fruits of *Citrullus colocynthis,* a potential crop of arid region are used in 'Ayurved' and 'Unani' medicine for wide range of ailments such as ulcers asthama, jaundice, bronchitis, dyspepsia, constipation and anaemia.

Adaptation in India

The crop has reached India in prehistoric times which resulted evolution of extensive variability and this designated India as the secondary centre of diversity of this crop. Cultivation of this crop began in ancient Egypt and India and then spread to different countries *via* Mediterranean, Near East and Asia. As a result of prolonged cultivation, and selection, new forms of table watermelon have evolved. The varieties grown today have little resemblance to the ancestral African forms. Watermelon cultivars grown in India comprise different locally adapted types under region specific names, improved varieties developed either through selection over indigenous types or combination breeding approaches and different introduced varieties.

Breeding behaviour

- Two sex forms occur in watermelon namely, monoecious and hermaphrodite and obviously monoecious sex form is predominant.
- Watermelon is highly cross-pollinated in nature due to the monoecious sex form.
- Hermaphrodite sex form favours a higher degree of natural self-pollination than in the monoecious condition.
- It is entomophillous and major pollinators are bees (Apis *florea, Apis dorsata, Nomioides sp., Halictine* sp., etc.).
- Beetles like *Conopophilus* sp., moths like *Planidia* sp., *Pygargonia* sp. also helps in natural cross-pollination.
- Although 100% natural cross-pollination occur in monoecious sex form, yet reasonable extent of self fertilization happens because of geitenogamy due to foraging of insect pollinators in the staminate flowers of the same vine.
- Watermelon like other members of the family Cucurbitaceae does not suffer much from inbreeding depression or do so only negligibly, in contrast to the cross-pollinated vegetable crops belonging to other families.
- Anthesis of both staminate and pistillate flowers generally starts in the morning hours from 5.30 to 8.00 a.m.
- Anther dehiscence in the staminate flower generally starts one hour before anthesis and continues till 8.00 a.m.
- Pollen viability is at its maximum at anthesis and it decreases drastically if kept at room temperature for five hours. Pollen from the staminate/ hermaphrodite flowers at the optimum stage can be stored for a few days in a desiccator.
- Stigma receptivity is at its maximum at the time of anthesis. At high temperature condition stigmatic fluid starts drying and the stigma becomes non-receptive.
- Best time of hybridization is between 6.00 and 9.00 a.m on the day of anthesis.
- Bagging the staminate and pistillate flowers prior to opening followed by hand pollination and subsequent bagging for two more days is the commercial method of hybrid seed production of watermelon.The stage of staminate flower development is very important. Young flowers do not provide pollens and pollens from the over-matured flowers do not produce viable seed.

- In monoecious sex form, harvesting the fully matured staminate flower buds in the afternoon hours that will open in the morning hours of the next day and keeping them covered with thin moist cloth over night for their use in cross-pollination in the morning hours is the conventional practice commercially.
- Presence of male sterility has somewhat simplified the hybrid seed production method.

Origin and taxonomy

Watermelons, which were originally domesticated in central and southern Africa, are an important part of the "most widespread and characteristic African agricultural complex adapted to savanna zones" in that they are not only a food plant but also a vital source of water in arid regions. Indeed, watermelons refer to as "botanical canteens." Watermelon emerged as an important cultigen in northern Africa and southwestern Asia prior to 6,000 years ago. Archaeological data suggest that they were cultivated in ancient Egypt more than 5,000 years ago, where representations of watermelons appeared on wall paintings and watermelon seeds and leaves were deposited in Egyptian tombs.

From their African origins, watermelons spread *via* trade routes throughout much of the world, reaching India by 800 and China by 1100. In both of these countries, as in Africa, the seeds are eaten and crushed for their edible oils. Watermelons became widely distributed along Mediterranean trade routes and were introduced into southern Europe by the Moorish conquerors of Spain, who left evidence of watermelon cultivation at Cordoba in 961 and Seville in 1158. Then watermelons spread slowly into other parts of Europe, perhaps largely because the summers are not generally hot enough for good yields. However, they began appearing in European herbals before 1600, and by 1625, the species was widely planted in Europe as a minor garden crop. Their first recorded appearance in Great Britain dates to 1597.

Watermelons reached the New World with European colonists and African slaves. Spanish settlers were producing watermelons in Florida by 1576, and by 1650 they were common in Panama, Peru, and Brazil, as well as in British and Dutch colonies throughout the New World. The first recorded cultivation in British colonial North America dates to 1629 in Massachusetts. Like cucumbers and melons, watermelons spread very rapidly among Native American groups. Prior to the beginning of the seventeenth century, they were being grown by tribes in the Ocmulgee region of Georgia, the Conchos nation of the Rio Grande valley, the Zuni and other Pueblo peoples of the Southwest, as well as by the Huron of eastern Canada and groups from the Great Lakes region. By the mid-seventeenth century, native Americans were cultivating them in Florida and the Mississippi valley, and in the eighteenth and early nineteenth centuries the western Apache of east-central and southeastern Arizona were producing maize and European-introduced crops including watermelons as they combined small scale horticulture with hunting and gathering in a low rainfall environment. This fact is ethnographically significant because other transitional foraging—farming groups, such as the San people of the Kalahari Desert of southern Africa, have parallel subsistence practices involving watermelons. Watermelons and melons were also rapidly adopted by Pacific Islanders in Hawaii and elsewhere as soon as the seeds were introduced by Captain James Cook (1778) and other European explorers.

The important role of watermelons in American popular culture in numerous areas including folk art, literature, advertising and merchandising, and the large number of annual summer watermelon festivals throughout the country with "parades, watermelon-eating contests, seed spitting contests, watermelon queens, sports events, and plenty of food and music". Growing and exhibiting large watermelons is an active pastime in some rural areas of the southern United States. Closely guarded family "secrets" for producing large watermelons and seeds from previous large fruit are carefully maintained. According to *The Guinness Book of Records,* the largest recorded watermelon in the United States was grown by B. Carson of Arrington, Tennessee, in 1990 which weighed a phenomenal 119 kg.

The somatic chromosome number is 2n=2x=22. From the results of morphological and cytogenetic studies, it was observed that the four species viz., *C. vulgaris*, *C. colocynthis*, *C. ecirrhosus* and *C. naudinianus* are related and cross-compatible with each other. *C. colocynthis* , the probable ancestor of watermelon which is found in the drought prone areas of Rajasthan and Gujrat are used in Ayurvedic medicine. It has similar morphological characters with watermelon but the fruits are bitter and seeds are small. However, African investigators considered bitter forms of *C. vulgaris* as the ancestor of cultivated watermelon.

Improved varieties

- Open pollinated varieties: Durgapura Meetha, Durgapura Kesar, Sugar Baby, Ashahi Yamato, Improved Shipper, Special No-1, PKM-1, Fuken.
- Public sector Hybrids: Arka Manik.

Cultural requirements

- Watermelon requires hot dry climate and a long growing season preferably with warmer days and cooler nights.
- It cannot withstand frost or very low temperatures.
- For seed germination, optimum moisture and a soil temperature between 25°–30°C is needed.
- Plant growth is optimum under 28°–30°C, while fruiting is better at 24°–27°C.
- Higher temperatures are beneficial during ripening.
- Arid regions of Rajasthan are best suited for production of quality fruits.

- Watermelon may be grown on a wide variety of soils.
- Sandy loams are best for early crop, while loams have high-yielding potential.
- Alluvial river-beds are also good for watermelon.
- Heavier soils do not permit perfect root growth and hence, only short duration varieties with smaller fruits are suitable.
- The soil should be well-drained and should have ample organic matter.
- It is generally cultivated in river-beds by making trenches and sowing in hills or pits.
- A soil pH of 6.5–7.0 is ideal.

Land preparation

- Prepare the field with 4-5 deep ploughings and make into a fine tilth before seed sowing.
- Remove all the weeds, stubbles and level the land by laddering.
- Manures should be mixed with the soil at the time of last ploughing.
- Generally furrows are opened at 2.0 to 3.0 m spacing and sowing is done on the top of the sides of furrows and the vines are allowed to trail on the ground.
- The crop is also grown on flat or raised beds, about 2.0 to 2.5 m wide and seed are sown on both sides of the bed.
- In rainy season, raised beds or mounds are prepared to facilitate drainage.
- Mounds in the furrow, flat bed method or in river bed system of planting are prepared with the mixture of 15-20 t FYM/ha and basal dose of fertilizers are mixed with the soil.

Sowing time

- In north Indian plains: late-February to mid-March
- In north-eastern and western India: November–January.
- In West Bengal : mid-November
- In central and south India: December–January.
- In Rajasthan: August–September.
- In north Indian hills: April–May.

Seed sowing

- Seeds are sown by hand dibbling in the mound.
- Sowings can be done following flat bed or furrow method or in river bed system depending on the season and convenience.

- The seed requirement for planting depends on spacing between plants and rows which again depend on vining habit of the variety.
- In general, a seed rate of 3.0–3.5kg/ha for small-seeded types and 5.0kg/ha for large-seeded types is sufficient.
- Hybrids need about one-third seed rate.
- Seeding can also be transplanted with ball of earth after raising in alkathene bags.
- For transplanting, 1–2 seeds/ bag are sown in perforated alkathene bags of 150–200 micron thickness and 8cm × 10cm size and seedlings at 2–3 leaves stage with ball of earth should be transplanted.
- Seeds should be soaked overnight and kept at moist warm place for 48hr for initiation of germination.
- Pre-germinated seeds can be sown in hills on the raised sides of furrows in upland and in trenches or in pits in the river-beds.
- Two plants/ hill in trenches or furrows and 4 plants/ pit are retained for growing.
- In pit sowing, pits of 60cm × 60cm size are dug up and filled with soil mixed with sufficient quantity of organic manure and fertilizer mixture before sowing.

Spacing

- Spring-summer crop: 2.0 – 3.0 m × 90 – 150 cm
- River bed system : 2.0 – 3.0 m × 100 – 150 cm
- Early autumn crop: 2.0 – 2.5 m × 60 – 90 cm

Nutrient management

- Watermelon responds well to manuring and fertilizer application.
- The dose of manures and fertilizers depend on the soil type, climate and system of cultivation.
- Well rotten farm yard manure at the rate of 15-20 t/ha should be applied at the time of land preparation.
- Apply a general fertilizer dose of 80 kg N, 60 kg P_2O_5 and 60 kg K_2O per hectare for open pollinated improved varieties.1/2 N along with full P and K fertilizers and farm yard manure should be given as basal and rest N should be top dressed in two split doses at the time of vining (25 - 30 days after sowing) and at blooming (45 - 50 days after sowing).
- For the hybrids apply 200 kg N, 100 kg P_2O_5 and 100kg K_2O per hectare. 1/3 N along with full P and K fertilizers and farm yard manure should be given as basal and rest N should be top dressed in two split doses at the time of vining (25 - 30 days after sowing) and at blooming (45-50 days after sowing).

Crop regulation

- In general, high N under high temperature conditions promotes maleness in flowering and lower number of pistilate flowers/ vine, resulting in low fruit-set and yield. The best way to control the vine growth within reasonable limits is by adjusting nitrogen fertilizer doses and frequency of irrigation and pruning.
- If apical shoot is pinched and 2–4 side shoots are allowed to grow it gives significantly higher fruit yield than un-pruned plants.
- Fruit thinning is useful and retaining of 2 fruits/ vine improves fruit size as well as fruit yield.
- Application of different growth regulators, secondary and micro-nutrients viz., 50 ppm 2,3,5 tri-iodobenzoic acid (TIBA), 50 ppm maleic hydrazide (MH), and 2 ppm boron twice, first at two true leaves of the plants i.e. 15 days after sowing and subsequently repeated after 7 days helps in increasing the yield.

Irrigation

- Interval and amount of irrigation depends upon the season and type of soils.
- Watermelon responds very much to irrigation but it cannot withstand waterlogged conditions.
- It is generally cultivated as a spring-summer crop in which frequency of irrigation is very important.
- Soil moisture is important for good germination.
- Frequent irrigations also promote excessive vegetative growth, especially in heavy soils which needs to be avoided.
- Soil moisture stress during pre-flowering, flowering and fruit development stages drastically reduces yield.
- Irrigation should be stopped during ripening as it adversely affects fruit quality and promotes fruit cracking.
- Application of water should be restricted to the base of the plant and root zone.
- Irrigation water should not wet the vines or vegetative parts, especially when flowering, fruit-set and fruit development is in progress.
- Frequent wetting of leaves, stems and developing fruits promote disease incidence.
- River-bed crop needs watering only in initial stages. Later when roots go as deep as 1.5m or up to the water table, no irrigation is given.
- The crop should be irrigated at 10-15 days intervals during summer.

Interculture and weed control

- Watermelons do not require much attention on interculture.
- In early stage, the beds and ridges should be kept weed-free.
- At the time of topdressing of nitrogenous fertilizer, weeding and earthing-up are done. When the vines start spreading, weeding in between rows or ridges becomes neither necessary not feasible since vine growth can smother the weeds.
- Vigorously growing weeds should be manually pulled out, without disturbing vines at later stages.
- Watermelon is also sensitive to weeds in initial stages of plant growth.
- Yield losses up to 30% have been observed due to weeds.
- Intercultural operations need to be started 15–20 days after sowing.
- Depending upon soil and environmental factors, 2–3 weedings are required.
- Pre-emergence herbicidal treatments of butachlor @ 2.0 kg a.i./ha and trifluralin @ 1–2 kg a.i./ha are effective.

Harvesting the crop

- Watermelons should be harvested at proper stage of maturity.
- Irrigation in preceding one week of harvest may give high turgidity to fruits and the skin of the fruit may crack while transporting them and hence this should not be practiced.
- Size of fruits and colour of the skin are not good indicators to know the proper stage.
- The crop is ready for harvesting 90–120 days after sowing depending upon cultivar and season.
- Change in the colour of the portion of the fruit which rests on the ground from white to creamy in case of light skin colour, or to yellow in case of dark skin colour, is a useful guide of maturity.
- A metallic sound when the fruit is tapped with the back of hand or with fingers denotes immaturity, whereas a heavy dull sound indicates ripeness.
- The drying of tendril at the base of the fruit is also a sign of maturity.
- However, the knack of recognizing a ripe melon comes with experience.
- The fruits become fit for harvesting 30–40 days after anthesis.
- Fruits should be separated from the vine with the help of a knife.

Post harvest handling and storage

- Improper harvesting, handling, transportation and distribution result in significant losses.

- Jerk to fruits while loading and transport should be avoided.
- They can be shipped to long distances, even without refrigeration.
- Flesh colour, sugar content and flavour do not improve/deteriorate appreciably after harvesting but there is a gradual softening of granulation of pulp around the seeds.
- Watermelons keep well for 1–3 weeks at 2.2°–4.4°C and 80% relative humidity.
- Watermelon is not processed commercially.

Plant protection measures

Physiological disorders

Preponderance of staminate flowers

The plants are monoecious in sex form hence, staminate and pistillate flowers are borne separately in the same plant and fruit yield depends on the number of pistillate flowers. Preponderance of staminate flowers is caused due to i) excessive nitrogen application, ii) high temperature condition , iii) long day length and iv) over irrigation.

Control measures

- Avoid rampant vegetative growth through control in fertilizer application, checking in irrigation and training and pruning.
- Apply nitrogenous fertilizer at a proper dose.
- Avoid excess irrigation
- Apply TIBA @ 50 ppm or MH @ 50 ppm twice, first at two true leaves of the plants i.e. 15 days after sowing and subsequently repeated after 7 days.

Delay in fruit ripening

Delay in fruit ripening is sometimes associated with less sweetness and cracking of fruits which occur due to high moisture level and temperature fluctuation at ripening stage.

Control measures

- Irrigation should be stopped at the ripening stage to hasten ripening.
- Sowing time should be adjusted in such a way that fruits ripe in hot and rainless condition which hastens ripening and at the same time improve sweetness of the fruits.

Blossom end rot

Brown, water-soaked discolouration appears at the blossom end of the fruit where senescent petals are attached at the immature stage of the fruit. The spots enlarge and darken rapidly and the affected portion becomes sunken, leathery and dark coloured. This disorder is caused due to i) sudden change in the rate of transpiration specially in moisture stress condition, ii) continuously high evapo-transpiration regime and a large leaf area, iii)increasing the level of nitrogen content in the fruits and iv) fall in the level of calcium content in the fruit.

Control measures

- Increase in the frequency of irrigation.
- Increase the level of phosphate which decreases the incidence of this disorder.
- Two foliar sparys of 0.2% calcium chloride at the time of fruit development.

Disease

Fusarium wilt (Fusarium oxysporm f. nivenum)

It is a common fungal disease. In young seedlings, colyledons wither and in older plants, leaves wilt suddenly and vascular bundles in the collar region become yellow or brown

Control measures

- Seed treatment with thiram or carbendazim @ 2.5g/kg of seed.
- Seed treatment ith *Trichoderma viride.*
- Spraying the crop with carbendazim (1g)+ mancozeb (2g) pen litre of water at weekly interval.

Anthracnose (*Colletotrichum lagenarium*)

Anthracnose is a destructive fungal disease occurring during warm and moist seasons. Significant damage can occur to cucumber, muskmelon, and watermelon unless resistant varieties are grown. All aboveground plant parts can be infected. Symptoms vary among the three principal cucurbits infected. Leaf lesions begin as water soaked and then become yellowish circular spots. On foliage the spots are irregular and turn dark brown or black.

Control measures

- Use disease-free seeds.
- Rotate vine crops with unrelated crops in a three-year rotation.
- Practice good sanitation by ploughing under fruits and vines at the end of the season.
- Spraying with carbendazim (0.15%) or chlorothalonil (0.2%) or mancozeb (0.25%) at 10 days interval for 2-3 times for the management of the disease.

Powdery mildew (*Sphaerotheca fuliginea* and *Erysiphe cichoracearum)*

White, powdery fungal growth develops on leaf surfaces, petioles, and stems. The growth is primarily asexual spores called conidia. It usually develops first on crown leaves, on shaded lower leaves, and on leaf undersurfaces. Yellow spots may form on upper leaf surfaces opposite powdery mildew colonies. Older plants are affected first. Infected leaves usually wither and die. Plants may senesce prematurely. Fruit infection is rare on watermelon and cucumber.

Control measures

- Spray the crop with Bavistin or Benlate or Topsin M @ 1g/l of water 3 to 4 times at 15 days interval.
- Two sprays of 0.1% Bavistin followed by two sprays of 0.1% karathane or tridemorph effectively control the disease.

Downy mildew (*Pseudoperonospora cubensis*)

This fungal disease is prevalent in high humidity, especially when summer rains occurs regularly.The diseases is characterized by formation of yellow, more or less angular spots on the upper surface of the leaves, white-purplish spores appear on the upper surface of the leaves. The disease spreads rapidly killing the plant quickly through rapid defoliation.

Control measures

- Removal of infected leaves and spray the crop with 0.25% mancozeb 3 to 4 times at 10 days interval effectively control the disease.
- Two sprays of 0.25% fosetyl Al or cyamoxanil- mancozeb or metalaxyl-mancozeb at 10 days interval effectively control the disease.

Insect pests

American bollworm (*Heliothes armigera*)

Some times it appears as a dengerous pest of watermelon causing devastating damage by feeding on the ovary of flower and developing fruits.

Control measures

- Collection and destruction of larvae mechanically at the early infestion can reduce the destructive pest population.
- Installation of bird perching site on the field bund helps in controlling invading population.
- Spraying with pyridalyl + fenpropathrin @ 1.25 ml/l is effective.

Thrips (*Scirtothrips dorsalis*)

The small insect attack on the leaves and young shoots causing silvery curling appearance. Severe population reduces the growth of the plants. They also indirectly damage by transmitting viral diseases in the plants.

Control measures

- Spraying the infested crops with acephate @ 1g/l or acetamiprid @ 0.2 g/l is very effective against the pest.

Yield

- 20 - 30 tonnes/hectare (open pollinated improved varieties).
- 50 – 60 tonnes/hectare (hybrids)

10. Muskmelon

Scientific Name: Cucumis melo
Family : Cucurbitaceae
Origin : Africa
Chromosome number: 2n = 24

Introduction

Muskmelon is one of the most commonly grown cucurbit in the tropical region. It is an important crop of Africa, Middleeast, Asia, Southeast Asia, Japan, USA, South and Central America, France, Spain, Israel and eastern European countries. Muskmelon is widely grown in Bihar, Uttar Pradesh, Punjab, Rajasthan, Andhra Pradesh and Tamil Nadu.

Importance and use

- Muskmelon is primarily used as fresh ripe fruit as dessert.
- The fruits are sweet and musky in flavour which are the main attractive qualities of muskmelon.
- Fruits are typically eaten uncooked, although immature fruits are cooked or pickled in some Asian cuisines.
- The flesh is used in ice-cream and preparation of candy.
- The seeds of some cultivars are roasted and consumed in parts of India.
- Dried and ground melon seeds are used as food in some African societies.
- Melon fruits, roots, leaves, and seeds play important roles in the treatment of a wide range of health problems in Chinese traditional medicine

Nutritional value

- The fruits contain 0.6% protein, 0.2% fat, 3.5% carbohydrates, 32 mg calcium, 14 mg phosphorus, 1.4 mg iron, 16 mg carotene and 26 mg vitamin C per 100 g fresh fruit.
- Muskmelon contains significant amount of dietary fiber, making it good for those suffering form constipation.

- The potassium present in muskmelons makes them quite helpful in the lowering of blood pressure.
- The fruit proves to be a good source of vitamin A and thus, helps maintain healthy skin.
- The folic acid present in muskmelons helps create healthy fetuses (in pregnant women) and can even prevent cervical cancer and osteoporosis.
- Muskmelon has a refreshing effect during tiresome hot summer periods.

Medicinal value

- The fruit has been associated with regulating heart beat and, possibly, preventing strokes.
- Muskmelons might reduce the risk of developing kidney stones and age-related bone loss.
- The fruits have a cooling effect on the body system.
- It is dimulscent, diuritic and aphrodisiac.
- Muskmelons have been found to have sedative properties, making them beneficial for the people who are suffering from insomnia.
- The fruit has almost zero cholesterol and thus, is safe for the people who suffer from the problem of high blood cholesterol.
- Regular consumption of muskmelon juice help treat lack of appetite, acidity, ulcer and urinary tract infections.
- It also serves as a mild antidepressant.
- Since it has high water content, the fruits help reduce the heat in the body and thus, prevent heat-related disorders.
- The seeds are diuretic, cooling, nutritive, and beneficial to the problem of enlarged prostate gland.

Adaptation in India

Muskmelon, a native of tropical Africa, is a polymorphic taxon comprising large number of botanical and horticultural groups, which includes both desserts as well as cooking and salad types. Muskmelon was introduced in India probably during Mughal invasion around 14th century from central Asian region (some parts of southern Russia, Iran, Afghanistan). Since then, it has spread to different parts of the country as far down to Andhra Pradesh and Karnataka. Existence of huge diversity in North-West India also suggests its probable secondary centre of origin here. Dessert type muskmelon grown in India comprises different local cultivars and improved varieties developed through selection from indigenous and combination breeding employing exotic introductions. However, deffernet local cultivars are much popular and cultivated under different names viz., Lucknow Safeda, Kanpuria, Jogia, Mathuria, Batti, Kajra, Jaunpuri, Mou, Bhagpat,

Jametha, Mahaban, Faizabadi (Uttar Pradesh), Tonk melon (Rajasthan), Sankheda, Galteshwar (Gujarat), Jalgaon, Kabri, Gurbeli, Peeli, Kavit (Madhya Pradesh), Sharbit-e-Anar, Bathesa (Andhra Pradesh), Kalapa (Karnataka), Kutana, Bhagpat (Haryana), Sharda melon, Haridhari (Punjab).

Breeding behaviour

- Five sex forms occur in muskmelon namely, monoecious, gynoecious, andromonoecious, gynomonoecious and hermaphrodite with monoecious and andro-monoecious sex form being predominant.
- Muskmelon is highly cross-pollinated in nature due to the monoecious sex form.
- Andromonoecious and hermaphrodite sex form favour a higher degree of natural self-pollination than in the monoecious or gynoecious condition.
- It is entomophillous and major pollinators are bees (*Apis florea, Apis dorsata, Nomioides sp., Halictine* sp., etc.).
- Beetles like *Conopophilus* sp, moths like *Planidia* sp, *Pygargonia* sp also help in natural cross pollination.
- Although 100% natural cross-pollination occur in monoecious and gynoecious sex form, yet reasonable extent of self fertilization happens because of geitenogamy due to foraging of insect pollinators in the staminate flowers of the same vine.
- Muskmelon like other members of the family Cucurbitaceae does not suffer much from inbreeding depression or do so only negligibly, in contrast to the cross-pollinated vegetable crops belonging to other families.
- Anthesis of both staminate and pistillate flowers generally starts in the morning hours from 5.30 to 7.00 a.m.
- Anther dehiscence in the staminate flower generally starts after anthesis.
- Pollens remain viable from 5.00 a.m. to 2.00 p.m. in ambient condition.
- Stigma receptivity is at its maximum at the time of anthesis although the stigma becomes receptive 2 hours earlier and 2-4 hours thereafter.
- Best time of hybridization is between 6.00 and 8.00 a.m. on the day of anthesis.
- Bagging the staminate and pistillate flowers prior to opening followed by hand pollination and subsequent bagging for two more days is the commercial method of hybrid seed production of monoecious form of muskmelon.
- In monoecious sex form, harvesting the fully matured staminate flower buds in the afternoon hours that will open in the morning hours of the next day and keeping them covered with thin moist cloth over night for their use in cross-pollination by gentle rubbing in the morning hours is the conventional practice commercially.

- In androemonocious and hermaphrodite flowers the flowers are emasculated a day prior to anthesis in the afternoon and subsequently bagged.
- Presence of male-sterility has made possible phenomenal progress in exploitation of F_1 hybrids.
- Muskmelon is crossable with *Cucumis melo* var. *momordica*, *Cucumis melo* var. *flexuosus*.

Origin and taxonomy

Melon is generally thought to have originated in western Africa, with China or India as possible secondary centers of diversity. Wild melons growing in natural habitats have been reported in desert and savanna zones of Africa, Arabia, southwestern Asia, and Australia. The true wild forms are found in eastern tropical Africa and the wild forms reported from India are feral escapes derived from local cultivars. However, it is unclear where melon was domesticated and it is conceivable that it was independently domesticated from different wild populations in Africa and southwestern Asia. Melon was an important food crop in ancient China, where archaeological data suggest that it has been cultivated for over 5,000 years. Archaeological evidence also suggests that melon was cultivated in Iran some 5,000 years ago and in Greece and Egypt about 4,000 years ago. As with cucumbers, melons were cultivated in the Roman Empire and diffused throughout Europe by the Middle Ages where the variety and quality of melon cultivars were evidently greatly increased by selection in Medieval gardens. As with cucumbers and watermelons, melons were introduced to the New World by Spanish colonial settlers in the late fifteenth and early sixteenth centuries and subsequently spread very rapidly among Native American horticultural groups. Later during the eighteenth century they reached the Pacific Islanders via British explorers.

Improved varieties

- **Improved cultivars:** Pusa Sharbati, Pusa Madhuras, Pusa Rasraj, Arka Jeet, Arka Rajhans, Durgapura Madhu, Hara Madhu, Hisar Madhur, Hisar Saras, MH-10, Punjab Sunheri, Punjab Rasila,
- **Public sector Hybrids:** Punjab Hybrid, Pusa Rasraj
- **Resistant to downy mildew:** Hisar Saras, Punjab Rasila

Cultural requirements

- Muskmelon flourishes well under warm climate and cannot tolerate frost.
- The optimum temperature for germination of seeds is 24°–29°C.
- With increase in temperature, the plants complete their vegetative growth earlier.
- Stormy weather particularly dust-storm during flowering reduces fruit setting.

- Dry weather with clear sunshine during ripening ensures a high sugar content, better flavour and high percentage of marketable fruits.
- High humidity increases the incidence of diseases, particularly those affecting foliage.
- Cool nights and warm days are ideal for accumulation of sugars in fruits.
- A well-drained, loamy soil is ideal.
- Lighter soils which warm up quickly in spring are usually utilized for early crop.
- In heavier soils, vine growth is more and fruit maturity is delayed.
- On sandy river-beds, alluvial substrata and subterranean moisture of river streams support its growth.
- In river-beds, its long tap root system is adapted to growth although, it is necessary that soils should be fertile well provided with organic matter.
- Muskmelon is sensitive to acid soils. The crop cannot be grown successfully below a pH of 5.5.
- A soil pH between 6.0 and 7.0 is ideal.
- Alkaline soils with high salt concentration are not suitable.

Land preparation

- Prepare the field with 4-5 deep ploughings and make into a fine tilth before seed sowing.
- Remove all the weeds, stubbles and level the land by laddering.
- Seeds should be sown on raised beds.
- The width of bed depends upon crop and the variety to be planted.
- Generally 3-4 m wide raised beds are prepared for muskmelon.
- Planting is done on both sides of the beds.
- The planting can also be successfully done in long water channels.
- The channels can be prepared either with a tractor or manualy.
- The distance between 2 channels should be according to width of the bed.
- Subsequently, the vines should be raised to grow in between the channels.

Sowing time

- In north India, early sowing is done in river-beds from November to January.
- In northern plains, sowing is done in February.
- In southern plains, sowing is done in December–January.

Seed sowing

- Muskmelon is basically a direct seeded crop and sowings can be done following flat bed or furrow method or in river bed system depending on the season and convenience.
- Optimum temperature for seed germination is 24°–29°C.
- To get an early crop, seeds can be sown in polythene bags under protected cover.
- The seedlings become ready for transplanting at 2–4 true leaf stage. This practice is prevalent in Punjab, where sowing is done in last week of January or first week of February. The seedlings become ready for transplanting by February end or first week of March.
- The seed rate of 3-4 kg/ha is optimum for sowing in field by dibbling method, or in polythene bags although, seed rate upto 7kg/ha is also given.
- Transplanted crop matures 15–20 days earlier than the direct-seeded one.
- For raising seedlings, take polythene bags of 15cm × 10cm size and of 100 gauge thickness.
- Fill them with a mixture of well-rotten farmyard manure, field soil and silt in equal proportion.
- For raising nursery for 1 hectare, 12.5–15 kg polythene bags are required.
- Seeds should be sown 1.5cm deep.
- Watering of the bags should be done daily with a sprinkling can.
- Stop watering the bags 2 days before transplanting.
- Cut and remove polythene bags before transplanting the seedlings.
- Take care that the earth-ball containing seedling does not break.
- Place the earth-balls in the pits in the field, put the soil all around and irrigate the channels immediately after transplanting.

Spacing

- Spring-summer crop: 3.0 – 3.5 m × 50 – 60 cm
- River bed system: 2.0 – 3.0 m × 1.0 – 1.5 m

Nutrient management

- Add 25–30 tonnes/ha farmyard manure in soil 10–15 days before sowing.
- Apply 125 kg N, 65kg P_2O_5 and 65kg K_2O/ha for improved varicties.
- Whole P and K and one-third of N should be applied in 2 parallel bands 45cm apart on both sides of the bed mark. Prepare channels in between fertilizer bands. Rest of the N should be applied to the vines 3–4 weeks after germination prior to earthing-up.

- For transplanted crop keep row-to-row and plant-to-plant distance same as for directly sown crop. Dig 15–20cm deep pits at the sites where seedlings are to be transplanted. Fill each pit with a mixture of 1 kg of farmyard manure, 10–15g of calcium ammonium nitrate, 40g of super phosphate and 10g muriate of potash before planting. Apply another dose of 10–15g of CAN (calcium ammonium nitrate) to each plant after one month.
- For the hybrids apply 200 kg N, 100 kg P_2O_5 and 100 kg K_2O per hectare. One third N along ith full P and K fertilizers and farm yard manure should be given as basal and rest N should be top dressed in to split doses at the lime of vining (28-30 days after sowing) and at flowering (45-50 days after sowiing).
- Spraying the plant twice at 10 days interval starting from 25 days after sowing with 20-30 ppm Ca and 8-10 ppm B is beneficial.

Crop regulation

- In general, high N under high temperature conditions promotes maleness in flowering and lower number of pistilate or perfect flowers/ vine, resulting in low fruits-set and yield. The best way to control the vine growth within reasonable limits is by adjusting nitrogen fertilizer doses and frequency of irrigation and pruning.
- Rampant vegetative growth should be avoided as it promotes maleness.
- Secondary branches are pinched off up to 6-7 nodes on emergence and tertiary branches are allowed to grow.
- Perfect/pistillate flowers are allowed to be borne on these branches to set fruits.
- Tertiary branches are detopped at 2 leaves above the first perfect/pistillate flowers.
- Application of different growth regulator like 200 ppm maleic hydrazide (MH) twice, first at two true leaves of the plants i.e. 15 days after sowing and subsequently repeated after 7 days helps in increasing the yield.

Irrigation

- Maintaining sufficient moisture is essential.
- Frequency of irrigation should be reduced during maturity period to get sweeter fruits.
- The irrigation should be as light as possible.
- The light sandy soils need more frequent irrigation than heavier ones.
- The flooding of the field should be avoided, particularly when fruits are reaching maturity.
- During dry summer, irrigate the crop at 5–7 days intervals.

Interculture and weed control

- Hoeing keeps soil loose and free from weeds.
- In the beginning, cultivation can be fairly close to the plants and shallow (5–10cm). The growth of lateral roots often equals or exceeds that of the above ground parts.
- Later hoeing should be relatively shallow.
- When vines cover the ground, cultivation should be stopped and only weeds should be pulled out by hand.
- During April–June severe dust storms not only hamper vegetative growth but also roll them up and in some cases even uproot the vines, if the fields are kept clear of weeds.
- It is, therefore, advisable particularly in the storm prone areas of Rajasthan, Punjab, Haryana, Uttar Pradesh to keep the area near the roots free from weeds but allow them to grow on the beds away from root zone, so that the vines during growth can get hold of them and are disturbed the least by fast winds.

Harvesting the Crop

- Fruits maturing on vine, without becoming over-ripe, are superior in quality to those harvested immature, or left on vine after they have become mature.
- There are two groups of cultivars which behave distinctly. Fruits of most of the cultivars slip out easily from the vine upon maturily with little pressure or jerk, called "Full slip stage".
- The other group do not separate from the vine at maturity which are harvested on the basis of external colour of the fruits.
- Harvesting at the proper stage is of major importance in marketing good quality produce. The fruits for distant markets should be harvested as they reach the half-slip maturity, to avoid losses from the over-ripeness and decay.
- For local markets, harvesting should be done at 'full-slip' stage. Hara Madhu Honey Dew, Casaba, etc.however, never reaches the full-slip stage and colour of the rind can be taken as the criteria of maturity.
- Fruits harvested when immature as yet, never attain these desirable characteristics.
- The soluble solids are considerably affected by the environment, the incidence of diseases and the vigour of the plant.

Post harvest handling and storage

- Improper harvesting, handling, transportation and distribution result in significant losses.

- Jerk to the fruits while loading and transport should be avoided.
- Muskmelon does not have good storage quality.
- Flavour and texture of the flesh improve for a few days after harvesting and attain highest quality if the fruits are harvested when they have developed their maximum sugar content.
- Muskmelons keep well for one week at 1.7°–3.3°C and 85-90 % relative humidity.

Plant protection measures

Physiological disorder

Preponderance of staminate flowers

In the vareties having monoceious sex form, staminate and pistillate flowers are borne separately in the same plant and fruit yield depends on the number of pistillate flowers. Preponderance of staminate flowers is caused due to i) excessive nitrogen application, ii) high temperature condition , iii) long day length and iv) over irrigation.

Control measures

- Avoid rampant vegetative growth through control in fertilizer application, checking in irrigation and training and pruning.
- Apply nitrogenous fertilizer at a proper dose.
- Avoid excess irrigation
- Apply 200 ppm MH twice, first at two true leaves of the plants i.e. 15 days after sowing and subsequently repeated after 7 days.

Delay in fruit ripening

Delay in fruit ripening is sometimes associated with less sweetness which occur due to high moisture level and temperature fluctuation at ripening stage.

- Irrigation should be stopped at the ripening stage to hasten ripening.
- Sowing time should be adjusted in such a way that fruits ripe in hot and rainless condition which hastens ripening and at the same time improve sweetness of the fruits.

Diseases

Fusarium wilt (Fusarium oxysporum f. melonis)

It is a common seed and soil borne fungal disease. In young seedlings, the colyledons droop and wither and in older plants the plants wilt suddenly with browning of vascular bundles at the collar region.

Control measures

- Same as mentioned in watermelon.

Anthracnose (*Colletotrichum lagenarium*)

Anthracnose is a destructive fungal disease of muskmelon and other cucurbits occurring during warm and moist seasons. All aboveground plant parts can be infected. Leaf lesions begin as water soaked and then become yellowish circular spots. On cucumber and muskmelon the spots turn brown and can enlarge considerably. Stem lesions on muskmelon can girdle the stem and cause vines to wilt.

Control measures

- Use disease-free seeds.
- Rotate cucurbits with unrelated crops in a three-year rotation.
- Practice good sanitation by ploughing under fruits and vines at the end of the season.
- Spraying with carbendazim (0.15%) or chlorothalonil (0.2%) or mancozeb (0.25%) at 10 days interval for 2-3 times for the management of the disease.

Powdery mildew (*Sphaerotheca fuliginea* and *Erysiphe cichoracearum)*

White, powdery fungal growth develops on leaf surfaces, petioles, and stems. The growth is primarily asexual spores called conidia. It usually develops first on crown leaves, on shaded lower leaves, and on leaf undersurfaces. Yellow spots may form on upper leaf surfaces opposite powdery mildew colonies. Older plants are affected first. Infected leaves usually wither and die. Plants may senesce prematurely.

Control measures

- Spray the crop with Bavistin or Benlate or Topsin M @ 1g/l of water 3 to 4 times at 15 days interval.
- Two sprays of 0.1% Bavistin followed by two sprays of 0.1% Karathane or tridemorph effectively control the disease.

Downy mildew (*Pseudoperonospora cubensis*)

This fungal disease is prevalent in high humidity, especially when summer rains occurs regularly.The diseases is characterized by formation of yellow, more or less angular spots on the upper surface of the leaves, white-purplish spores appear on the upper surface of the leaves. The disease spreads rapidly killing the plant quickly through rapid defoliation.

Control measures

- Removal of infected leaves and spray the crop with 0.25% mancozeb 3 to 4 times at 10 days interval effectively control the disease.
- Two sprays of 0.25% fosetyl Al or cyamoxanil- mancozeb or metalaxyl-mancozeb at 10 days interval effectively control the disease.

Insect pests

Fruit fly (Dacus cucurbita, Dacus dorsalis)

Adult fly of this polyphagous pest having a wide host range is 4 – 5 mm long, ferruginous-brown in colour with hyaline wings. On hatching, the maggots feed inside the fruits and infested fruits can be identified by the presence of resinous liquid which oozes out of the punctures made by the flies for oviposition. The infested fruits start rotting due to secondary infection of different micro-organisms.

Control measures

- Collect and destroy the infested fruits along with the maggots to prevent the carry over of the pest.
- Stir the soil under the vines frequently and plough the infested field after harvest the crop to kill pupae.
- Spray the crop with 0.05% endosulfan.
- Apply poison bait. Bait is prepared by mixing 20 g Malathion 50% WP or 50 ml Diazinon 20% EC with 500 g molasses + 20 g yeast hydrolysate. This mixture is mixed with 2 litres of water for poison baiting and 20 liters of water for bait spray.
- Adopt male annihilation technique using 0.1% methyl eugenol + 0.025% Malathion in a bait trap during fruiting season. Methyl eugenol is a sex attractant and males are attracted towards the bait and killed by malathion. Apply this trap solution at 7-10 days interval.

Red pumpkin beetle (Aulacophora foveicollis)

Shiny yellowish-red adults of this polyphagous pest lay eggs in moist soils around the plant. On hatching, the grubs feed on the roots and underground portion of the plant and also on the fruits touching the soils. Adult beetles feed voraciously on leaf lamina making irregular holes. Young seedlings and tender leaves are mainly preferred by the adults and damage at this stage may kill the seedlings.

Control measures

- Clean cultivation.
- Collect and destroy the beetles during cool hours of early morning when they remain sluggish.
- Deep ploughing of infested fields just after harvesting to kill the grubs in the soil.
- Dust the crop with 5% carbaryl or spray with 0.2% carbaryl.
- Repellent dusting of ash mixed with kerosene oil but heavy dusting should be avoided as it hampers plant growth.

Thrips (*Scirtothrips dorsalis*)

The small insect attack on the leaves and young shoots causing silvery curling appearance. Severe population reduces the growth of the plants. They also indirectly damage by transmitting viral diseases in the plants.

Control measures

- Spraying the infested crops with acephate @ 1g/l or acetamiprid @ 0.2 g/l is very effective against the pest.

Yield

- Improved varieties: 15-20 tonnes/ha
- Hybrids: 30-35 tonnes/ha

11. Pumpkin

Scientific Name: Cucurbita moschata Duch. ex Poir.
Family : Cucurbitaceae
Origin : North-western South America
Chromosome No. : 2n = 40

Introduction

The cultivated species of *Cucurbita* were selected by the American Indians long before 1492 and they were staple items of their diet. The genus *Cucurbita* giving pumpkin, squashes or vegetable marrows is the most economically viable genus of the family Cucurbitaceae with 27 species, five of which (*C. moschata, C. pepo, C. maxima, C. mixta* and *C. ficifolia*) are cultivated all over the tropical and sub-tropical world. Pumpkin is the most important species and extensively cultivated in India, Africa, Latin America, Southern Asia and United States. The English name of this species "Tropical Pumpkin" is an appropriate one because the greatest diversity lies in the neotropics where the vines are grown under a wide range of ecological conditions including the hotter conditions than are tolerated by the other cultivated *Cucurbita* species. It is one of the important and popular vegetable crops extensively grown during spring-summer, summer-rainy and early autumn seasons in all parts of the country particularly in Assam, West Bengal, Tamil Nadu, Karnataka, Kerala, Madhya Pradesh, Uttar Pradesh, Orissa and Bihar.

Importance and use

- Pumpkin occupies a prominent place among the vegetables owing to its high productivity, nutritive value, good storability, long period of availability and better transport potentialities.

- It is an important cucurbitaceous vegetable extensively grown in different parts of the country.
- It is the natural source of β-carotene, the precursor of vitamin A.
- During summer it is used commonly in North India in marriage parties.
- Nearly all above grown parts of pumpkin like leaves, staminate flowers and both mature and immature fruits are utilized as vegetable in different forms.
- The flesh is delicious when stewed, boiled or baked.
- Fully matured fruits, apart from utilization as cooked vegetable can be used in preparing sweets, candy or fermented into beverages.
- Delicate sweet items like "halwa", other sweets and jams are prepared from meshed flesh of fully matured fruit.
- Fully matured ripe fruits are also mixed with tomato in the preparation of sauce and ketchup.
- Pumpkin pulp is also used for the preparation of GTF (glucose tolerance factor) pumpkin milk powder.
- Matured fruits are also used as industrial raw material for carotene production.
- The "Yerusseri" prepared from immature fruits is very popular in Kerala.
- The seeds after removing the seed coats are used in confectionary.
- Seed oil (50 % of the seed) is a common salad oil in the southern part of Austria, Slovenia and Hungary.

Nutritional value

- Average nutritive value of pumpkin (2.68) is higher than brinjal (2.41), tomato (2.09) and cucumber (1.69) mainly because of stored carbohydrate (mainly glucose) and carotenoids (74 % β-carotene) along with moderate quantities of ascorbic, nicotinic, pantothenic and folic acids and different minerals.
- Edible portion of pumpkin contains 1.4% protein, 4.6% carbohydrates, and 0.1% fat and 0.6 % minerals.
- The young leaves, flowers and fruits are rich in β-carotene, a precursor of vitamin-A and provide on an average 50 mg β-carotene per 100 g edible portion of fruit.
- It provides about 30 mg of P, 10 mg of Ca and 0.7 mg of Fe per 100 g edible portion of the fruit.

Medicinal value

- Pumpkin is particularly important for the supply of antioxidants and especially β-carotene in the foods.

- It helps to alleviate cardiovascular diseases and constipation.
- It is also a good diuretic food because of having high K : Na ratio.
- Its combination with sugar and honey has soothing effect to ulcers and blisters.
- It is used in the treatment of diabetis, rheumatism, eczema and burns and against worms and other parasites.
- The fruit pulp is also used for poultice for boils, carbuncles and ulcers etc.
- Its seeds along with sugar are taken to kill tape worms and are useful in urinary diseases.

Adaptation in India

Pumpkin was introduced to India from South America long back by foreign navigators and emissaries. Its early introduction and subsequently wide range acceptability and cultivation have favoured the development of huge genetic variability of pumpkin in India. A number of vernacular names of pumpkin *viz.,* Kumra, Misti kumra, Sitaphal, Kashiphal, Kaddu, Lalbhopla, Lalkumbhara, Meetha kaddu, Pilum kohalum, Poashani, Gumbala, Kannada, Duddini, Red gourd, Saphurii kumra, etc. also suggests its wide range cultivation in India. Wide range of diversity of pumpkin genotypes is found in India and all the improved varieties have been developed through selection followed by inbreeding from different local cultivars. In India mostly different local cultivars are grown under different name mainly after the name of the place where the particular local types are available however, a good number of improved varieties have also been developed in India from the indigenous diversity.

Breeding behaviour

- It is highly cross pollinated in nature due to monoecious sex form and entomophilies in pollination.
- Major pollinators are honey bees (*Apis* sp.).
- Although it is 100% cross pollinated in nature yet high extent of self fertilization happens because of bee foraging in the staminate flowers of the same vine.
- Pumpkin does not suffer much from inbreeding depression or do so only negligibly, in contrast to the cross pollinated vegetable crops belonging to other families.
- Anthesis starts in the early morning at 3:30 a.m. and maximum number of flowers opens between 4:00 and 5.00 a.m.
- Flowers remain open for about 8 to 10 hours after which they start losing their yellow colour and wither.
- Pollen dehiscent generally starts from 5:30 am. in summer months and it is some what delayed with the advent of low temperature.

- Stigma receptivity is maximum at the time of anthesis although stigma becomes receptive one day prior to and remains one day after anthesis also.
- Best time of hybridization is within 6.00 am on the day of anthesis.
- Harvesting the fully matured staminate flowers in the afternoon hours that will open in the next day morning and keeping them covered with thin moist cloth overnight for their use in cross pollination in the early morning hours is the conventional practice commercially.
- Bagging the staminate and pistillate flowers prior to opening followed by hand pollination and subsequent bagging for two more days is the commercial method of hybrid seed production.
- Pumpkin occupies a central position among the cultivated *Cucurbita* spp. and can be crossed with difficulty with *C. pepo, C. maxima* and *C. mixta.*
- Embryo culture and or amphidiploidy is generally employed to overcome the problem of poorly developed embryo and F_1 sterility usually encountered in these interspecific crosses.

Origin and taxonomy

Pumpkin and squashes have been associated with man in the America for at least 10,000 years. American Indians selected the cultivated species long before and they were staple items of their diet. It was cultivated in southern Mexico in 5000 B.C. and in Peru in 3000 B.C. Archaeological evidences showed that pumpkin was widely distributed in both North and South America with the primary centre of origin being Mexico and Peru. Considering its seed colour, pumpkin is also believed to be the native of Guatemala and Columbia. Wild forms of pumpkin have not yet been described but primitive-appearing land races are known from Central America to northern Peru. This coupled with archaelogical evidence recently suggests that the centre of origin of pumkin is in the north-western South America.

The genus *Cucurbita* belongs to the family Cucurbitaceae, sub-family Cucurbitoideae, Tribe Melothrieae and sub-tribe Cucumerinae. There are 26 *Cucurbita* species including five cultivated species namely, *C. moschata* – Pumpkin, *C. Pepo* – Summer squash, *C.maxima* – Winter Squash, *C. mixta* (Syn. *C. angyrosperma)* – Winter Squash and *C. ficifolia* – Fig leaf gourd or Malabar gourd. The wild form of *Cucurbita* are classified into

(a) Xerophytic form: *C. digitata*, *C. palmate*, *C. cylindiata*, *C. foetodissima.*

(b) Mesophytic form: *C. ficifolia*, *C. sororia*, *C.lundelliana*

Pumpkin appears to be more closely related to xerophytic forms and is considered as the axis through which the species under xerophytic forms are related to each other.

The genomic structure of *Cucurbita moschata* is AABB indicate that it is an amphidiploid. Isozyme studies in pumpkin revealed that there are 28 loci conditioning the isoenzymic phenotypes and this coupled with high chromosome number (2n = 40) suggest its allotetraploid origin.

Open pollinated varieties

Pumpkin is cultivated in India mostly using different local cultivars under different name and a number of improved varieties have also been developed form these diverse genotypes.

- Improved varieties: Arka Suryamukhi, Solan Badami, Arka Chandan, CO-1, CO-2, Ambili, Pusa Biswas, Pusa Vikas.
- Local cultivars: Large Red, Large Round, Red Flesh, etc.

Cultural requirements

- Pumpkin is basically a warm season crop and susceptible to frost.
- The favourable temperature range for the growth and development is 18.0 - 30°C and optimum being 20°C-25°C.
- Short days, comparatively low-night temperature and high relative humidity is the best for pumpkin cultivation as this condition increases the propensity of pistillate flowers in the vine.
- High temperature and long days help to increase the number of staminate flowers and reduce the number of pistillate flowers.
- It can also be grown at high temperature range of 30 -35°C.
- Ideal range of soil temperature for seed germination is 20°C-25°C.
- Seed germination is hampered at soil temperature below 10°C and above 30°C.
- Deep well drained loamy soils with pH range 6.0 – 7.0 are ideal.
- Sandy loam soils are ideal for early crop but clay loam soils are suitable for getting high yield.
- Soil moisture should be at least 10-15% above the permanent witting point for proper growth and yield of the crop.
- Pumpkin is sensitive to water logging condition.
- Conventionally pumpkin in India is grown after "Aman" paddy (sowing in November-December) and in relay cropping with potato (sowing in January-February).

Land preparation

- Prepare the field with 2-3 ploughing and make into a good tilth before seed sowing.
- Remove all the weeds, stubbles and level the land by laddering.
- The plant has extensive root system.
- In spring-summer or early autumn season, prepare pits of 60 cm diameter and 45 cm depth at a spacing of 90 to 150 cm in rows spaced 2.0-3.0 metre.
- Furrow can also be opened at 2.0 to 3.0 m spacing and mounds are prepared just at the side of the furrow top.
- In rainy season, raised beds or mounds are prepared to facilitate drainage.
- Soil of the pits is filled and mounds are prepared with the mixture of 20-25 tonnes of FYM per hectare and basal dose of fertilizers are mixed with the soil.

Sowing time

Pumpkin grows successfully in tropical and subtropical climate hence, can be grown year round in the southern states due to prevailing mild winter there. However, the crop is grown in three main seasons: spring-summer, early autumn and rainy season.

- Late January – March: Spring-summer crop in eastern and northern plain.
- May-July: Rainy season crop in eastern and northern plain.
- August - September: Early autumn crop in eastern India.
- December - January: Winter – spring crop in southern and western regions.
- March – April: Hill regions.

Seed sowing

- Pumpkin gives little success on direct transplanting and , in general, seeds are sown directly in the field.
- Seeds are sown by hand dibbling in the mound.
- Sowings can be done following pit, mound or furrow method depending on the season and convenience.
- Seedlings raised in plastic packets particularly under polyethylene shed in winter months can be transplanted after 35 – 40 days with the advent of spring without disturbing the root system.
- 3.0 – 5.0 kg seed/ha is sufficient depending on the plant population.
- It is better to do seed sowing in pre-irrigated moist soil than irrigate the field after sowing.

- Pre-sowing soil application of 20-22 kg Furadon 3G helps in protecting the plants from root knot nematodes and other pests during initial 4-5 weeks.
- On an average 4 to 5 thousand plants can be accommodated in one hectare land.

Spacing

- Spring-summer crop: 2.0 – 3.0 m × 90 – 150 cm
- Rainy season crop: 3.5 m × 1.5 – 2.0 m
- Early autumn crop: 2.0 – 2.5 m × 60 – 90 cm

Nutrient management

- The dose of manures and fertilizers depend on the soil type, climate and system of cultivation.
- Apply a general fertilizer dose of 100 kg N, 60 kg P_2O_5 and 60 kg K_2O per hectare for open pollinated improved varieties. 1/2 N along with full P and K fertilizers and farm yard manure should be given as basal and rest N should be top dressed in two split doses at the time of vining (25 - 30 days after sowing) and at full blooming (55 – 60 days after sowing).
- High nitrogen under high temperature condition promotes staminate flowers resulting in low fruit set and yield so, fertilizer dose should be adjusted according to season of cultivation.

Crop regulation

- Excessive vine growth due to high nitrogen nutrition coupled with high temperature and high soil moisture condition promote staminate flowers in the vine resulting low fruit set and low yield. The best way to control the vine growth within reasonable limits is by adjusting nitrogen fertilizer doses and frequency of irrigation.
- Application of ethylene releasing growth regulator, Ethrel @ 250 ppm concentration (2.5 ml/10 l of water) four times, first at two true leaves of the plants i.e. 15 days after sowing and subsequently repeated three times at 7 days interval induces production of pistillate flowers in the vine which helps in increasing the yield.

Irrigation

- Interval and amount of irrigation depends upon the season and type of soils.
- Apply water at the initiation of first true leaf during spring-summer and at its expansion during rainy season.
- Irrigate the field subsequently 5-6 days intervals during summer and early autumn crop and as and when required for the rainy season crop depending on the rain.
- In light soils irrigate the crop at frequent intervals.

- The crop must be irrigated during its most critical stages i.e. flowering and fruit setting stages.
- Over irrigation during vegetative and early flowering stages may cause excessive vine growth resulting induction of more staminate flowers in the vine.
- Excessive irrigation at fruit maturity stage adversely affects the storability of the fruits.
- Ridge and furrow or ring methods of irrigation is the best for pumpkin.

Interculture and weed control

- The crop may be trained over low trellis of 1.5 m high or trailed over the dried twigs spread over the ground for proper growth.
- Soil at the base of the plant should be loosened before top dressing with nitrogen fertilizers followed by irrigation.
- At the initial stages of weed growth hand weeding and hoeing are sufficient to check the weed growth.
- Application of herbicides effectively controls weeds.
- Pre-emergence application of Besulide @ 4-6 kg a.i. /ha or Alachlor @ 2.5 kg a.i. /ha checks weed population in the fields.
- Application of Gramaxone @ 0.2% controls the weeds during early stage of crop growth.

Harvesting the crop

- Fruits generally reach full maturity 160 to 220 days after sowing depending on the variety and season of cultivation.
- Generally fruits are harvested at full maturity or complete ripe stage when the harvested fruits are stored or seeds are extracted from it.
- At full maturity generally rind colour of the fruits turn brown or reddish-brown from green and peduncle either dries up or separates from the fruits.
- Periodical harvests of tender fruits increases the fruit number and thereby yield.

Post harvest handling and storage

- For long storage life and long distance transportation fully matured fruits have to be harvested.
- Fruits can be kept for 2 to 4 months without any damage at room temperature however, the fruit should not be kept in heaps.
- It is better to keep the fruits in shelves where they can be spread out in single layer with a space between the fruits.
- The fruits can be kept for 6 months at 10 to 12°C temperature with 70-75% relative humidity condition.

Plant protection measures

Physiological disorder

Preponderance of staminate flowers

The plants are monoecious in sex form hence, staminate and pistillate flowers are borne separately in the same plant and fruit yield depends on the number of pistillate flowers. Preponderance of staminate flowers is caused due to i) excessive nitrogen application, ii) high temperature condition , iii) long day length and iv) over irrigation.

Control measures

- Avoid rampant vegetative growth through control in fertilizer application and checking in irrigation.
- Apply nitrogenous fertilizer at a proper dose.
- Avoid excess irrigation.
- Apply ethephon / ethrel (2-chloroethyl phosphonic acid) @ 250 ppm four times, first at two true leaves of the plants i.e. 15 days after sowing and subsequently repeated three times at 7 days interval

Diseases

Powdery mildew (*Sphaerotheca fuliginea* and *Erysiphe cichoracearum*)

This fungal disease appears under prolonged dry conditions. White floury or powdery patches or coatings appear on the lower surface of leaf extending to upper surface causing yellowing and death of the leaves.

Control measures

- Spray the crop with Bavistin or Benlate or Topsin M @ 1g/liter of water 3 to 4 times at 15 days interval.
- Two sprays of 0.1% Bavistin followed by two sprays of 0.1% Karathane or tridemorph effectively control the disease.

Downy mildew (*Pseudoperonospora cubensis*)

This fungal disease is prevalent in high humidity, especially when summer rains occurs regularly.The diseases is characterized by formation of yellow, more or less angular spots on the upper surface of the leaves, white-purplish spores appear on the upper surface of the leaves. The disease spreads rapidly killing the plant quickly through rapid defoliation.

Control measures

- Removal of infected leaves and spray the crop with 0.25% mancozeb 3 to 4 times at 10 days interval effectively control the disease.
- Two sprays of 0.25% fosetyl Al or cyamoxanil- mancozeb or metalaxyl-mancozeb at 10 days interval effectively control the disease.

Mosaic (Pumpkin mild mosaic virus, cucumber mosaic virus (CMV), Watermelon Mosaic Virus (WMV).

These viruses are transmitted by sap, seed and by several species of aphids.The young leaves develop small greenish-yellow areas which are more translucent than the remaining parts of the leaf. Leaves show characteristics mottling with yellowish-green and blistering and in severe infection, growth of the plant is stunted.

Control measures

- Spray the crop with 0.05% dimethoate or imidachloprid(3.5 ml/10 l) or 0.05% phosphamidon at weekly interval to control aphid vectors.
- Eliminate weed hosts and use of barrier crops like sorghum, maize around the field.
- Crop rotation with non-cucurbitaceous crops.
- Remove infected plants from the field.

Insect Pest

Fruit fly (*Dacus cucurbita, Dacus dorsalis*)

Adult fly of this polyphagous pest having a wide host range is 4 – 5 mm long, ferruginous-brown in colour with hyaline wings. On hatching, the maggots feed inside the fruits and infested fruits can be identified by the presence of resinous liquid which oozes out from the punctures made by the flies for oviposition. The infested fruits start rotting due to secondary infection of different micro-organisms.

Control measures

- Collect and destroy the infested fruits along with the maggots to prevent the carry over of the pest.
- Stir the soil under the vines frequently and plough the infested field after harvest the crop to kill pupae.
- Spray the crop with 0.05% endosulfan.
- Apply poison bait. Bait is prepared by mixing 20 g malathion 50% WP or 50 ml diazinon 20% EC with 500 g molasses + 20 g yeast hydrolysate. This mixture is mixed with 2 litres of water for poison baiting and 20 liters of water for bait spray.
- Adopt male annihilation technique using 0.1% methyl eugenol + 0.025% malathion in a bait trap during fruiting season. Methyl eugenol is a sex attractant and males are attracted towards the bait and killed by malathion. Apply this trap solution at 7-10 days interval.

Red pumpkin beetle (Aulacophora foveicollis)

Shiny yellowish-red adults of this polyphagous pest lay eggs in moist soils around the plant. On hatching, the grubs feed on the roots and underground portion of host plant and also on the fruits touching the soils. Adult beetles feed voraciously on leaf lamina making irregular holes. Young seedlings and tender leaves are mainly preferred by the adults and damage at this stage may kill the seedlings.

Control measures

- Clean cultivation.
- Collect and destroy the beetles during cool hours of early morning when they remain sluggish.
- Deep ploughing of the infested fields just after harvesting to kill the grubs in the soil.
- Dust the crop with 5% carbaryl or spray with 0.2% carbaryl.
- Repellent dusting of ash mixed with kerosene oil but heavy dusting should be avoided as it hampers plant growth.

Red spider mite (Tetranychus urticae)

Spider mite causes significant damage attacking on the foliages of the crop. Numerous populations continuously suck cell sap from both the surface of mature leaf which is expressed in the form of numerous whitish specks on the dorsal surface of leaf. Hot and dry weather is very congenial for growth and development of red spider mite species. The infested plants loss vigour and dry up quickly.

- Prophylactic application of azadirectin 1% @ 1ml/l of water reduces the pest population.
- Conserve and release of predatory mite, *Amblyseius longispinosus* reduce the destructive form of spider mite population.
- Spray acaricide like, dicofol @ 2.5 ml/l or propergite @ 1.5 ml/l.

Root knot nematode (Meloidogyne incognita)

These plant parasitic eelworms incite root galls causing poor and stunted growth of the infested plants.

Control measures

- Long crop rotation with non-host crops like, cereals, marigold, etc.
- Ploughing the fields in the summer months.
- Apply neem cake @ 12-15 t/ha.
- Apply granular insecticides like Aldicarb 10 G @ 25 kg/ha or Thimet 10G @ 20 kg/ha or Furadon 3G @ 40 kg/ha.

Yield

- 20 to 30 tonnes/ hectare (open pollinated improved varieties)

12. Bottle gourd

Scientific Name: Lagenaria siceraria
Family : Cucurbitaceae
Origin : Tropical Africa
Chromosome No: 2n = 22

Introduction

Bottle gourd or Calabash gourd might be the only plant species that was known to man in both the New and Old world from very early pre-historic times. Archaeological evidence estimates a time depth of at least 12000 years for *Lageneria* remains, both in new and old worlds. The bottle like shape of fruit and its use as a container of wines and spirits in the past gave it the common name of bottle gourd. It is a commonly grown cucurbit vegetable crop of India, Africa, Central America and other warmer regions of the world. In India it is cultivated in almost all parts, including the plains and hills. It is one of the important and popular vegetable crops extensively grown during spring-summer, summer-rainy and early autumn seasons in all parts of the country particularly in Uttar Pradesh, Punjab, Haryana, Gujarat, West Bengal, Tamil Nadu, Karnataka and Bihar

Importance and use

- Bottle gourd is also called Calabash gourd or Zucca melon.
- Bottle gourd is an important cucurbitaceous vegetable extensively grown in different parts of the country.
- It is also cultivated in different river beds of the country.
- It is suitable for cultivation in dry areas.
- It has great export potential to South-East Asia and Gulf countries.
- Tender fruits of 30 – 40 cm in length are generally exported.
- The fruits may be long, oblong or round depending on the variety.
- Tender fruits are widely used as cooked vegetables in a number of ways.
- In northern India "Kofta" is most popular vegetable preparation.
- Immature fruits are used with milk for the preparation of different sweets like, "Kheer", "Payesh" and "Halwa".
- Matured fruits are also used for the preparation of different sweets like, "Petha" and "Burfi".
- Immature fruits are also used with curd for the preparation of " Rayata" and in the preparation of pickles, especially in the hills.
- The green leaves are used as potherbs.
- Seed and seed oils are also edible.

- The hard shell of the fruit is used for making wide range of articles of common use including bowls, bottles, ladles, different containers, floats for fishing nets, pipes and musical instruments.

Nutritional value

- Edible portion of bottle gourd contains 0.2% protein, 2.9% carbohydrates and 0.5% fat and 11.0 mg vitamin C per 100 g fresh edible portion..
- Tender fruits provide about 10 mg of P, 20mg of Ca per 100 g edible portion.
- The young leaves are rich in carbohydrate (6.1%) and protein (2.3%).

Medicinal value

- Fruits are easily digestible and have cooling effect.
- It is good for people suffering from biliousness and indigestion.
- Fruits are cardiatonic and diuretic.
- The pulp is good for overcoming constipation and cough.
- The pulp also acts as an antidote against certain poisons.
- The plant extract is used as a cathartic.
- The seeds are used for treating dropsy.
- Decoction of leaves mixed with sugar is given for treating jaundice.
- The bitter fruits are poisonous and are used as a strong purgative.

Adaptation in India

Bottle gourd was introduced in India in very early times because of its independent domestication in Africa and Asia. Sanskrit language equivalents signifying Vedic period and Aryan usage is available for bottle gourd as "alabu". The bottle gourd was probably used by Proto-Australoides and cooking of this fruit "alabu" or "iksvaku" have been mentioned in "Yajurveda". Such mentioning of bottle gourd in ancient Indian literatures signifies that it was introduced into India in very early times and has been undergone significant diversification in this country mainly in plant type, fruit shape, fruit colour and weight. It has been found wild on the coast of Malabar and in the humid forests of Dehra Dun. It is grown as spring-summer, summer-rainy and early autumn crop in the field or grown in different riverbeds of the country.

Bottle gourd is cultivated in India mostly using different local cultivars under different name and some of them are Doodhi, Long Green, Three Feet Long Green, Jora Bonta, Kali Lau, etc. A number of improved varieties have been developed through selection followed by inbreeding from different local cultivars.

Breeding behaviour

- Bottle gourd is highly cross pollinated in nature due to monoecious sex form and entomophilies.
- Major pollinators are honey bees (*Apis* sp.).
- Although it is 100% cross pollinated in nature yet reasonable extent of self fertilization happens because of bee foraging in the staminate flowers of the same vine.
- Bottle gourd like other members of the family Cucurbitaceae does not suffer much from inbreeding depression or do so only negligibly, in contrast to the cross pollinated vegetable crops belonging to other families.
- Anthesis generally starts in the afternoon hours from 2.30 p.m. in north and eastern Indian condition. In Bangalore condition, anthesis also occurs in the morning hours.
- Anther dehiscence generally starts in the afternoon hours.
- Stigma receptivity is maximum at the time of anthesis although stigma becomes receptive one day prior to and remains one day after anthesis also.
- Best time of hybridization is between 2.00 and 5.00 p.m. on the day of anthesis.
- Harvesting the fully matured staminate flowers in the morning hours that will open in the afternoon hours on that day itself and keeping them covered with thin moist cloth for their use in cross pollination in the afternoon morning hours is the conventional practice for commercial hybrid seed production.
- Bagging the staminate and pistillate flowers prior to opening followed by hand pollination and subsequent bagging for two more days is the commercial method of hybrid seed production.
- Hybrids of bottle gourd and one of the wild *Lagenaria*, *L. sphaerica* is sterile.

Origin and taxonomy

Bottle gourd is believed to be indigenous to tropical Africa on the basis variability present in fruits and seeds. However, it is found in wild form in both South Africa and India which amply suggest its independent domestication in Africa, Asia although its progenitor is unknown. In fact, it has bihemespheric distribution in Africa, the Old world and America, the New world. It has now been concluded that the bottle gourd might have originated in Africa and been carried to America by trans-oceanic drift or human transport. It had pre-historic existence in India and was cultivated during ancient times. Archaeological evidences indicated its presence in Egyptian tombs in 3000 to 5000 B.C., Spirit caves of Thailand in 6000 to 10,000 B.C., Mexico in 5000 to 7000

B.C. and Peru in 3000 to 4000 B.C. Its utilization by man is about 16,000 years old in the Old world and 12,000 years old in the New world.

The bottle gourd, *Lagenaria siceraria* (Mol.) Standl. belongs to the family Cucurbitaceae, sub-family Cucurbitoideae, tribe Benincaseae and sub-tribe Benincasinae. The genus *Lagenaria* includes six species having pantropical distribution in Africa, Madagascar, Indo-Malaysia and the neotropics. Of these only *Lagenaria siceraria* is cultivated which is annual and monoecious while the other five species are wild, perennial and dioecious in sex form. The wild *Lagenaria* occur only in east Africa and Madagascar.

The modern forms of bottle gourd are secondary polyploids derived from its progenitor with a basic chromosome number, x = 5 and n = 5+5+1 (2n = 22). Hybrids of bottle gourd and one of the wild *Lagenaria*, *L. sphaerica* is sterile.

Improved varieties

- Long fruited varieties: Pusa Summer Priolific Long, Pusa Naveen, Arka Bahar, Kalyanpur Long Green, Samrat, NDBG-1,Phule BTG-1, Punjab Long, Rajendra Chamatkar, Azad Nutan.
- Round fruited varieties: Pusa Summer Prolific Round, Punjab Round, Punjab Komal, Pusa Sandesh.
- Public sector Hybrids: Pusa Hybrid-3 (club shaped fruit), Pant Sankar Lauki (long fruited), NDBG-4 (long fruited)

Local cultivars

Doodhi, Long Green, Three Feet Long Green, Jora Bonta, etc.

Cultural requirements

- Bottle gourd is basically a warm season crop and susceptible to cold temperature and frost.
- Low temperature causes stunting of growth and poor germination of seeds.
- It can be grown in warm and humid and slightly cool climate.
- Optimum temperature range for the growth and development is 24°C-27°C.
- Soil temperature of 18-22 °C promotes good growth and ensures better yield.
- Flowering is highly sensitive to photoperiod.
- Short days, comparatively low-night temperature and high relative humidity increase the propensity of pistillate flowers in the vine.
- Long days and high temperature increase the number of staminate flowers and reduce the number of pistillate flowers.
- High rainfall coupled with prolonged cloudiness promote high incidence of disease and in turn drastically reduce yield.

- Soil moisture is important for rapid growth and it should be at least 10 – 15 % above the permanent wilting point.
- It can also be grown in dry areas.
- It is susceptible to water logging.
- Deep well drained fertile, loamy or sandy-loam soils with pH range 6.0 – 7.0 are ideal.
- Sandy loam soils are ideal for early crop.

Land preparation

- Prepare the field with 4-5 deep ploughing and make into a fine tilth before seed sowing.
- Remove all the weeds, stubbles and level the land by laddering.
- Manures should be mixed with the soil at the time of last ploughing.
- The plant has comparatively shallow root system and is susceptible to water logging so land should be well drained.
- Generally furrows are opened at 2.0 to 3.0 m spacing and sowing is done on the top of the sides of furrows and the vines are allowed to trail on the ground.
- The crop is also grown on flat or raised beds, about 2.0 to 2.5 m wide and seed are sown on both sides of the bed.
- In rainy season, raised beds or mounds are prepared to facilitate drainage.
- Mounds in the furrow, flat bed method or in river bed system of planting are prepared with the mixture of 15-20 tonnes of FYM per hectare and basal dose of fertilizers are mixed with the soil.

Sowing time

Bottle gourd grows successfully in tropical and subtropical climate hence, can be grown year round in the southern and central Indian states due to prevailing mild winter there. However, the crop is grown in three main seasons: spring-summer, early autumn and rainy season.

- November-December: Spring-summer crop in eastern and northern plain (river bed).
- February-March: Summer crop in eastern and northern plain.
- June-July: Rainy season crop in eastern and northern plain.
- August - September: Early autumn crop in eastern India.
- December - January: Winter – spring crop in southern and western regions.
- March – April: Hill regions.

Seed sowing

- Bottle gourd gives little success on direct transplanting and, in general, seeds are directly sown in the field.

- Seeds are sown by hand dibbling in the mound.
- Sowings can be done following flat bed or furrow method or in river bed system depending on the season and convenience.
- Pre-sowing treatment of seeds with 600 ppm succinic acid for 12 hours improves germination and seedling growth.
- Seed soaking in water for 12 hours also improves germination.
- Seedlings raised in plastic packets particularly under polyethylene shed in winter month can be transplanted after 35 – 40 days with the advent of spring without disturbing the root system.
- 4.0 – 6.0 kg seed/ha is sufficient depending on the plant population.
- It is better to do seed sowing in pre-irrigated moist soil than irrigate the field after sowing.
- Pre-sowing soil application of 20-22 kg Furadon 3G helps in protecting the plants from root knot nematodes and other pests during initial 4-5 weeks.
- On an average 4 to 5 thousand plants can be accommodated in one hectare land.

Spacing

- Spring-summer crop: 2.0 – 3.0 m × 90 – 150 cm
- River bed system: 2.0 – 3.0 m × 100 – 150 cm
- Rainy season crop: 3.5 m × 1.5 – 2.0 m
- Early autumn crop: 2.0 – 2.5 m × 60 – 90 cm
- Crop on the bower: 2.5 – 3.0 m × 1.0 m

Nutrient management

- Bottle gourd responds well to manuring and fertilizer application.
- The dose of manures and fertilizers depend on the soil type, climate and system of cultivation.
- Well rotten farm yard manure at the rate of 15-20 t/ha should be applied at the time of land preparation.
- Apply a general fertilizer dose of 90 kg N, 60 kg P_2O_5 and 60 kg K_2O per hectare for open pollinated improved varieties.1/2 N along with full P and K fertilizers and farm yard manure should be given as basal and rest N should be top dressed in two split doses at the time of vining (25 - 30 days after sowing) and at full blooming (50 – 60 days after sowing).
- For the hybrids apply 200 kg N, 100 kg P_2O_5 and 100kg K_2O per hectare. 1/4 N along with full P and K fertilizers and farm yard manure should be given as basal and rest N should be top dressed in three split doses at the time of vining (25 - 30 days after sowing) , at full blooming (50 – 60 days after sowing) and after first harvest (85-90days after sowing).

- Bottle gourd is a comparatively shallow rotted crop hence, responses well in top-dressing.
- High nitrogen under high temperature condition promotes staminate flowers resulting in low fruit set and yield so, fertilizer dose should be adjusted according to season of cultivation.

Crop regulation

- Excessive vine growth due to high nitrogen nutrition coupled with high temperature and high soil moisture condition promote staminate flowers in the vine resulting low fruit set and low yield. The best way to control the vine growth within reasonable limits is by adjusting nitrogen fertilizer doses and frequency of irrigation and pruning.
- Application of different growth regulators, secondary and micro-nutrients viz., 150 ppm Ethrel (1.5 ml/10 l of water), 50 ppm tri-iodobenzoic acid, 3-4 ppm boron or 20 ppm calcium twice, first at two true leaves of the plants i.e. 15 days after sowing and subsequently repeated 7 days after helps in increasing the yield.

Irrigation

- Interval and amount of irrigation depends upon the season and type of soils.
- The crop requires frequent irrigation as high humidity is favourable for the crop.
- Irrigation after sowing improves seed germination because seed germination require high water supply.
- Apply irrigation at the initiation of first true leaf during spring-summer and at its expansion during rainy season.
- Irrigate the field subsequently 5-6 days intervals during summer and early autumn crop and as and when required for the rainy season crop depending on the rain.
- In light soils irrigate the crop at frequent intervals.
- The crop must be irrigated during its most critical stages i.e. flowering and fruit setting stages.
- Over irrigation during vegetative and early flowering stages may cause excessive vine growth resulting induction of more staminate flowers in the vine.
- Ridge and furrow or ring methods of irrigation are the best for bottle gourd.
- Drip irrigation gives 48% more fruit yield than furrow irrigation.

Interculture and weed control

- Plant grown on trellis gives better crop stand and fruit yield mainly because of better use of sunlight by maximum number of leaves and higher number of side branches.
- The crop needs to be trained over low trellis of 1.5 m high above the ground.
- Training the vines over the trellis or bower coupled with some pruning is very effective.
- The auxillary buds are to be removed at weekly interval till the vine attains the height of the trellis and finally the growing point of the vine is pruned 15 cm below the trellis allowing two side branches to grow over the trellis or bower.
- After 85 to 90 days of sowing older leaves are near the bottom of the vine are pruned.
- Hand pollination increases yield both during summer and monsoon seasons because of enhanced fruit set.
- Soil at the base of the plant should be loosened before top dressing with nitrogen fertilizers followed by irrigation.
- In early stage of growth, the beds, ridges etc. required to be kept free from weeds.
- At the initial stages of weed growth hand weeding and hoeing are sufficient to check the weed growth.
- Deep intercultivation should be avoided as it may damage the shallow roots of the crop.
- Application of herbicides effectively controls weeds.
- Pre-emergence application of linuron @ 0.5 kg/ha, alachlor 2.5 kg/ha, dichlormate 0.4kg/ha, duiron 1.25kg/ha checks weed population in the fields.

Harvesting the crop

- The tenderness and edible maturity are judged by pressing the skin and little pubescence persisting on the skins.
- In the early variety or hybrid, first picking of fruits may be done even 60 to 65 days after sowing.
- Fruits generally take 12-15 days after fruit set to reach marketable stage when the fruits are tender.
- In some big fruited variety it takes 20 days to reach marketable maturity.
- In small fruited hybrids fruits are harvested at 3-4 days intervals.
- Fruits harvested with knife by cutting the peduncle.

Post harvest handling and storage

- Improper harvesting, handling, transportation and distribution result in significant losses.
- Proper cultural operations, harvesting, transportation, storage and pre and post harvest treatments minimize post harvest losses.
- Under normal cool and shady conditions fruits can be kept for 3-5 days.
- In cold storage at 8.0 to 10°C temperature the fruits can be stored for 2 to 3 weeks.
- Fruits are packed in polythene bags and kept in small boxes for distance transportation and export.
- For export, the fruits should be picked at edible stage and then packed in polythene pipe bags and these bags are kept in boxes of convenient capacity.

Plant protection measures

Physiological disorder

Preponderance of staminate flowers

The plants are monoecious in sex form hence, staminate and pistillate flowers are borne separately in the same plant and fruit yield depends on the number of pistillate flowers. Preponderance of staminate flowers is caused due to i) excessive nitrogen application, ii) high temperature condition , iii) long day length and iv) over irrigation.

Control measures

- Avoid rampant vegetative growth through control in fertilizer application, checking in irrigation and training and pruning.
- Apply nitrogenous fertilizer at a proper dose.
- Avoid excess irrigation
- Apply ethephon / ethrel (2-chloroethyl phosphonic acid) @ 150 ppm twice, first at two true leaves of the plants i.e. 15 days after sowing and subsequently repeated after 7 days.

Diseases

Fusariums wilt (*Fusarium solani* and *Fusarium oxysporum* f.sp. *lagenariae*)

This soil-borne as well as seed-borne fungus attacks the young seedlings causing withering of cotyledons. Older plants wilt suddenly and vascular bundles in the collar region become yellow or brown.

Control measures

- Seed treatment with thiram or carbendazim @ 2.5 g /kg of seed.
- Seed treatment with *Trichoderma viride* along with neem cake application is effective to control the disease.

- Application of lime to raise the soil pH to 7.0- 7.4
- Spraying the crop with 0.2% captan or Bavistin at weekly interval checks further spread of the disease.

Gummy stemblight and leaf blight (*Phoma sp.*)

Dark brown to black irregular spots sppear in the leaf through the margin due to the inflection of this fungus. In the stem lesion starts as water-soaked areas, later gummy exudation and stem cracking occur leading to death in severe case.

Control measures

- Seed treatment with carbendazim (2g/kg) or thiram (3g/kg seed)
- Spraying with carbendazim 1g+chlorothalonil 2g/l at 10 days interval.

Anthracnose (*Colletotrichum legenarium*)

This seed borne fungal disease appears more during the period of heavy summer rains i.e. in high humidity and warm conditions. Light brown spots appear on the leaves which later turn to deep brown. The lesions are dotted with pink conidia in moist condition.In severe cases, fruits detach from the pedicel.

Control measures

- Use disease free seeds.
- Seed treatment with thiram or carbendazim@ 2.5 g /kg .
- Spray the crop with 0.1% Bavistin or 0.1% hexaconazole or 0.2% chlorothalonil, 3-4 times at 10 days interval.
- Crop rotation excluding cucurbits.

Downy mildew (*Pseudoperonospora cubensis*)

This fungal disease is prevalent in high humidity, especially when summer rains occurs regularly. The diseases is characterized by formation of yellow, more or less angular spots on the upper surface of the leaves, white-purplish spores appear on the upper surface of the leaves. The disease spreads rapidly killing the plant quickly through rapid defoliation.

Control measures

- Management schedules as mentioned in pumpkin.

Powdery mildew (*Sphaerotheca fuliginea* and *Erysiphe cichoracearum*)

This fungal disease appears under prolonged dry conditions. White floury or powdery patches or coatings appear on the lower surface of leaf extending to upper surface causing yellowing and death of the leaves.

Control measures

- Management schedules as mentioned in pumpkin.

Fruit Rot (*Pythium, Phytophthora, Rhizoctonia*, etc.)

The fungi can infect the seedlings, young plants and even young fruits.Initial infection is noticed as small spots on the leaves, which rapidly increases in number and size. After coalescing of adjoining spots and on severely affected leaves, a burning effect or blight symptoms is seen. Fruit rot may occur in the field or it may start after harvest.

Control measures

- Use of disease free seeds and its treatment with thiram @ 2.5 g/kg of seeds or carbendazim@ 2.5 g/kg of seeds .
- Crop rotation and proper drainage in the field.
- Spraying the crop with 0.25% mancozeb or 0.25 % metalaxyl-mancozeb or 0.25 % cyamoxanil-mancozeb effectively check fruit rot in the field.
- Borax wash (2.5%) at 45°C for 30 seconds or at 40°C for 2 minutes before packing of the fruits prevents fruit rots in transit.

Mosaic (Bottle gourd mosaic virus, cucumber mosaic virus (CMV), Watermelon Mosaic Virus (WMV).

These viruses are transmitted by sap, seed and by several species of aphids.The young leaves develop small greenish-yellow areas which are more translucent than the remaining parts of the leaf. Leaves show characteristics mottling with yellowish-green and blistering and in severe infection, growth of the plant is stunted.

Control measures

- Spray the crop with 0.05% dimethoate or 0.05% monocrotophos or 0.05% phosphamidon at weekly interval to control aphid vectors.
- Eliminate weed hosts and use of barrier crops like sorghum, maize around the field.
- Crop rotation with non-cucurbitaceous crops
- Remove infected plants from the field.

Insect Pest

Fruit fly (*Dacus cucurbita, Dacus dorsalis*)

Adult fly of this polyphagous pest having a wide host range is 4 – 5 mm long, ferruginous-brown in colour with hyaline wings. On hatching, the maggots feed inside the fruits and infested fruits can be identified by the presence of resinous liquid which oozes out of the punctures made by the flies for oviposition. The infested fruits start rotting due to secondary infection of different micro-organisms.

Control measures

- Collect and destroy the infested fruits along with the maggots to prevent the carry over of the pest.

- Stir the soil under the vines frequently and plough the infested field after harvest the crop to kill pupae.
- Spray the crop with 0.05% endosulfan.
- Apply poison bait. Bait is prepared by mixing 20 g mkalathion 50% WP or 50 ml diazinon 20% EC with 500 g molasses + 20 g yeast hydrolysate. This mixture is mixed with 2 litres of water for poison baiting and 20 liters of water for bait spray.
- Adopt male annihilation technique using 0.1% methyl eugenol + 0.025% Malathion in a bait trap during fruiting season. Methyl eugenol is a sex attractant and males are attracted towards the bait and killed by malathion. Apply this trap solution at 7-10 days interval.

Red Pumpkin Beetle (*Aulacophora foveicollis*)

Shiny yellowish-red adults of this polyphagous pest lay eggs in moist soils around the plant. On hatching, the grubs feed on the roots and underground portion of host plant and also on the fruits touching the soils. Adult beetles feed voraciously on leaf lamina making irregular holes. Young seedlings and tender leaves are mainly preferred by the adults and damage at this stage may kill the seedlings.

Control measures

- Clean cultivation.
- Collect and destroy the beetles during cool hours of early morning when they remain sluggish.
- Deep ploughing of infested fields just after harvesting to kill the grubs in the soil.
- Dust the crop with 5% carbaryl or spray with 0.2% carbaryl.
- Repellent dusting of ash mixed with kerosene oil but heavy dusting should be avoided as it hampers plant growth.

Root knot nematode (*Meloidogyne incognita*)

These plant parasitic eelworms incite root galls causing poor and stunted growth of the infested plants.

Control measures

- Long crop rotation with non-host crops like, cereals, marigold, etc.
- Ploughing the fields in the summer months.
- Apply neem cake @ 12-15 t/ha.
- Apply granular insecticides like Aldicarb 10 G @ 25 kg/ha or Thimet 10G @ 20 kg/ha or Furadon 3G @ 40 kg/ha.

Yield

- 20 - 25 tonnes/hectare (open pollinated improved varieties).
- 40 - 50 tonnes/hectare (hybrids).

13. Bitter gourd

Scientific Name: Momordica charantia L.
Family: Cucurbitaceae
Origin: Indo-Burma region
Chromosome No: 2n = 22

Introduction

Bitter gourd or Balsam pear or Bitter cucumber is an important and commonly grown vegetable in India, Indonesia, Malaysia, Singapore, Thailand, China, Japan, tropical Africa and South America. Recording of bitter gourd as "Kara vellika" in two medical treatises of India, "Charaka Samhita" and Yajurveda" is the testimony of its antiquity in India. It is one of the important and popular vegetable crops extensively grown during spring-summer, summer-rainy and early autumn seasons in all parts of the country. The leading bitter gourd growing states are Uttar Pradesh, Orissa, West Bengal, Maharashtra, Gujarat, Andhra Pradesh, Tamil Nadu and Kerala.

Importance and use

- Bitter gourd is also called Balsam pear or Bitter cucumber or Bitter melon.
- Bitter gourd is an important cucurbitaceous vegetable extensively grown for tender fruits in different parts of the country.
- It has high export potential to South-East Asia, Gulf and also European countries.
- Complete green, lustrous, pricklier, medium long and spindle shaped fruits are suitable for export.
- Tender fruits are widely used as cooked vegetables in a number of ways.
- The fruits are more commonly used as fried, boiled and stuffed.
- White fruited varieties are less bitter in taste and particularly preferred in South India.
- It can be canned, pickled and stored.
- The fruit slices can be dried and used as a vegetable as and when required and the ratio of fresh fruit: dehydrated product, in general, is 16:1.
- Cooked vegetable of bitter gourd remain quite fit for consumption for 2-3 days.

- The bitter glucoside, cucurbitacin may help in preventing spoilage of cooked vegetable of bitter gourd.
- Extracts of fresh fruits contain alkaloids, highly aromatic essential oil and saponin alkaloid "momordicine".
- Young leaves are also used as pot herb preparations.
- It is also grown as ornamental crop in the USA, Japan and some other countries.

Nutritional value

- The tender fruits are rich in iron, vitamin A, B, C and are good source of proteins and minerals.
- Edible portion of bottle gourd contains 2.0% protein, 5.0-10.0% carbohydrates and 0.2 – 1.0 % fat and 96.0 mg vitamin C and 210 IU vitamin A per 100 g fresh edible portion..
- Tender fruits also provide about 2.0 mg Fe, 40-70 mg of P, 23mg of Ca per 100 g edible portion.
- It is a rich source of phosphorus and regular use of bitter gourd invigorates human body.
- Cultivars with small to medium sized fruits contain higher carbohydrate and fat and lower phosphorus contents than the big fruited ones.

Medicinal value

- Fruits are antidotal, antipyretic tonic, appetizing, stomachic, antibilious and laxative.
- Several researches showed that it contains a hypoglycaemic or insulin like principle, designated as 'plant–insulin' which bas been found highly beneficial in lowering the blood and urine sugar levels.
- It is highly beneficial in the treatment of blood disorders like blood boil, scabies, itching, psoriasis, ring worm and other fungal diseases.
- It is also good for curing rheumatism, diabetes and asthma.
- Fruits and leaves are considered anthelmentic and vermifuge and useful in piles, leprosy and jaundice.
- Fresh leaf juice is an effective medicine in early stages of cholera and other types of diarrhoea during summer.
- Leaf juice is used as emetic, purgative, given in bilious problems and rubbed in burning of the soles of the feet.
- Roots are astringent and useful in haemorrhoids.

Adaptation in India

Bitter gourd was found mentioned in Sanskrit language equivalent "Kara vellika" in two ancient medical treatises of India, "Charaka Samhita" and Yajurveda". Bitter gourd has also been listed as a vegetable in the "Ain-i-Akbari", the detailed account of the Mughal rulers in India. Such mentioning of bitter gourd in ancient Indian scriptures signifies its very early history in India. Its cultivation throughout India has resulted significant diversification in the genotypes. Bitter gourd has been growing in this country from ancient period and undergone significant diversification mainly in fruit size, fruit shape, fruit colour and fruit weight. A number of improved varieties have been developed through selection followed by inbreeding from different local cultivars. Large number of indigenous cultivars are mostly grown in India under different names like, Small Green, Large White, Karela Guchhedar, Karela Long, Baramasi, Pride of Surat, Pride of Gujarat (both are small, roundish, whitish green and cultivated widely in Gujarat), Meghna, Gangajali (medium in size, green, spindle shaped with smooth warts and cultivated widely in West Bengal), etc. Improved varieties have been developed in India chiefly through selection from the wide genetic diversity present in different states and subsequent purification of the genotype by inbreeding because bitter gourd like other cucurbits does not show inbreeding depression

Breeding behaviour

- Bitter gourd is highly cross pollinated in nature due to predominantly monoecious sex form and entomophilies.
- Gynoceious sex form controlled by single recessive gene has also been recorded in bitter gourd.
- Major pollinators are honey bees (*Apis* sp.).
- Bitter gourd like other members of the family Cucurbitaceae does not suffer much from inbreeding depression or do so only negligibly, in contrast to the cross pollinated vegetable crops belonging to other families. Hence, inbred selection becomes the chief breeding method for this crop.
- Anthesis generally starts by 4.00 a.m and gets completed by 9.00 a.m.
- Anther dehiscence starts by 5.00 am and gets completed by 7.30 am.
- Stigma receptivity is maximum at the time of anthesis although stigma becomes receptive one day prior to and remains one day after anthesis also.
- Best time of hybridization is between 5.00 and 7.00 a.m. on the day of anthesis.
- Harvesting the fully matured staminate flowers in the afternoon hours that will open in the morning hours next day and keeping them covered with thin moist cloth overnight for their use in cross pollination in the next day morning hours is the conventional practice for commercial hybrid seed production.

- Bagging both the staminate and pistillate flowers prior to opening followed by hand pollination and subsequent bagging for two more days is the commercial method of hybrid seed production.
- Hand pollination is carried out by rubbing the anthers of the freshly opened or prior harvested matured staminate flowers against the stigma of the protected pistillate flowers.
- Bitter gourd does not cross with *M. dioica*, *M. balsamina* and *M. cymbalaria*.

Origin and taxonomy

The origin of bitter gourd remains obscure. However, it is considered native of tropical Asia, particularly eastern India and southern China. Indo-Burma centre of origin of this crop is widely accepted. This proposition is amply supported by long history of its cultivation in India. It is widely distributed in India, China, Malayasia, Thailand and tropical Africa. Bitter gourd from its native place was taken to Brazil early in the slave trade and is now wide spread throughout he tropics.

Bitter gourd, *Momordica charantia* L. belongs to the family Cucurbitaceae, sub-family Cucurbitoideae, tribe Joliffieae and sub-tribe Thladianthinae. The genus *Momordica* has about 80 species, among which *M. charantia* is the most widely cultivated species. The other species grown for edible fruits are *M. cochinchinensis (*giant spine gourd), *M. dioica* (spine gourd), *M. subangulata* ssp. *renigera* (teasle gourd or kakrol), *M. tuberosa* and *M. balsamina* (balsam apple). The wild species *M. charantia* var. *abbreviata* Ser. of Asia is perhaps the progenitor of the cultivated bitter gourd.

The cytological investigations have indicated that perhaps aneuploidy is involved in the evolution of the genus *Momordica*. Chromosome number of *Momordica* species ranges between 2n = 22 (*M. charantia, M. balsamina* and *M. tuberosa*) and 2n = 44 (*M. foetida*) and two dioecious and vegetatively propagated species, *M. cochinchinensis* and *M. dioica* remain in between in this respect (2n = 28). Natural triploid (2n = 33) has also been reported in *M. charantia.*

Improved open pollinated varieties

Pusa Do Mousumi, Pride of Surat, Pride of Gujarat, Small Green, Arka Harit, Pusa Vishesh, Priyanka, Konkan Tara, Gangajali, Hirkani, Coimbatore Long White, MC-23, MDU-1, Coimbatore Green, Phule Green, Coimbatore Long, Preethi, VK-1 Priya, Punjab 14, Pant Karela 1, Kalyanpur Baramasi, Kalyanpur Sona, C-96, NDB-1, Phule Green Gold, MDU-1, Meghna, etc.

- Small fruited variety (7.5 to 10.0 cm long): Pusa Do Mousumi, Pride of Surat, Pride of Gujarat, Small Green.
- Medium long fruited variety (10 to 15 cm long): Arka Harit, Pusa Vishesh, Priyanka, Konkan Tara, Gangajali, Meghna.

- Long fruited variety (15 to 20 cm ling): Hirkani
- Extra long fruited variety (20 to 40 cm long): Coimbatore Long White, MC-23, MDU-1, Coimbatore Green, Phule Green, Coimbatore Long, Preethi, and VK-1 Priya.
- Spindle shaped variety: Konkan Tara, Hirkani , Arka, Harit, Priyanka, Gangajali.
- Green and dark green fruited variety: Pusa Do Mousumi, Arka Harit, VK-1 Priya, CO-1, Pusa Vishesh, Coimbatore Green, Hirkani, Phule Green, Konkan Tara, Punjab 14, Meghna.
- White fruited variety: Preethi (MC84), Priyanka (Sel.1010), Coimbatore Long White.
- Red pumpkin beetle resistant variety: Pant Karela 1.
- Varieties suitable for export: Arka Harit, Pusa Vishesh, MBTH –1, Hirkani, Konkan Tara, Preethi, Meghna.
- Public sector hybrids: Pusa Hybrid-1, RHR BGH-1

Cultural requirements

- Bitter gourd is basically a warm season crop but has a wide range of adaptability and can be grown well in subtropical climate.
- It is also grown in the hills in the summer season.
- It is susceptible to frost and does not grow well under cool conditions.
- Temperature below 18°C causes poor germination of seeds, stunting of growth resulting poor yield.
- Optimum temperature range for seed germination and growth and development is 25°C-30°C.
- Temperature above 36°C causes poor development of pistillate flowers leading to poor yield.
- Seed germination is inhibited at and below 8°C and above 40°C.
- Short days and comparatively low-night temperature increase the propensity of pistillate flowers in the vine as well as lowers the node number at which the first pistillate flowers appear.
- Long days and high temperature increase the number of staminate flowers and reduce the number of pistillate flowers.
- High rainfall coupled with prolonged cloudiness promote high incidence of disease particularly downy mildew and in turn drastically reduce yield.

- It is susceptible to water logging.
- It can be grown in all types of soil however, well drained fertile, sandy loam and silt loam soils with pH range 6.5– 7.0 are ideal.
- Sandy loam soils are ideal for early crop.
- It is also suitable for cultivation in the partial shady condition in the coconut plantation as practiced in Kerala.

Land preparation

- Prepare the field with 4-5 deep ploughing and make into a good tilth before seed sowing.
- Remove all the weeds, stubbles and level the land by laddering.
- Farm yard manures @ 20-25 t/ha should be mixed with the soil at the time of last ploughing.
- The plant has comparatively shallow root system and is susceptible to water logging so land should be well drained.
- Generally long furrows or channels of 60 cm width are opened at 2.0 to 2.5 m spacing and mounds are prepared at 1.5 m spacing along the channel.
- Seed sowing is done on the mounds by the sides of furrows and the vines are allowed to trail on the ground and it is the best method of growing bitter gourd.
- The crop is also grown on flat or raised beds; about 2.0 to 2.5 m wide and seed are sown on both sides of the bed.
- In rainy season, raised beds or mounds are prepared to facilitate drainage.
- Mounds in the furrow, flat bed method or in river bed system of planting are prepared with the mixture of 20-25 tonnes of FYM per hectare and basal dose of fertilizers are mixed with the soil.

Sowing time

Bitter gourd grows successfully in tropical and subtropical climate hence, can be grown year round in the southern and central Indian states due to prevailing mild winter there. However, the crop is grown in three main seasons: spring-summer, early autumn and rainy season.

- November-December: Spring-summer crop in eastern and northern plain.
- January- March: Summer crop in eastern and northern plain.
- June-July: Rainy season crop in eastern and northern plain.
- August - September: Early autumn crop in eastern India.
- December - January: Winter – spring crop in southern and western regions.
- April- June: Hill regions.

Seed sowing

- Bitter gourd gives little success on direct transplanting hence, are direct seeded in the field.
- Seeds are sown by hand dibbling in the mound.
- Sowings can be done following flat bed or furrow method on the season and convenience.
- Pre-sowing treatment of seeds with thiram @ 3g per kg of seeds checks damping off of the seedlings.
- Seed soaking in water for 6 hours also improves germination.
- The seed has hard seed coat and germinate slowly due to slow absorption of water.
- Germination takes longer time at low temperatures.
- Seedlings raised in plastic packets particularly under polyethylene shed in winter month can be transplanted after 35 – 40 days with the advent of spring (January – February) without disturbing the root system.
- 4.0 – 6.0 kg seed/ha is sufficient depending on the plant population.
- It is better to do seed sowing in pre-irrigated moist soil than irrigate the field after sowing.
- Pre-sowing soil application of 20-22 kg Furadon 3G helps in protecting the plants from root knot nematodes and other pests during initial 4-5 weeks.

Spacing

- Spring-summer crop: 1.2 – 1.5 m × 45 - 60 cm
- Rainy season crop: 1.5 – 2.0 m × 50 - 60 m
- Early autumn crop: 1.2 – 1.5 m × 45 - 60 cm
- Crop on the bower: 1.5 – 2.0 m × 75 - 80 cm

Nutrient management

- Bitter gourd responds well to manuring and fertilizer application.
- The dose of manures and fertilizers depend on the soil type, climate and system of cultivation.
- Well rotten farm yard manure at the rate of 20-25 t/ha should be applied at the time of land preparation.
- Apply a general fertilizer dose of 100 kg N, 60 kg P_2O_5 and 60 kg K_2O per hectare for open pollinated improved varieties. 1/2 N along with full P and K fertilizers and farm yard manure should be given as basal and rest N should be top dressed in two split doses at the time of vining (25 - 30 days after sowing) and at full blooming (50 – 55 days after sowing).

- For the hybrids apply 200 kg N, 100 kg P_2O_5 and 100kg K_2O per hectare. 1/4 N along with full P and K fertilizers and farm yard manure should be given as basal and rest N should be top dressed in three split doses at the time of vining (25 - 30 days after sowing) , at full blooming (50 – 55 days after sowing) and after first harvest (60-75 days after sowing).
- Bitter gourd is a comparatively shallow rotted crop hence, responses well in top-dressing.
- High nitrogen under high temperature condition promotes staminate flowers resulting in low fruit set and yield so, fertilizer dose should be adjusted according to season of cultivation.

Crop regulation

- Like other monoecious cucurbits, excessive vine growth due to high nitrogen nutrition coupled with high temperature and high soil moisture condition promote staminate flowers in the vine resulting low fruit set and low yield. The best way to control the vine growth within reasonable limits is by adjusting nitrogen fertilizer doses and frequency of irrigation and pruning.
- High levels of endogenous GA like substances occur in the plants between 45 and 60 days when the ratio of staminate: pistillate flower is low.
- Application of different growth regulators and micro-nutrients viz., 150 ppm ethrel, 50 ppm maleic hydrazide (0.5ml/10 l of water), 50-100 ppm cycocel, 60 ppm gibberelic acid or 3-4 ppm boron twice, first at two true leaves of the plants i.e. 15 days after sowing and subsequently repeated 7 days after helps in increasing the yield.
- Soaking of seeds in 20 ppm ethrel, 3-4 ppm silver nitrate, 3-4 ppm B9 or 3-4 ppm boron induce significantly higher number of pistillate flowers in the vine and produce higher fruit yield.

Irrigation

- Bitter gourd is a shallow rooted crop and roots are mostly concentrated at top 60 cm soil layer.
- Give first irrigation immediately after sowing.
- Irrigation after sowing improves seed germination because seed germination require high water supply.
- Apply irrigation at the initiation of first true leaf during spring-summer and at its expansion during rainy season.
- The crop should be irrigated at 3-4 days interval in summer until it flowers.
- Critical stages of irrigations are flower bud development and early fruit development stages when irrigation is mandatory.

- Over irrigation during vegetative and early flowering stages may cause excessive vine growth resulting induction of more staminate flowers in the vine.
- Ridge and furrow method of irrigation are the best for bitter gourd.
- Drip irrigation gives more fruit yield than furrow irrigation.

Interculture and weed control

- In the bed system of cultivation the vines are allowed to grow on the bed with out any support.
- Plant grown on trellis or pandal or bower gives better crop stand and fruit yield mainly because of better use of sunlight by maximum number of leaves and higher number of side branches.
- Bower system of bitter gourd cultivation is very much practiced in Kerala, Tamil Nadu, Karnataka, Maharashtra, Andhra Pradesh and West Bengal.
- Soil at the base of the plant should be loosened before top dressing with nitrogen fertilizers followed by irrigation.
- In early stage of growth, the beds, ridges etc. required to be kept free from weeds.
- At the initial stages of weed growth hand weeding and hoeing are sufficient to check the weed growth.
- Deep intercultivation should be avoided as it may damage the shallow roots of the crop.
- Application of herbicides effectively controls weeds.
- Pre-emergence application of glyphosphate at 4.5 kg a.i. /ha checks weed population in the fields.

Harvesting the Crop

- Flowering starts generally 40-45 days after sowing and the first picking can be done 65-70 days after sowing depending on the variety, sowing time, soil type and management practices.
- Regular harvesting at short intervals increase the number of fruits in the vine and irregular harvesting may delay the formation of successive fruit production and affect the yield adversely.
- The colour of the tender fruits may be light green, dark green or whitish green depending on the variety.
- Fruits generally take 10-15 days after fruit set to reach marketable stage when the fruits are tender.
- In some big fruited variety it takes 20-25 days to reach marketable maturity.

Post harvest handling and storage

- After harvesting remove all fruits affected with insect-pests diseases or deformed ones.
- The harvested fruits can not be kept for long time in ambient condition so need to be sent to the market as soon as possible.
- Sprinkle water over the fruits to maintain the freshness for sometime at the initial stage.
- Keep the fruits in polypropylene bags for extending the shelf life.
- Dipping the fruits in solution of wax emulsion (6-12%) plus sodium phenyl phenate for 30-60 seconds prolong the shelf life up to 3 days at room temp.
- Fruits can be stored in cold storage at 0.6 –1.7^{0}C and 85–90% relative humidity for 4 weeks.

Plant protection measures

Physiological disorder

Preponderance of staminate flowers

The plants are monoecious in sex form hence, staminate and pistillate flowers are borne separately in the same plant and fruit yield depends on the number of pistillate flowers. Preponderance of staminate flowers is caused due to i) excessive nitrogen application, ii) high temperature condition , iii) long day length and iv) over irrigation.

Control measures

- Avoid rampant vegetative growth through control in fertilizer application, checking in irrigation and training and pruning.
- Apply nitrogenous fertilizer at a proper dose.
- Avoid excess irrigation.
- Apply ethephon / ethrel (2-chloroethyl phosphonic acid) @ 150 ppm or 6-benzylaminopurine (BA) @ 25 ppm twice, first at two true leaves of the plants i.e. 15 days after sowing and subsequently repeated after 7 days.

Disease

Downy mildew (*Pseudoperonospora cubensis*)

Bitter gourd is particularly susceptible to this fungal disease which is prevalent in high humidity, especially when summer rains occurs regularly. The diseases is characterized by formation of yellow, more or less angular spots on the upper surface of the leaves, white-purplish spores appear on the upper surface of the leaves. The disease spreads rapidly killing the plant quickly through rapid defoliation.

Control measures

- Management schedules as mentioned in pumpkin.

Damping off and Fruit rot (*Fusarium, Rhizoctonia and Phytophthora* spp.)

The fungi can infect the young seedlings causing the rot of the seedlings. However, fruit rot is very much prevalent which may occur in the field itself as well as in after harvest in storage or transit. Infected fruits are enveloped by luxuriant cottony mycelial growth of fungus.

Control measures

- Use of disease free seeds and its treatment with thirum @ 2.5 g/kg of seeds or carbendazim@ 2.5 g/kg.
- Crop rotation and proper drainage in the field.
- Spraying the crop with 0.25% mancozeb or carbendazim (1g) + mancozeb(2g) or metalaxyl-mancozeb (0.25%) effectively check fruit rot in the field.

Mosaic (Bottle gourd mosaic virus, cucumber mosaic virus (CMV), Watermelon Mosaic Virus (WMV).

These viruses are transmitted by sap, seed and by several species of aphids. The young leaves develop small greenish-yellow areas which are more translucent than the remaining parts of the leaf. Leaves show characteristics mottling with yellowish-green and blistering and in severe infection, growth of the plant is stunted.

Control measures

- Spray the crop with 0.05% dimethoate or 0.05% monocrotophos or 0.05% phosphamidon at weekly interval to control aphid vectors.
- Eliminate weed hosts and use of barrier crops like sorghum, maize around the field.
- Crop rotation with non-cucurbitaceous crops.
- Growing 5-6 rows of border crops like sunflower, sorghum and pearl millet all round the crop sown 50-60 days before sowing the cucurbits.
- Remove infected plants from the field.

Insect Pest

Fruit fly (*Dacus cucurbita, Dacus dorsalis*)

It is the most serious insect pest of bitter gourd.Adult fly of this polyphagous pest having a wide host range is 4 – 5 mm long, ferruginous-brown in colour with hyaline wings. On hatching, the maggots feed inside the fruits and infested fruits can be identified by the presence of resinous liquid which oozes out of the punctures made by the flies for oviposition. The infested fruits start rotting due to secondary infection of different micro-organisms.

Control measures

- Collect and destroy the infested fruits along with the maggots to prevent the carry over of the pest.
- Stir the soil under the vines frequently and plough the infested field after harvest the crop to kill pupae.
- Spray the crop with 0.05% endosulfan.
- Apply poison bait. Bait is prepared by mixing 20 g malathion 50% WP or 50 ml diazinon 20% EC with 500 g molasses + 20 g yeast hydrolysate. This mixture is mixed with 2 litres of water for poison baiting and 20 liters of water for bait spray.
- Adopt male annihilation technique using 0.1% methyle eugenol + 0.025% malathion in a bait trap during fruiting season. Methyl eugenol is a sex attractant and males are attracted towards the bait and killed by malathion. Apply this trap solution at 7-10 days interval.

Epilachna beetle or Hadda beetle (*Henosepilachna vigintioctopunctata*)

It is a very destructive pest of bitter gourd causing damage by defoliating the entire crop of the field. Both adult and grabs are destructive, feed by scraping the leaf tissue, leaving behind the skeletonized leaf. The late season crops are badly affected.

Control measures

- Collection and destruction of yellow egg masses, grabs and the invading adult population is very effective.
- Application of carbaryl @ 2g/l is very effective against this pest.

Red pumpkin beetle (*Aulacophora foveicollis*)

Shiny yellowish-red adults of this polyphagous pest lay eggs in moist soils around the plant. On hatching, the grubs feed on the roots and underground portion of host plant and also on the fruits touching the soils. Adult beetles feed voraciously on leaf lamina making irregular holes. Young seedlings and tender leaves are mainly preferred by the adults and damage at this stage may kill the seedlings.

Control measures

- Clean cultivation.
- Collect and destroy the beetles during cool hours of early morning when they remain sluggish.
- Deep ploughing of infested fields just after harvesting to kill the grubs in the soil.
- Dust the crop with 5% carbaryl dust or spray the crop with 0.2% carbaryl or 0.05% endosulfan.
- Light repellent dusting of ash mixed with kerosene.

Aphids (Aphis gossypii, Myzus persicae)

Both nymph and adults suck sap from tender twigs and ventral surface of the leaves, affected parts turn yellow, get curled, wrinkled and ultimately die away. The aphids also transmit several viral diseases.

Control measures

- Clip off and destroy the affected plant parts along with the crowded aphids thereon in the initial stage of attack.
- Spray the crop with 0.03% phosphamidon or 0.05% malathion or 0.03% methyl demeton.

Yield

- 10-15 tonnes/hectare (open pollinated improved varieties).
- 25-30 tonnes/hectare (hybrids)

14. Cucumber

Scientific Name: Cucumis sativus
Family : Cucurbitaceae
Origin : India
Chromosome No: 2n = 14

Introduction

Cucumber is one of the oldest cultivated vegetable crops in India. It has been known in history for about 3000 years and was probably originated in India from where it seems to have spread eastwards to China and westwards to Asia Minor, North Africa and Southern Europe and subsequently to entire Europe. It was introduced to the new world by Columbus who planted it in Haiti in 1494 and perhaps soon afterwards it was brought to the USA. It was grown by the ancient Greeks and Romans in about 300 B.C. Cucumber was cultivated almost throughout Europe in the middle ages and was common in France in the ninth century in England in 1327 A.D. Cucumber is the second most widely grown cultivated cucurbit in the world after watermelon. In the temperate countries it is extensively grown in the green house. Cucumber particularly gherkin has great export potential to South-East Asia and Gulf and European countries. It is one of the important and popular vegetable crops extensively grown during spring-summer, summer-rainy and early autumn seasons in all parts of the country including the hills of the southern and northern regions. Gherkin (*Cucumis anguria*) and small sized pickling cucumbers have become important processed vegetables for export from Karnataka, Andhra Pradesh and Tamil Nadu.

Importance and use

- Cucumber is the second most widely grown cultivated cucurbit in the world after watermelon.
- Immature fruits are universally used as salad and for pickling.
- Off-season cucumber particularly during winter is sold at premium price.
- Mature fruits are also used as cooked vegetable in India.
- Cucumber particularly gherkin has great export potential to South-East Asia and Gulf and European countries.
- Gherkin varieties or hybrids having green, small sized (7.5 cm × 2.4 cm) fruits of about 200 in number per kg is considered as premium.
- Immature fruits are also used with curd for the preparation of " Rayata".
- The green leaves are used as potherbs.
- Seeds and seed oils are also edible.

Nutritional value

- Edible portion of cucumber contains 0.4% protein, 2.5% carbohydrates and 0.1% fat and 7.0 mg vitamin C per 100 g fresh edible portion.
- Tender fruits provide about 25 mg phosphorus, 10mg cacium and 1.5 mg iron per 100 g edible portion.

Medicinal value

- The fruits and seeds have cooling effect and prevent constipation.
- Fruits are good for people suffering form constipation, jaundice and indigestion.
- Fruits are demulcent and diuretic.
- Fruits contain photolytic enzymes, ascorbic acid oxidase, succinic and malic dehydrogenases.
- Fruits are also used as an astringent and antipyretic.
- Seed oil also used as antipyretic.

Adaptation in India

Cucumber is believed to be indigenous to India although it is never found in wild state in India. Mentioning of cucumber as Sanskrit language equivalents "Urvaruka" in Rigveda and "Ervaruka" in Caraka Samhita signifies its usage by the Aryan in the Vedic period. Cucumber has wide distribution in India showing large genetic variability. It is grown as spring-summer, summer-rainy and early autumn crop in the field both as sole and mixed crop or grown in different riverbeds of the country. Huge genetic diversity in cucumber exists in India because of its probable primary gene center in Indian subcontinent (Hindustani centre). Large numbers of indigenous cultivars are mostly grown in India under different names like, Balam Khira, Khira Poona, Long Spined Green, Long Spine less Green, Baro Pata, Beldanga Local, Sikkim Local, etc. A number of improved varieties have also been developed through purification and subsequent improvement over the indigenous cultivars.

Breeding behaviour

- Four sex forms occur in cucumber namely, monoecious, gynoecious, andromonoecious, and hermaphrodite and obviously monoecious sex form is predominant.
- Cucumber is highly cross pollinated in nature due to the monoecious and gynoecious sex form.
- Andromonoecious and hermaphrodite sex form favour a higher degree of natural self-pollination than in the monoecious or gynoecious condition.
- It is entomophillous and major pollinators are bees (*Apis florea, Apis dorsata, Nomioides* sp., *Halictine* sp., etc.).
- Beetles like *Conopophilus* sp., moths like *Planidia* sp., *Pygargonia* sp. also helps in natural cross pollination.
- Although 100% natural cross pollination occur in monoecious sex form, yet reasonable extent of self fertilization happens because of foraging of insect pollinators in the staminate flowers of the same vine.
- Cucumber like other members of the family Cucurbitaceae does not suffer much from inbreeding depression or do so only negligibly, in contrast to the cross pollinated vegetable crops belonging to other families.
- Anthesis generally starts in the morning hours from 5:30 to 7.00 a.m.
- Pollen dehiscent generally starts after anthesis and pollens remain fertile in ambient condition remain upto 2.00 p.m.
- Stigma receptivity is the maximum at the time of anthesis.
- Best time of hybridization is between 7.00 and 10.00 a.m on the day of anthesis.
- Harvesting the fully matured staminate flowers in the afternoon hours that will open in the morning hours of the next day and keeping them covered with thin moist cloth over night for their use in cross pollination in the morning hours is the conventional practice commercially.
- Bagging the staminate and pistillate flowers prior to opening followed by hand pollination and subsequent bagging for two more days is the commercial method of hybrid seed production of cucumber.
- Presence of gynoecious sex forms has made possible phenomenal progress in exploitation of F_1 hybrids under green house condition.
- The temperate gynoecious lines are unstable for gynoecy under high temperature and long photoperiodic conditions.

- Tropical gynoecious lines with parthenocarpic gene in it have also been developed.
- Cucumber is crossable with *C. sativus* var. *hardwickii.*

Origin and taxonomy

The genus *Cucumis* comprises about 30 species distributed over two distinct geographic areas like South east of Himalayan group and African group. The Asiatic species which are mostly annuals are distributed in India, China, Myanmar and Korea and the African species which include annual, perennial, monoecious and dioecious are found in different parts of Africa, Middle Esat, Sudan, Egypt, Ethiopia and Pakistan. The melons belong to the *Cucumis* species of the African group and cucumber belongs to the *Cucumis* species of the Asiatic group. So, cucumber (*C. sativus*) and melon (*C. melo*) are considered to be separate and unrelated species.

Cucumber is thought to be indigenous to India although it is never found in wild state. The other Cucumis species of the Asiatic group occurring in India as wild include *C. sativus* var. *hardwickii, C. callosus, C. hystrix* and *C. setosus. C. sativus* var. *hardwickii* found along the southern foothills of the Himalayas and having cross compatibility with cucumber is the feral or progenitor form of cultivated cucumber. From India, cucumber spread to China and westwards to Asia Minor, North Africa and Southern Europe and later to other parts of Europe.

Cucumber, *Cucumis sativus* L. belongs to the family Cucurbitaceae, sub-family Cucurbitoideae, tribe Melothrieae and sub-tribe Cucumerinae. The cucumber along with other Indian *Cucumis* species are diploid (2n= 14) with a basic chromosome number, x = 7. The *Cucumis* species of the African group to which melns belog ar also diloid (2n = 24) but with a different basic chromosome number (x = 12).

Improved varieties

- Slicing type varieties: Japanese Long Green, Straight Eight, Poinsett, China, Himangi, Phule Shubhangi, Sheetal, Khira-90, Khira 75, CO-1, Punjab-1, Pusa Uday, Swarna Ageti, Swarna Sheetal, Pant Khira-1, Kalyanpur Ageti.
- Public sector hybrids: Pusa Sanyog, AAUC-1, AAUC-2, PCUC- F1, Solan Hybrid.

Cultural requirements

- Cucumber is basically a warm season crop but is successfully grown in tropical, sub-tropical and temperate regions.
- It prefers slightly milder temperature than watermelon and muskmelon.
- It is very susceptible to frost.
- Optimum temperature range for the growth and development is 18°C-24°C.

- The minimum temperature should not go below 10 ^{0}C and maximum above 30°C.
- Seeds germinate well at 25°C day temperature.
- Low temperature causes stunting of growth and poor germination of seeds.
- Sex ratio in monoecious form is sensitive to photoperiod and temperature condition.
- Short days, comparatively low-night temperature and high relative humidity increase the propensity of pistillate flowers in the vine.
- Long days and high temperature above 30°C increase the number of staminate flowers and reduce the number of pistillate flowers in the vine.
- Excess humidity promotes some important diseases like, downey mildew and anthracnose and increases the attack of fruit fly.
- Well-drained loamy soil is preferred, lighter soils that warm quickly in spring are usually used for early yields.
- In heavier soils vine growth is greater and fruits mature late.
- The soils should not crack in summer and should not be water-logged in rainy season.
- Cucumber prefers a soil pH between 6.0 and 7.0 and below pH 5.5 cucumber can not be grown.
- Sandy river beds have moisture beneath and warm up quickly in spring, that is the reason why cucumber in river-beds survives low temperature periods of winter months and produce early crops in spring.
- In riverbed cucumber is grown usually as a mixed cropping.

Land preparation

- Prepare the field with 4-5 deep ploughing and make into a good tilth before seed sowing.
- Remove all the weeds, stubbles and level the land by laddering.
- Manures should be mixed with the soil at the time of last ploughing.
- The plant has comparatively shallow root system and is susceptible to water logging so land should be well drained.
- Generally furrows are opened at 1.5 to 2.5 m spacing and sowing is done on the top of the sides of furrows at 60-90 cm spacing and the vines are allowed to trail on the ground, especially during summer.
- The crop is also grown on flat or raised beds, about 1.5 to 2.5 m wide and seed are sown on the periphery of the beds.
- In rainy season, raised beds or mounds are prepared to facilitate drainage.

- Mounds in the furrow, flat bed method or in river bed system of planting are prepared with the mixture of 15-20 tonnes of FYM per hectare and basal dose of fertilizers are mixed with the soil.
- In river bed system of sowing, trenches of 30 cm wide, 60 cm deep and of convenient length and spaced by 2.0 to 3.0 m are made followed by filling the trenches with farm yard manure and soil.

Sowing time

Cucumber grows successfully in tropical and subtropical climate hence, can be grown year round in the southern and central Indian states due to prevailing mild winter there. However, the crop is grown in three main seasons: spring-summer, early autumn and rainy season.

- November - December: Spring-summer crop in eastern plain.
- February - March: Summer crop in eastern and northern plain.
- June - July: Rainy season crop in eastern, southern and northern plain.
- August - September: Early autumn crop in eastern and nonthern India.
- September - October : Autumn crop of southern India.
- December - January: Winter – spring crop in southern and western regions.
- November - January: River bed cultivation in northern India.
- April - May: Hill regions.

Seed sowing

- Cucumber is a seed propagated crop and *in situ* sowing is generally practiced.
- Seeds are sown by hand dibbling in the mound or bed.
- Sowings can be done following flat bed or furrow method or in river bed system depending on the season and convenience.
- In river bed system, pre-germinated seeds are sown in the mound.
- Seed soaking in water for 12 hours before sowing improves germination.
- Seedlings raised in plastic bags (15cm x10 cm) particularly under polyethylene shed in winter month can be transplanted at 2-4 true leaf stage after 35 – 40 days with the advent of spring without disturbing the root system.
- This is basically practiced in north Indian condition during winter to get early harvest in spring.
- This seedling transplanting can be practiced in the hills to get early crop in July.
- 2.5 – 3.0 kg seed/ha is sufficient depending on the plant population.
- It is better to do seed sowing in pre-irrigated moist soil than irrigate the field after sowing.

- Pre-sowing soil application of 20-22 kg Furadon 3G per hectare helps in protecting the plants from root knot nematodes and other pests during initial 4-5 weeks.

Spacing

- Spring-summer crop: 1.5 – 2.0 m × 60 - 90 cm
- River bed system : 2.0 – 3.0 m × 60 - 90 cm
- Rainy season crop: 2.0 – 2.5 m × 80 - 90 cm
- Early autumn crop: 1.5 – 2.0 m × 60 - 90 cm
- Crop on the bower: 2.0 – 2.5 m × 80 - 90 cm

Nutrient management

- Cucumber responds well to manuring and fertilizer application.
- The dose of manures and fertilizers depend on the soil type, climate and system of cultivation.
- Well rotten farm yard manure at the rate of 15-20 t/ha should be applied at the time of land preparation.
- Apply a general fertilizer dose of 100 kg N, 60 kg P_2O_5 and 60 kg K_2O per hectare for open pollinated improved varieties.1/2 N along with full P and K fertilizers and farm yard manure should be given as basal and rest N should be top dressed in two split doses at the time of vining (25-30 days after sowing) and at full blooming (50 – 55 days after sowing).
- For the hybrids apply150 kg N, 90 kg P_2O_5 and 90 kg K_2O per hectare. 1/4 N along with full P and K fertilizers and farm yard manure should be given as basal and rest N should be top dressed in three split doses at the time of vining (25 - 30 days after sowing) , at full blooming (50 – 55 days after sowing) and after first harvest (75-80 days after sowing).
- Cucumber is a comparatively shallow rotted crop hence, responses well in top-dressing.

Crop regulation

- Excessive vine growth due to high nitrogen nutrition coupled with high temperature and high soil moisture condition promote staminate flowers in the vine resulting low fruit set and low yield. The best way to control the vine growth within reasonable limits is by adjusting nitrogen fertilizer doses and frequency of irrigation.
- Application of different growth regulators viz., 100 - 200 ppm Ethrel (100-200 ml/ l of water) twice, first at two true leaves of the plants i.e. 15 days after sowing and subsequently repeated 7 days after helps in increasing the number of pistillate flowers in the vine and subsequently yield.

- Growth regulators like GA @1500-2000 ppm or silver nitrate @ 200-300 ppm induce staminate or hermaphrodite flower in the gynoecious line which enables maintenance of the gynoecious line in pure form.

Irrigation

- Pits, ridges or beds are lightly irrigated a day or two days prior to sowing, preferably sprouted seeds planted and subsequent irrigation is given 4-5 days after seed sowing.
- Apply irrigation at the initiation of first true leaf during spring-summer and at its expansion during rainy season.
- Irrigation at regular interval of 4-5 days after is very important for spring – summer crop.
- Irrigation may not be necessary at all, if rainfall is well distributed between July- September for rainy season crop.
- Irrigate the field subsequently 5-6 days intervals during summer and early autumn crop and as and when required for the rainy season crop depending on the rain.
- In light soils irrigate the crop at frequent intervals.
- Water stressed plants generally leach out large amount of amino acids and sugars from the roots during recovery upon irrigation.
- The crop must be irrigated during its most critical stages i.e. flower bud development and early fruit development stages.
- Over irrigation during vegetative and early flowering stages may cause excessive vine growth resulting induction of more staminate flowers in the vine.
- Ridge and furrow method of irrigation is the best for cucumber.
- Drip irrigation gives more fruit yield than furrow irrigation.

Interculture and weed control

- Plant grown on trellis gives better crop stand and fruit yield mainly because of better use of sunlight by maximum number of leaves and higher number of side branches.
- The crop needs to be trained over low trellis of 1.5 m high above the ground.
- Training the vines over the trellis or bower coupled with some pruning is very effective.
- Cucumber needs interculture operations for successful cultivation.
- During the early stage of the crops, beds, ridges etc need to be kept free from weeds.

- At the time of top dressing with nitrogenous fertilizers, weeding and earthing up are done.
- Pruning all the primary branches after 2 nodes gives highest yield in green house.
- When the vines start spreading, weeding may not be needed in between the rows or ridges since vine growth can smother the weeds
- Deep intercultivation should be avoided as it may damage the shallow roots of the crop.
- Application of herbicides effectively controls weeds.
- Pre-emergence application of fluchloralin @ 1.20kg a.i/ha or fluchloralin 0.48 kg a.i. + Nitrofen @ 0.50kg a.i/ha checks weed population in the fields.

Harvesting the crop

- Cucumber should be harvested when they are tender.
- Right edible maturity is dependent on the individual kinds (slicing or pickling type) and varieties.
- Fruits of the pickling or slicing types generally reach optimum edible maturity stage within a week or slightly later of anthesis.
- If harvesting is delayed dark green skin colour of the slicing type fruits turn into brownish-yellow or russeting.
- In pickling type having pale-green fruit colour, turning to yellow also indicate over-maturity.
- White spine colour is also an useful indication for edible maturity.
- In small fruited types, length of fruit at edible maturity stage is around 8-12 cm.
- Over mature fruits show carpel separation in transverse section.

Post harvest handling and storage

- Crop damaged by summer rains, fruit fly and diseased fruit should be culled out before sending market.
- Crook neck and wasp shaped and other misshapen fruits should also be separated from the normal fruits before sending the market.
- Cucumber can not stand much too long transportation.
- Cucumber can be stored for 2 week at 10.0-11.7°C temperature with 92% relative humidity in storage.
- Chilling injury as a physiological storage disorder occurs when the fruits are exposed to the temperature below 0°C for prolonged period.
- It is better to use the fruits immediately after removal from the storage.

Plant protection measures

Physiological disorder

Preponderance of staminate flowers

The plants are monoecious in sex form hence, staminate and pistillate flowers are borne separately in the same plant and fruit yield depends on the number of pistillate flowers. Preponderance of staminate flowers is caused due to i) excessive nitrogen application, ii) high temperature condition , iii) long day length and iv) over irrigation.

Control measures

- Avoid rampant vegetative growth through control in fertilizer application, checking in irrigation and training and pruning.
- Apply nitrogenous fertilizer at a proper dose.
- Avoid excess irrigation
- Apply ethephon / ethrel (2-chloroethyl phosphonic acid) @ 150 ppm twice, first at two true leaves of the plants i.e. 15 days after sowing and subsequently repeated after 7 days.

Diseases

Fusarium wilt (*Fusarium solani* and *Fusarium oxysporum* f.sp. *lagenariae*)

This soil-borne as well as seed-borne fungus attacks the young seedlings causing withering of cotyledons. Older plants wilt suddenly and vascular bundles in the collar region become yellow or brown.

Control measures

- Seed treatment with thiram @ 2.5 g /kg of seed or carbendazim @ 2.5 g /kg of seed.
- Seed treatment with *Trichoderma viride* along with neem cake application is effective to control the disease.
- Application of lime to raise the soil pH to 7.0 -7.4 is effective.
- Spraying the crop with 0.2% captan or carbendazim (1g) + mancozeb (2g) per litre at weekly interval checks further spread of the disease.

Anthracnose (*Colletotrichum legenarium*)

This seed borne fungal disease appears more during the period of heavy summer rains i.e. in high humidity and warm conditions. Light brown spots appear on the leaves which later turn to deep brown. The lesions are dotted with pink conidia in moist condition.In severe cases, fruits detach from the pedicel.

Control measures

- Use disease free seeds.
- Treat seeds with thiram @ 2.5 g /kg of seed or carbedazim @ 2.5 g /kg of seed.
- Spray the crop with 0.1% Bavistin or 0.2% chlrothalonil or 0.1% hexaconazole 3-4 times at 10 days interval.
- Crop rotation excluding cucurbits.

Downy mildew (*Pseudoperonospora cubensis*)

This fungal disease is prevalent in high humidity, especially when summer rains occurs regularly. The diseases is characterized by formation of yellow, more or less angular spots on the upper surface of the leaves, white-purplish spores appear on the upper surface of the leaves. The disease spreads rapidly killing the plant quickly through rapid defoliation.

Control measures

- Management schedules as mentioned in pumpkin.

Powdery mildew (*Sphaerotheca fuliginea* and *Erysiphe cichoracearum*)

This fungal disease appears under prolonged dry conditions. White floury or powdery patches or coatings appear on the lower surface of leaf extending to upper surface causing yellowing and death of the leaves.

Control measures

- Management schedules as mentioned in pumpkin.

Fruit Rot (*Pythium, Phytophthora,* etc.)

The fungi can infect the seedlings, young plants and even young fruits. Initial infection is noticed as small spots on the leaves, which rapidly increases in number and size. After coalescing of adjoining spots and on severely affected leaves, a burning effect or blight symptoms is seen. Fruit rot may occur in the field or it may start after harvest.

Control measures

- Use of disease free seeds and its treatment with thiram @ 2.5 g /kg of seed or carbedazim @ 2.5 g /kg of seed.
- Crop rotation and proper drainage in the field.
- Spraying the crop with 0.25% mancozeb or metalaxyl-mancozeb (0.25%) effectively checks fruit rot in the field.
- Borax wash (2.5%) at 45°C for 30 seconds or at 40°C for 2 minutes before packing of the fruits prevents fruit rots in transit.

Angular leaf spot (*Pseudomonas syringae pv, lachrymans*)

It is an important bacterial disease where angular brown spots appear on the leaves as first symptom. On these spots bacterial ooze dries out to a white incrustation. The center of such spot on leaves later dry and fall out giving ragged appearance. Spots on fruits are circular which crack open to exude amber- coloured liquid containing bacteria that subsequently infect the fruits and cause severe rotting.

Control measures

- Use disease free seeds.
- Hot water treatment of seeds at 50°C for 30 minutes.
- Seed treatment with 0.01% streptocycline solution for 30 minutes.
- Spraying the crop with 0.01% streptocycline or Agromycin -100 immediately after the appearance of the disease checks the secondary spreads.

Cucumber mosaic (*Cucumber mosaic virus (CMV*)

This virus disease are transmitted by sap, seed and by several species of aphids .The young leaves develop small greenish- yellow areas which are more translucent than the remaining parts of the leaf. Yellow mottling in the leaf is followed by leaf distortion and stunting of plants. Infected fruits are mottled with yellowish green colour and blistered.

Control measures

- Spray the crop with 0.05% dimethoate or 0.05% monocrotophos or 0.05% phosphamidon at weekly interval to control aphid vectors.
- Eliminate weed hosts and use of barrier crops like sorghum, maize around the field.
- Crop rotation with non-cucurbitaceous crops.
- Remove infected plants from the field.

Insect Pest

Fruit fly (*Dacus cucurbita, Dacus dorsalis*)

Adult fly of this polyphagous pest having a wide host range is 4 – 5 mm long, ferruginous-brown in colour with hyaline wings. On hatching, the maggots feed inside the fruits and infested fruits can be identified by the presence of resinous liquid which oozes out of the punctures made by the flies for oviposition. The infested fruits start rotting due to secondary infection of different micro-organisms.

Control measures

- Collect and destroy the infested fruits along with the maggots to prevent the carry over of the pest.
- Stir the soil under the vines frequently and plough the infested field after harvest the crop to kill pupae.

- Spray the crop with 0.05% endosulfan.
- Apply poison bait. Bait is prepared by mixing 20 g malathion 50% WP or 50 ml diazinon 20% EC with 500 g molasses + 20 g yeast hydrolysate. This mixture is mixed with 2 litres of water for poison baiting and 20 liters of water for bait spray.
- Adopt male annihilation technique using 0.1% methyle eugenol + 0.025% malathion in a bait trap during fruiting season. Methyl eugenol is a sex attractant and males are attracted towards the bait and killed by malathion. Apply this trap solution at 7-10 days interval.

Red pumpkin beetle (*Aulacophora foveicollis*)

Shiny yellowish-red adults of this polyphagous pest lay eggs in moist soils around the plant. On hatching, the grubs feed on the roots and underground portion of host plant and also on the fruits touching the soils. Adult beetles feed voraciously on leaf lamina making irregular holes. Young seedlings and tender leaves are mainly preferred by the adults and damage at this stage may kill the seedlings.

Control measures

- Clean cultivation.
- Collect and destroy the beetles during cool hours of early morning when they remain sluggish.
- Deep ploughing of infested fields just after harvesting to kill the grubs in the soil.
- Dust the crop with 5% carbaryl or spray with 0.2% carbaryl.
- Repellent dusting of ash mixed with kerosene oil but heavy dusting should be avoided as it hampers plant growth.

Root knot nematode (*Meloidogyne incognita*)

These plant parasitic eelworms incite root galls causing poor and stunted growth of the infested plants.

Control measures

- Long crop rotation with non-host crops like, cereals, marigold, etc.
- Ploughing the fields in the summer months.
- Apply neem cake @ 12-15 t/ha.
- Apply granular insecticides like Aldicarb 10 G @ 25 kg/ha or Thimet 10G @ 20 kg/ha or Furadon 3G @ 40 kg/ha.

Yield

- 15 - 20tonnes/hectare (open pollinated monoecious varieties).
- 30 – 40 tonnes/hectare (hybrids).
- 150-200 tonnes/ha (gynoecious hybrids in green house).

15. Ridge gourd

Scientific Name : Luffa acutangula
Family : Cucurbitaceae
Origin : Sub-tropical Asia particularly India
Chromosome number: 2n = 26

Introduction

Fruits of ridge gourd are ribbed and that of sponge gourd under the same genus are non-ribbed and smooth. It is cultivated both on a commercial scale and in kitchen-gardens throughout India as spring summer and rainy season crops. This species contain a gelatinous compound called luffein. This crop is cultivated in India, Southeast Asia, China, Japan, Egypt and other parts of Africa.

Importance and use

- The immature fruits are cooked in many ways and are quite commonly used fried, boiled, stuffed and cooked and also in curries.
- The genus derives its name from the netted material of the dried fruit "loofah" which is used in bathing sponges, scrubber pads, doormats, pillows, mattresses and for cleaning utensils.
- Young leaves, blossoms and immature seeds are also edible. They can be simmered until tender and added to the vegetable dishes.
- Cultivation cost of this crop is comparatively less.
- The crop has got fair export demand.

Nutritional value

- Ridge gourd is not only inexpensive but better source of β-carotene.
- Ridge gourd contains 0.5% protein, 3.4% carbohydrate, 37mg carotene and 18 mg vitamin C per 100g edible portion.

Medicinal value

- Fruit is demulcent, diuretic and nutritive.
- The seed possesses purgative and emetic properties.
- The pounded leaves are applied locally to splenitis, haemorrhoides and leprosy.
- The juice of the fresh leaves is dropped in to the eyes of children in granular conjunctivitis, also to prevent the lids adhering at night from excessive meibomian secretion.
- The oil from the seeds is known to cure cutaneous problems.
- The roots have laxative effects.

Adaptation in India

The genus *Luffa* under which both ridge and sponge gourd belongs has been originated in subtropical Asia including India. Both ridge and sponge gourd were domesticated and known in very early times. Luffa was mentioned in Kautilya's Arthashastra, c. 350-300 B.C. in India.Expectedly wide genetic variability in this crop is present in India. A large number of local cultivars varying in shape and size (one metre to a few centimetres long) are grown in India. Some of them are even slightly bitter in taste. Improved varieties have been developed in India mainly by single plant selection from the existing genetic variability followed by inbreeding to stabilize the genotype.

Breeding behaviour

- Different sex forms occur in ridge gourd namely, monoecious, gynoecious, andromonoecious, androecious, gynomonoecious and hermaphrodite and obviously monoecious sex form is predominant.
- The cultivar "Satputia" found in Bihar is hermaphrodite in sex form and small fruits are borne in cluster.
- It is is highly cross-pollinated in nature due to the monoecious sex form.
- Andromonoecious and hermaphrodite sex form favour a higher degree of natural self-pollination than in the monoecious or gynoecious condition.
- It is entomophillous and major pollinators are bees (*Apis florea, Apis dorsata, Nomioides sp., Halictine* sp., etc.).
- Although 100% natural cross-pollination occur in monoecious sex form, yet reasonable extent of self fertilization happens because of geitenogamy due to foraging of insect pollinators in the staminate flowers of the same vine.
- Ridge gourd like other members of the family Cucurbitaceae does not suffer much from inbreeding depression or do so only negligibly, in contrast to the cross-pollinated vegetable crops belonging to other families.
- Anthesis of both staminate and pistillate flowers generally starts in the afternoon hours from 4.00 to 8.00 p.m.
- Anther dehiscence in the staminate flower generally coincides with the anthesis.
- Pollen fertility is the maximum on the day of anthesis which may remain viable up to 1 to 3 days after anthesis in ambient condition depending on the season.
- Stigma receptivity is at its maximum at the time of anthesis although the stigma becomes receptive 6 hours earlier to 48 hours after anthesis.
- Best time of hybridization is between 4.00 and 6.00 p.m. on the day of anthesis however, hybridieation can effectively be done in the next day morning also.

- Bagging the staminate and pistillate flowers prior to opening followed by hand pollination and subsequent bagging for two more days is the commercial method of hybrid seed production of ridge gourd.
- In monoecious sex form, harvesting the fully matured staminate flower buds in the morning hours that will open in the afternoon hours on that day and keeping them covered with thin moist cloth for their use in cross-pollination in the afternoon hours or next day morning hours is the conventional practice commercially.
- Ridge gourd is crossable with sponge gourd, *Luffa cylindrica.*

Origin and taxonomy

The Loofah gourds (*Luffa acutangula* and *Luffa cylindrica*) originated from the subtropical regions of Asia, most probably from India as a primary centre. *Luffa* is an essentially old world genus, consisting of two cultivated species, *Luffa acutangula* and *Luffa cylindrica,* and two wild species, *Luffa graveolens* and *Luffa echinata. Luffa graveolens* is considered as the progenitor species of the two cultivated species. *Luffa graveolens* is found in north-eastern plains, extending south to Tamil Nadu and sporadic in Eastern Himalayas. The advanced sex form of dioecism has been found only in wild primitive species *Luffa echinata*. Wild forms of the ridge gourd have extremely bitter in taste but the domesticated types are less bitter. The chromosome number of both cultivated and wild species is same (2n=26).

Improved open pollinatedvarieties

- Pusa Nasdar, Pusa Sada Bahar, Co-1, Kankan Harita, Punjab Sadabahar, IIHR-8, PKM-1, Arka Sujat, Arka Sumeet.
- Hermaphrodite flower bearing cultivar: Satputia

Cultural requirements

- It requires long warm season for best production.
- It grows best at a temperature of 25-30° C.
- Rain during flowering and fruiting reduces yield considerably.
- Well drained loamy soil with good amount of organic matter is preferable for ridge gourd cultivation.
- Soil having P^H 6.0-7.5 in suitable for ridge gourd cultivation.
- Soil should have good water-holding capacity especially in summer season.

Land preparation

- The land should be prepared to obtain good tilth to facilitate rapid and better germination.

- The raised bed, furrows or pits are prepared and field is kept readily for planting.
- In riverbeds trenches or pits are prepared for sowing seeds.
- Mounds in the furrow, flat bed method or river bed system are prepared with addition of 15-20t FYM/ha and basal dose of fertilizers mixed with the soil.

Sowing time

- For summer season crop: January-February
- For rainy season crop:June-July
- In Areas having mild winter : October-November (for early crop in March).
- In river beds in northern India: November- January.
- In hills : April-May.

Seed sowing

- 3.5-5.0 kg seeds are sufficient for one hectare.
- Two seeds per hill in case of raised beds at the side.
- Three to four seeds per pit in core pit system planting.
- 12-24 hours soaking of seeds in water before sowing hasten the germination.
- Transplanting of seedlings raised in small polybags is also sometimes practiced as it economise the seed quantity, ensures better plant stand and early crop.

Spacing

- Row to row distance : 1.5-2.5 m
- Hill to hill distance: 60-120 cm

Nutrient management

- Apply a general fertilizer dose of 15-20 tonnes FYM, 100kg N, 60 kg P_2O_5 and 60 kg K_2O per hectare
- For hybrids, apply 150 kg N, 80 kg P_2O_5 and 80 kg K_2O per hectare.
- Farmyard manure should be mixed in the soil at the time of preparation of field.
- Full doses of super phosphate, potash and one third nitrogen should be applied as basal dose and the rest nitrogen should be given as top dressing in two splits, 30 and 60 days after planting.

Crop regulation

- Application of elhrel (200 ppm) first 15 days after sowing and subsequently repeated after 7 days increases the pistillate flower production and in turn increases the yield significantly.

Irrigation

- The first irrigation should be given immediately after the sowing.
- Summer crop requires more frequent irrigation than rainy season crop. It should be irrigated once in a week depending on soil moisture.
- During rainy season, irrigation may be required during the early growth period.

Interculture and weed control

- In general, vines are trained over trellises or arbours or pandals at 1.5-2.0 m height.
- The crops are trained when the seedlings are about 10-15 cm tall.
- In some areas the vines are not generally trained and allowed to trail on the ground.
- For the rainy season crop, it is advisable to trail the vine on a pandal so that the fruits are prevented from direct contact with soil.
- Hand weeding should be done 15-20 days after sowing.
- Second weeding may be done 20-25 days after the first one.
- Glyphosate @ 4.5 kg a.i. /ha applied after weed emergence and before sowing effectively control weeds.

Harvesting the crop

- Fruits are harvested when they are still immature, the flesh should not turn fibrous and picking should be done earlier.
- Picking is done in about 8-10 days after anthesis of pistillate flower.
- The crop is ready for harvest (first picking) in about 60-80 days after seed sowing depending upon the season and variety.
- Picking should be done every 3-4 days, if there is delay in harvesting, the fruits became more fibrous and unfit for human consumption.

Post harvest handling and storage

- Grading should be done according to size.
- The fruits are packed in bamboo baskets with proper cushionings to prevent injury during transit.
- The plastic crates are also used for packing of fruits.
- Fruits harvested at the marketable stage can stand 3-4 days in a cool place without any adverse effects.
- The fruits can be stored for 1-2 weeks at 7-10°C with 90-95% relative humidity.

Plant protection measures

Physiological disorder

Preponderance of staminate flowers

The plants are monoecious in sex form hence, staminate and pistillate flowers are borne separately in the same plant and fruit yield depends on the number of pistillate flowers. Preponderance of staminate flowers is caused due to i) excessive nitrogen application, ii) high temperature condition , iii) long day length and iv) over irrigation.

Control measures

- Avoid rampant vegetative growth through control in fertilizer application, checking in irrigation and training and pruning.
- Apply nitrogenous fertilizer at a proper dose.
- Avoid excess irrigation.
- Apply ethephon / ethrel (2-chloroethyl phosphonic acid) @ 200 ppm twice, first at two true leaves of the plants i.e. 15 days after sowing and subsequently repeated after 7 days.

Diseases

Powdery mildew (Sphaerotheca fuliginea and *Erysiphe cichoracearum)*

The disease is particularly severe in dry rainless weather. White floury patches or coating appears first as the under surface of the leaves which later spread to both the leaf surfaces, stem, petiole and other succulent plant parts. Severely attacked leaves become brown and shriveled leading to premature defoliation. Fruits of the affected plant do not develop fully.

Control measure

- Spray the crop with Bavistin or Benlate or Topsin M @ 1g/liter of water 3 to 4 times at 15 days interval.
- Two sprays of 0.1% Bavistin followed by two sprays of 0.1% karathane or tridemorph effectively control the disease.

Downy mildew (Pseudoperonospora cubensis)

This fungal disease is prevalent in high humidity, especially when summer rains occurs regularly.The diseases is characterized by formation of yellow, more or less angular spots on the upper surface of the leaves, white-purplish spores appear on the upper surface of the leaves. The disease spreads rapidly killing the plant quickly through rapid defoliation.

Control measures

- Removal of infected leaves and spray the crop with 0.25% mancozeb 3 to 4 times at 10 days interval effectively control the disease.

- Two sprays of 0.25% Fosetyl Al or cyamoxanil- mancozeb or metalaxyl-mancozeb at 10 days interval effectively control the disease.

Insect pests

Fruit fly (*Dacus cucurbita, Dacus dorsalis*)

Adult fly of this polyphagous pest having a wide host range is 4 – 5 mm long, ferruginous-brown in colour with hyaline wings. On hatching, the maggots feed inside the fruits and infested fruits can be identified by the presence of resinous liquid which oozes out of the punctures made by the flies for oviposition. The infested fruits start rotting due to secondary infection of different micro-organisms.

Control measures

- Collect and destroy the infested fruits along with the maggots to prevent the carry over of the pest.
- Stir the soil under the vines frequently and plough the infested field after harvest the crop to kill pupae.
- Spray the crop with 0.05% endosulfan.
- Apply poison bait. Bait is prepared by mixing 20 g malathion 50% WP or 50 ml diazinon 20% EC with 500 g molasses + 20 g yeast hydrolysate. This mixture is mixed with 2 litres of water for poison baiting and 20 liters of water for bait spray.
- Adopt male annihilation technique using 0.1% methyle eugenol + 0.025% malathion in a bait trap during fruiting season. Methyl eugenol is a sex attractant and males are attracted towards the bait and killed by malathion. Apply this trap solution at 7-10 days interval.

Red pumpkin beetles (*Aulacophora foveicollis*)

Pumpkin beetles feed on the foliage. The grubs are soil dweller, feeding on the root system, twigs and tender fruit that touch ground. Infestation is more severe during summer months and the young seedlings may even be killed.

Control measures

- Clean cultivation.
- Collect and destroy the beetles during cool hours of early morning when they remain sluggish.
- Deep ploughing of infested fields just after harvesting to kill the grubs in the soil.
- Dust the crop with 5% carbaryl dust or spray the crop with 0.2% carbaryl or 0.05% endosulfan.
- Light repellent dusting of ash mixed with kerosene.

Epilachna beetle (*Henosepilachna vigintioctopunctata*)

Both the adults and grubs feed on the leaves by scarping in a characteristic manner producing lace like symptoms and often cause defoliation. The bellow cigar shaped eggs and oval spiny grubs are seen on the foliage. They remain active throughout the year, though their activity is reduced during winter.

Control measures

- Clean cultivation, mechanical collection and destruction of adult beetles during cool hours of early morning when they remain sluggish
- Spraying the crop with 1.5ml endosulfan, or 2.5ml chloropyriphos per litre of water 15 days interval effectively controlled the pest.

Yield

- Open pollinated improved varieties: 8-12 tonnes /ha.
- Hybrids: 20-25 tonnes/ha

Root Crops

16. Radish

Scientific Name: Raphanus sativus
Family: Brassicaceae (Cruciferae)
Origin: Europe, Western Asia
Chromosome No.: 2n = 2x =18

Introduction

Radish, one of the most ancient vegetables, is grown for fleshy roots and leaves in both tropical and temperate climate. It is cultivated widely in Asia (India, China, Japan, and Korea), Europe and USA. Inscriptions on the walls of pyramids show that radish was an important vegetable in Egypt about 2000 B.C. It was also cultivated by the ancient Babylonians, Greeks and Romans. It was spread to china about 500 B.C. and to Japan A.D. 700. Radish was also cultivated in India from ancient times (300-800 B.C.) and it is grown through out India and more extensively in Uttar Pradesh, Bihar, West Bengal, Punjab, Haryana, Assam, Himachal Pradesh and Gujarat.

Importance and use

- Radish is the crop of tropical, sub-tropical and temperate climate.
- The tropical, high temperature tolerant types mainly developed in India are more pungent than both Oriental and European types.
- The European radishes are basically early, juicy and mild type with crispy roots.

- True Oriental radishes are also mildly pungent and crispy in nature.
- It is grown extensively in India.
- Large number of tropicalized cultivars with heat tolerance has been evolved in India.
- Edible portion of radish is the swollen primary root.
- Fleshy edible roots are eaten raw as salad or cooked as a vegetable.
- Radish roots are used for preparation of pickles.
- Young leaves are cooked as leafy vegetable particularly in northern and eastern India.
- The rat tail radish (*Raphanus sativa var. caudatus*) is extensively grown for its long slender pods and is cooked as a vegetable or eaten raw as salad.
- The radish *Raphanus sativa var. oleifer* is used for fodder, vegetable oil and green manure.
- Radish has been tried successfully as a fodder crop in some countries like U.K.
- The characteristic pungent flavour of radish is due to the presence of volatile isothiocyanates (trans-4-methyle thiobutenyl isothiocyanate)
- The colour of pink cultivar due to the presence of anthocyanin pigments.

Nutritional value

- Radish contains fair amount of vitamin C (ascorbic acid), 15-40 mg/100 g of edible portion.
- Edible portion of radish contains 0.7% protein, 3.4-6.8% carbohydrates and 0.2% fat and 50 IU vitamin A per 100 g fresh edible portion..
- Tender roots provide about 50 mg calcium, 22 mg phosphorus and 0.5 mg iron per 100 g edible portion.
- It also contains some trace elements including aluminum, barium, lithium, manganese, silicon, titanium, fluorine and iodine upto 18 mg/100 g fresh.
- Pink skinned radish is generally richer in carbohydrate, ascorbic acid than the white skinned varieties.
- Radish contain glucose as major sugar and smaller quality of fructose and sucrose

Medicinal value

- Radish was considered as a medicinal plant in the Puranas like, Ramayana and medical treatise like "Susutra Samhita".
- In the Sutra period (800 B.C to 300 B.C) the mention of radish root was made as aiding digestion when munched after meals.
- Radish roots are good for liver and gallbladder trouble.
- The roots are good appetizer.

- In homeopathy they are used for neuralgic headaches, sleeplessness and chronic diarrhoea.
- The roots are useful in urinary complaints, piles and problems of spleen.
- The seeds are said to be peptic, expectorant, diuretic and carminative.
- The juice of fresh leaves is used as diuretic and laxative.

Adaptation in India

The introduction of radish into India has not been chronicled. However, mentioning of radish in different ancient literatures points its introduction before Christian era and "Mulaka" was the Sanskrit equivalent of radish in ancient India. Radish was considered as a medicinal plant in the Puranas like, Ramayana. In the Sutra period (800 B.C. to 300 B.C), the mention of radish root was made as aiding digestion when munched after meals. In "Susutra Samhita", the medical trreatise of 3rd to 4th century A.D. radish had been recorded having medicinal properties. Development of large range of tropicalised cultivars suitable for growing in subtropical and tropical condition with heat tolerance and capable of forming roots even under day temperature of even 35°C is significant feature of adaptation of Asiatic races in India. These adaptable Asiatic types of India (Desi type) have been utilized in breeding either in straight selection or in combination breeding scheme.

The cultivars grown in India can be categorized under two broad groups viz., European or temperate type and tropicalized Asiatic type depending on their temperature sensitivity, root characters and seed production behaviour. The European radish maintains its distinctness as temperate race and most of the cultivars of temperate European radish race grown in India are exotic introductions mainly at the time of British regime. The tropicalized Asiatic types flower and produce seeds freely in the plains with out vernalization or low temperature requirement. On the other hand, temperate European radish cultivars require vernalization for induction of flowering.

In India there are many local cultivars of Asiatic radish grown by the farmers under different names. Some of such cultivars are Jaunpuri Giant (very big, about 1 metre long, 50-60 cm in girth and 10-15 kg in weight) in Jaunpur, Uttar Pradesh; Kontai Long (big, red skinned, pungent, juicy ,2-4 kg), Bombay Long Red (big, red skinned, pungent, juicy,2-4 kg), Bombay Long White (big, white skinned, pungent, juicy ,1-3 kg), Aus Mula, Baramasi (small rooted grown at very high temperature condition) in West Bengal; Newari, Kannauji, Baramasi, Bombay Red, Nadauni in Himachal Pradesh, etc.

Breeding behaviour

- The tropicalized Asiatic cultivars mainly developed in India are high temperature tolerant and behave like annual and set seeds freely in the plains with the advent of long day condition.
- The Oriental (Chinese and Japanese) radishes are not true biennial but require low temperature for bolting and seed set.

- The European types behave like biennial crop which require 6-8 weeks of vernalization at mean temperature range of 4° to 8° C after they have passed the full root development phase for the production of seed stalks. Seeds of these types of radish can only be produced in the high hill situations of Himachal Pradesh and Kashmir.
- Radish bear hermaphrodite flowers but varying levels of cross fertilization takes place even upto 80 percent in different open pollinated varieties due to manifestation of self-incompatibility of different intensities.
- It shows considerable inbreeding depression on selfing.
- Self-incompatibilty system which is sporophytic in nature and governed by multiallelic single gene (*S* gene) with generally dominant effect. Intensities of self-incompatibility depend on the presence of *S* allele in the genotype.
- In the early bud stage, 2-3 days before anthesis when the stigma has already become receptive (protogyny), self-fertilization can be done by self-pollinating with the pollens from other opened flowers of the same plant.
- The "S" alleles are active in the stigma in flowering stage but remain inactive in the bud stage, i.e., 2-4 days before anthesis.
- In radish cross pollination is generally done by honey bees, wild honey bees and wild flower flies.
- Bees and other insect pollinators frequently visit the flowers of radish because of the presence of nectar in the flowers.
- To maintain the purity of the varieties, isolation distance of at least 1000 m is safe for seed production.
- Cytoplasmic male sterility referred to as "Ogura and McCollum" type have been identified in radish.
- Hybrids of radish have been developed mainly in Japan, Korea and Taiwan utilizing self-incompatibility and cytoplasmic male sterility mechanisms.
- Anthesis starts early in the morning and continues up to 11.30 am depending on the prevailing weather conditions.
- Dehiscence of anther is longitudinal and it generally occurs few hours after anthesis.
- The stigma becomes receptive at least two days before anthesis although receptivity is maximum at anthesis and it continues at least four days after.

Origin and taxonomy

Europe is considered as original home of radish however, wide variability in the cultivated forms for root morphology, ecology and temperature sensitivity amply indicate multicentre origin of this crop particularly western Asia. So, two distinct types of radish are available on the basis of their origin viz., European or Occidental radish and Oriental

(Chinese and Japanese) radish. Maximum diversity in *Raphanus* species is available in the Mediterranean region. A few important wild species viz., *Raphanus raphanistrum, R.maritimus, R. landra, R. microcarpus and R. rostratus* available in the Mediterranean region are considered to be the probable progenitors of the small rooted European radish. This European radish maintains its distinctness as temperate race.

Asiatic races of radish of widely varying root size and suiting to sub-tropical to temperate conditions have been evolved in western Asia. It was introduced into China more than 2400 years ago from the eastern Mediterranean region through ancient silk route and then to Japan more than 1250 years ago. The Chinese radishes are native to western region of Central Asia and south-west China or Central Asia. The Japanese radish has close relationship with *Raphanus sativus* var. *raphanistroides* which occurs wild in the coastal regions of Japan, Korea and South China. Even though China and Japan have a large range of radish varieties suiting to sub-tropical to temperate conditions, the tropical radish with heat tolerance and capable of forming roots even under day temperature of 35°C is the significant feature of adaptation of Asiatic races in India. The Indian group of radishes including the rat-tail radish evolved in the area of their present distribution in north-western region or west Asia but their ancestors do not exist.

Radish, *Raphanus sativus* belongs to the section Raphanis and family Brassicaceae (Cruciferae). There are five botanical varieties of radish having the same chromosome number ($2n = 2x = 18$) which are commonly cultivated in various regions of the world.

- *R sativus* var *radicula*: Small, cool season radish
- *R sativus* var. *niger*: Black or Spanish radish; large radish with wider range of temperature adaptation.
- *R sativus* var. *caudatus* : Rat tail radish or "Mougri" radish forming no fleshy roots but produce large slender (20-60cm) pods.
- *R sativus* var. *oleifera* : Oilseed radish; fodder radish also producing no fleshy roots.
- *R sativus* var. *raphanistroides* : Japanese and Chinese winter radish.

All the five botanical varieties intercross freely among each other and also with related wild species.

Improved varieties

- Tropicalized Asiatic varieties: Pusa Desi, Pusa Reshmi, Pusa Chetki, Japanese White, Punjab Safed, Punjab Ageti, CO-1, Kalyanpur No-1, Kashi Sweta, Arka Nishant, Chinese Pink.
- European or Temperate varieties: White Icicle, Pusa Himani, Rapid Red White Tipped, Scarlet Globe, Scarlet long.

Cultural requirements

- Radish is predominantly a cool season crop and grows best in mild and cool climate.
- These temperature requirements is of higher side for tropicalized Asiatic cultivars and lower side for European cultivars.
- The tropicalized Asiatic types can tolerate high temperature than the temperate types although it performs best in cool growing condition.
- The high temperature tolerant Asiatic cultivars when grown at high temperature condition generally produce small root and comparatively profuse top.
- The roots become tough during hot weather condition.

  High temperature during 20-30 days after sowing inhibits the formation of secondary meristem in the centre of roots leading to development of intercellular speaces which would otherwise have been filled with parenchyma.
- Relatively high temperature of 20-25°C for early root growth and low temperature of 10-18°C in the later part of root development is the best for best flavour, texture and size of the roots.
- It can be grown all type of soil, but best results are obtained as light, friable loam soil that contain ample humus.
- For early crop, sandy or sandy loam soils are preferred.
- For summer crop a cool, moist soil gives best result.
- Heavy soil or those with rocks, stones or pebbles below the surface produce rough, misshapen roots with a number of small fibrous laterals, and such soils should be avoided.
- Radish can be grown in slightly acidic soil also with soil pH ranging between 5.5 and 6.8.

Land preparation

- The soil should be thoroughly prepared so that there are no clods to interfere with the development of the root.
- The soil should not contain any undecomposed organic matter, because they may result in forking or misshapen roots.
- Remove all stubbles, weeds, etc. from the land.
- Add well rotten organic manure preferably farm yard manure or compost @ 10-15 t/ha during land preparation and level the land properly.
- It is better to prepare 20-25 cm high ridges as it facilitates irrigation in furrows made along the ridges.

Sowing time

The time of sowing depends on the type, variety and location. The tropical types when sown later than November in the plains of India bolt early. On the other hand, the European types when sown before October in the plains produce poorly developed roots. In places where the climate is mild, radish can be grown all the year round.

Northern, eastern and central Indian plains

Asiatic types: June-January

European types: October – February

South Indian plains

Asiatic types: July-January

European types: October- December

Northern eastern and north-eastern hills

First fortnight of March till late October or beginning of November

South Indian Hills

April- June

Sowing of seeds

- Radish is direct seeded crop because swelled primary roots are economic plant parts.
- Roots may be misshaped in the transplanted seedlings.
- The seeds generally take 5-10 days for germination under optimum environmental conditions.
- Seed germination in field can be improved when seeds are treated with 5-10 ppm GA_3 before sowing
- The seeds are sown on ridges or on flat beds.
- The seeds are sown as far as possible on the surface of the soil and covered by thin layer of soil.
- For long rooted cultivars the seeds have to be sown at a depth of 1.5 – 3cm.
- 10-12 kg seeds of the varieties of Asiatic types are sufficient for one hectare land.
- 15-18 kg seeds of the varieties of European types are sufficient for one hectare land.
- Seeds are sown either by broadcasting or line sowing.
- After a fortnight of sowing, the seedlings are thinned out maintaining the requisite spacing.

Spacing

- Asiatic cultivars: 30-45cm × 8-10cm
- European cultivar: 20-25cm × 5-6 cm

Nutrient management

- Adequate nutrition is necessary for obtaining higher yield and better root quality.
- Well decomposed FYM @ 10-15 t/ha should be applied at the time of land preparation, 20 – 30 days before sowing of seeds.
- The FYM to be applied should be well decomposed because fresh or undecomposed manure induce forking of roots with many fibrous roots.
- Apply a general fertilizer dose of 100 kg N, 60 kg P_2O_5 and 80 kg K_2O per hectare for open pollinated improved varieties. FYM along with the whole fertilizers should be given as basal before sowing seeds.
- For better growth and yield, the fertilizer should be ploughed to a depth of 25 cm.
- Two foliar sprays of 0.3% borax after 25, 40 days after sowing is beneficial for enhancing root growth.

Water management

- Irrigation immediately after sowing is the best to get good seed germination and good establishment of the crop.
- Next irrigation should be on the 3rd day and subsequently once in 5-7 days.
- Moist condition was found better in increasing the yield than wet condition
- During summer month frequent irrigation is necessary otherwise the growth will be checked and the roots will be more pungent and tough.
- The irrigation requirement for radish were found to be about 210-250 mm.

Interculture and weed control

- The plants should be thinned as soon as the seedlings establish.
- Shallow weeding help for better root development.
- Regular hand weeding is necessary to check the growth of weeds.
- Radish normally tend to bulge out as it grows, so proper earthing up after 20-25 days of sowing is necessary to get quality roots.
- Root yield is high from the plants raised from seeds treated with GA_3 at 10 ppm.
- Pre-sowing application of Tok E-25 (nitrofen 25%) control weeds effectively in the field.

Harvesting the crop

- Radish roots become ready for harvesting in 25-55 days after sowing depending upon the type and variety.
- Varieties of the European types are quite early and become ready 25-30 days after sowing
- Roots should be harvested at tender stage as delay in harvesting cause pithiness of roots.
- Varieties of the European types become bitter and pithy much quickly if left unharvested.
- A light irrigation need to be given before harvest of roots to facilitate easy lifting of roots.
- Roots are pulled out by hand ensuring that the roots do not break or get damaged during harvest.

Post harvest handling and storage

- The harvested roots are washed, graded and tied in bunches before transported to market
- In developed countries commercial radish growers use a single row harvester that pulls the plants from the soil, cuts the roots from the tops then places them in burlap bags for transportation to a packing shed where the roots are washed, graded and bagged in plastic bags for market.
- Radish roots can not be stored for more than 2-3 days under room temperature without impairing its quality.
- Roots can be stored for about 2 months in cold storage at 0° – 1°C and 90-95 percent relative humidity.

Plant protection measures

Physiological Disorder

Hollow root

High temperature during 16-30 days of sowing inhibits the formation of secondary meristem in the centre of roots leading to development of intercellular spaces which would otherwise have been filled with parenchyma. This environmental condition and/or boron deficiency results in the hollow root formation.

Wart

Wart is a physiological disorder which is a white protrusion of white inner root tissue through splits in skin, mainly occur due to soil moisture deficiency.

Akashin

This physiological disorder is caused due to boron deficiency coupled with exposure of high day and night temperature of 30ºC/20ºC.

Pithyness

It occurs more in summer than spring or autumn crop. Excess application of fertilizers, soil moisture stress and high temperature condition 3 weeks before harvest results pithiness.

Control measures

- Keeping proper moisture condition in the field.
- Avoid growing sensitive varieties during summer.
- Avoid applying excess fertilizers.
- Two foliar sprays of 0.3% borax after 25, 40 days after sowing.

Root forking

Roots get forked in poorly prepared soil having clods. Roots in touch of the lumps of undecomposed organic matter also get forked.

Control measure

- Land should be prepared finly so that the primary root may grow freely.

Diseases

Alternaria blight (*Alternaria raphani*)

It is an important fungal disease particularly for the seed crop. Small, yellowish, slightly raised lesion first appears on the leaves of seed stem. These enlarge many times as they become older. Lesions appear later on the stems and seed pods. Infection spreads rapidly during rainy weather, and the entire pod may be so infected that the style end becomes black and shriveled. The fungus also penetrates the pod tissues and ultimately infects the seeds.

Control measures

- Use of disease free seed.
- Seed treatment with hot water at 50ºC for 30 minutes.
- Seed treatment with captan @ 2g/kg of seed or thirum @ 2.5 g/kg of seed.
- Spraying with mancozeb (0.25%) or propiconazole (0.1%) or difenconazole (0.05%).

White rust (*Albugo candida*)

Disease symptoms due to infection of this fungus appear on the leaves and flowering shoots which become deformed and bear only malformed flowers. It produces white powdery substances in patches on the undersurfaces of the leaves.

Control measures

- Spraying the crop with 0.2-0.3% zineb or 0.25% metalaxyl-mancozeb.

Insect pests

Aphids (*Myzus persicae, Brevicoryne brassicae*)

Aphids are the most serious pests of radish. It attacks both seedlings and mature crops. Cloudy and humid conditions favour the spread of their infestation. In case of heavy infestation the plants are completely devitalized, leaves and shoot curl up, become yellowish and finally die.

Control measure

- Spray the crop with Rogor @ 1.5ml/l of water.

Mustard saw fly (*Athalia lugens proxima*)

The insect attacks the crop in both vegetative and flowering stage. The blackish larvae are damaging which feed on the foliage. The leaves of severely infested crops are eaten up completely. The larvae are very shy in nature which roll and fall down with little disturbance.

Control measures

- Spray the crop with Nuvacron @ 1 ml/l, or Dursban (chlorpyriphos) @ 2.5 ml/l or malathion @ 2 ml/l or Sevin W. P. @ 4 g/l of water.

Yield

- Tropicalized Asiatic types: 15-20 tonnes/ha
- Temperate types : 5-7 tonne/ha.

17. Carrot

Scientific Name : Daucus carota L.
Family : Umbelliferae
Origin : Afganistan
Chromosome number: 2n = 18

Introduction

Carrot, being a cool season crop is grown all over the world in spring, summer and autumn in temperate countries and during winter in tropical and sub-tropical climate. It is an important root crop of the European countries, USA, China, Japan, Konea and India. It is one of the most important root vegetables taken raw as well as in cooked form. Main carrot growing states of India are Uttar Pradesh, Assam, Karnataka, Andhra Pradesh, Punjab and Haryana.

Importance and use

- Undoubtedly carrot is one of the most ancient vegetables.
- Swollen primary roots are used as vegetable for soups, stews, curries and pies.
- Carrot *Halwa* is a delicious and popular dish in India.
- Carrot jam is also popular and the roots in the form of disc, and slices can be dehydrated.
- Grated roots are used as salad, tender roots as pickles and sweetmeat.
- Carrot juice is a rich source of carotene and is sometimes used for colouring butter and other foods.
- Black carrot is used for the preparation of a beverage called "kanji" considered to be a good appetizer.
- Carrot seed which is the source of an essential oil, used for flavouring liquors, and foods.

Nutritional value

- Orange-coloured carrots are rich in β carotene, a precursor of vitamin A and contain appreciable quantity of thiamine and riboflavin.
- The Asiatic types have more of anthocyanin pigments and less of β carotene and may be less nutritive.
- Green carrot leaves are highly nutritive, rich in protein, minerals and vitamins and used as fodder and also for preparation of poultry feed.

Medicinal value

- Carrot being the rich source of provitamin A prevents night blindness.
- An infusion of carrot has been used as folk medicine for thread worms.
- Essential oil from carrot seeds has antibacterial properties.
- Consumption of carrot increases the quantity of urine and helps the elimination of uric acid.
- Addition of large amount of carrot to the diet has a favourable effect on the nitrogen balance in our body.

Adaptation in India

Carrot (*Daucus carota* L.) was introduced into India early in its history. The black and red types might have been introduced from the central Asian region of its origin, probably Afganistan (considered to be the main center of origin), Persia and southern Russia. These types with low carotenoid content were mostly annual, juicy and forming roots under comparatively higher temperature condition. These types showing wide variation in colour, shape and size of roots form the distinct tropical or oriental type. Colour of these types vary from absolute colourless to light lemon, light orange, deep orange, light purple, deep purple and almost black. Himachal hills and Kashmir valley still abound in its wild form where wild animals especially brown bear feed on its roots. In some parts of Kashmir, people still eat the wild carrots. Adapted cultivars of this type have been utilized in combination breeding programmes to develop tropical varieties in India. Carrot was derived from wild *Dancus carota* sp. *carota,* commonly knon as "Queen Ann's lace" which is still found in Afganistan, southeastern Asia, southern Europe, month Africa and month America. The orange coloured temperate types having higher carotenoid content, believed to have originated in Europe, was introduced in India much later with the European traders. Most of the varieties under this group grown in India are exotic and these varieties have also been used in combination breeding to develop temperate carrot varieties in India.

Breeding behaviour

- Carrot is a long day crop and produce seeds with the advent of spring.
- The tropical or oriental cultivars are annual and adapted to high temperature condition. Cultivars of this type do not require low temperature exposure for flowering. They flower and set seeds freely in the plains of India after winter months with the advent of long day condition in the spring.
- The European types behave like biennial crop which require 6-8 weeks of vernalization at mean temperature range of 4.8^{o} to 10° C after they have passed the full root development phase for the production of seed stalks. Seeds of these types of radish can only be produced in the high hill situations of Himachal Pradesh and Kashmir.
- Carrot bear hermaphrodite flowers but varying levels of cross-fertilization takes place even upto 80 per cent in different open pollinated varieties due to protandry, high foraging of insects for nectars and strong smell and prolonged receptivity of stigma.
- It shows very high inbreeding depression on selfing.
- Cross-pollination is generally done by honey bees and wild bees.
- Bees and other insect pollinators frequently visit the flowers of carrot because of the presence of nectar in the flowers.
- To maintain the purity of the varieties, isolation distance of at least 1000 m is safe for seed production.
- Anthesis starts in the morning at about 7.00 a.m. and continues up to 11.30 a.m. depending on the prevailing weather conditions.

- Anthesis in a single umbel is completed by 7-9 days.
- Anthers deshise shortly after anthesis.
- The stigma becomes receptive five days after anthesis and it remains so for 8 days. Better fruit set is on 6 – 11 days after anthesis.
- Pollen becomes viable 3-4 days before stigma becomes receptive.
- Male-sterility in carrot, cytoplasmic-genic in nature is determined by an interaction between a cytoplasmic factor (*N* for normal cytoplasm and *S* for sterile cytoplasm) and a nuclear gene (dominant gene *Ms* for sterility and recessive gene *ms* for fertility).
- There are two systems of cytoplasmic-genic male sterility namely, brown anther type *(ba)* and petaloid type *(pt)*.
- The male-sterile line has the genetic constitution *SMsMs* or *SMsms* and the male fertiles are *NMsMs*, *NMsms*, *Nmsms*
- Male-sterility can be maintained by crossing with any male fertile genotype.
- For experimental hybridization purpose the umbel having maximum buds at the time of emasculation are selected and opened flowers along with young buds are removed. Maximum number of buds are emasculated and then bagged. The emasculated umbel is pollinated by enclosing an umbel from the intended male parent or by rubbing the umbel from the male parent on the emasculated umbel of the female parent.
- Commercially hybrid seed production of carrot is done utilizing the cytoplasmic-genic male-sterile lines.

Origin and taxonomy

Carrot was originated in Afganistan in Western Asia which is the primary centre of origin having a wide diversity. The secondary centres of diversity are perhaps Ethiopia and north America. From its centre of origin in Afganistan, it was taken by the Arabs to eastern Mediterranean and Spain in 12th century and north-western Europe in 14th century. Carrots were cultivated in Asia Minor in 10th and 11th centuries. It was introduced in India from Persia in 13th or 14th century. Carrot was first introduced into China in 13-14th century and Japan around 17th century. Cultivated carrot belongs to the family Apiaceae (=Umbelliferae), subfamily Apioideae, genus *Daucus*, section *Daucus*, Species *Daucus carota* L. Based on morphological and anatomical features, species *Daucus* is divided into five sections: Daucus L. (12 species), Anisactis DC. (three species), Platyspermum DC. (three spcies), Chrysodaucus Thell. (one species) and Meoides Lange (one species). All cultivated carrots, together with the wild ones which are known in Europe, belong to the species *Daucus carota* L. Most wild *Daucus* forms are found in the South-Western Asia and the Mediterranean, a few in Africa, Australia and America.

All wild forms found in Asia, Asia Minor, Japan and U.S.A. have the same chromosome number (2n=2x=18) as their European relatives. Neither polyploidy nor structural changes in the chromosome seem to have played any role in the differentiation of the species. Mutation plus selection rather than hybridization has been responsible for the development of the cultivated carrots. The western carotene-rich carrot has been originated from the eastern anthocyanin carrot as a mutant and the latter has probably been derived from anthocyanin containing forms of *D. carota* ssp. *carota*. Anthocyanin carrots are still grown in eastern countries together with the western orange cultivars.

Improved varieties

Oriental or tropical and European or temperate carrot varieties both are cultivated in India. Oriental or tropical varieties are mainly cultivated using different adaptable local cultivars under different name.

- Tropical or Oriental varieties: Pusa Kesar, Pusa Meghali, Sel- 233, HC-1, HC-2 and Hisar Gairic
- European or Temperate varieties: Chantaney, Early Nantes, Imperator, Nantes Half Long, Pusa Yamdagni, Zeno, Early Horn, Early Gem Imperator and Ooty-1

Cultural requirements

- The carrot is a cool-season crop though some of the tropical types tolerate quite high temperatures.
- Seeds germinate under a wide temperature range of 7.2 to 23.9^{0}C.
- The colour development and growth of the roots are affected by temperature.
- The top growth is optimum at a mean temperature of 16- 18^{0}C and the root yield and colour development is more at a temperature range of 20-22^{0}C.
- Top growth is reduced at mean temperature of 28^{0}C and roots become very strong flavored.
- The tropical types can form roots even at a higher temperature of 25°C.
- Cool night temperature is essential for carrot production in the tropics.
- Carrot can be grown on all types of soils.
- It thrives best on a deep, loose and loamy soil.
- For early crop, a sandy loam soil is preferred but for high yield, deep, loose, loamy soil is desirable.
- The long, smooth, slender roots desired for fresh market can successfully be grown on deep well-drained light soils.
- Carrots grown on heavy soils are more rough and coarse than those grown on light sandy soils.
- The pH of 6.7 is ideal and yield is extremely low at pH below this.

Land preparation

- Prepare the land in well advance with repeated ploughings (at least 4-5 ploughings) to a fine tilth.
- Remove all stubbles, weeds, etc from the land.
- Add well rotten organic manure preferably farm yard manure or compost @ 25-30 t/ha during land preparation and level the land properly.

Sowing time

- In the northern and eastern Indian plains, the tropical types should be sown from August to early-October and temperate types in October–December.
- In hills, sowing is done during March–July
- In south and central India, seeds are sown during January–February, June–July and October–November.

Sowing of seeds

- Carrot is direct seeded crop because swelled primary roots are economic plant parts.
- Roots may be misshaped in the transplanted seedlings.
- Seeds should be rubbed to remove the fine hairs before sowing.
- The seeds generally take 8-10 days for germination under optimum environmental conditions.
- Seeds are soaked in water for 12-24 hours prior to sowing to hasten germination.
- The seeds are sown on ridges or on flat beds.
- The seeds are sown as far as possible on the surface of the soil and covered by thin layer of soil.
- 6-8 kg seeds of the varieties of Asiatic types and 5-6 kg for the European varieties are sufficient for one hectare land.
- Seeds are generally sown by line sowing.
- Seeds should be sown not deeper than 1cm.
- After a fortnight of sowing, the seedlings are thinned out maintaining the requisite spacing.

Spacing

- Asiatic type : 30-45 cm × 22-25 cm
- Temperate type: 30 cm × 22-25 cm

Nutrient management

- Application of 25-30 tonnes/ha of well-rotten farmyard manure 10-15 days before sowing.
- Fresh cowdung manure should not be applied to a carrot field as it may result in branching of roots.
- Application of 90 kg N, 50 kg P_2O_5 and 80 kg K_2O/ha is optimum.

- Apply half of N and full dose of P_2O_5 and K_2O before sowing. Rest of N should be top dressed 30-35 days after sowing.
- Excess N application reduce root quality (decrease in sugar, dry matter, β carotene and vitamin C contents) and storability.

Irrigation

- After sowing, keep the ridges moist till the germination is completed.
- Irrigate at 8–10 days intervals before wilting of leaves starts.
- If pre-sowing irrigation is not given, a light irrigation soon after sowing is beneficial.
- The crop requires frequent irrigation at 4-6 days interval during autumn and spring and 8-10 days interval during winter.
- Insufficient moisture results in low yields.
- Excessive irrigation may decrease yield.
- Carrots should be irrigated before any wilting of leaves takes place.

Interculture and weed control

- Thinning should be done within 15 days of sowing for proper root development.
- The seedlings are thinned to a distance of 22-25cm.
- The seedlings grow slowly at first, and cannot compete with weeds.
- Remove weeds especially in the early stages.
- The soil should be hoed time-to-time to allow proper aeration.
- To control the weeds, spray Stomp @ 3.5 litres/ha immediately after sowing.
- Care should be taken that there is proper moisture at the time of application of Stomp and if moisture is less light irrigation can be given.
- Pre-emergent application of propazin 1.12 kg ai or Amiben 2.24 kg ai/ha is beneficial.
- During weeding and hoeing, the crown of the roots is often exposed to light which causes greening and lowers the quality of roots.
- For proper development of roots one earthing-up may be done during root formation.

Harvesting the crop

- The Asiatic (tropical types) carrots attain marketable stage of maturity when these are 2.5–4cm in diameter at the upper end.
- Delay in harvesting cause firm and splitling of roofs.
- Early carrots for market are pulled out manually when partly developed. Roots are normally dug out with a spade or khurpi when the soil is sufficiently moist.
- Give light irrigation before harvest.

Post harvest handling and storage

- After harvesting, the roots are washed, cleaned, graded and tied in bunches.
- Carrots are sometimes marketed with their tops attached to indicate firmness.
- Fresh carrots cannot be stored for more than 3–4 days under ambient condition.
- Long-term storage is possible without appreciable change in quality in cold storage.
- At temperature of 0°–4.5°C with 93–98% relative humidity carrots can be stored for 6 months.
- Storage at 98–100% relative humidity results in less decay than at 90–95%.
- At higher humidity moisture loss is considerably less and the roots remain crisper, firmer and of better colour.
- The optimum temperature for minimizing decay is 0°–1°C.
- There is no change in the nutritive and organoleptic properties of carrots stored in crates covered with perforated plastic films at 0°C and 93–96% relative humidity for 7–8 months. The stored carrots yield juice of acceptable commercial quality.

Plant protection measures

Physiological disorders

Splitting or cracking of roots

It is a major problem in many carrot-growing areas. Although the tendency of splitting seems to be controlled by genetic factors, a number of other factors may be involved.

Control measures

- The splitting is reduced by low N and increases with the amount of N in the soil increases.
- High concentration of ammonium compounds causes more serious splitting than by other forms of N.
- Carrot splitting is not affected by time of planting or variety.
- Wider the spacing, the greater is the propensity of splitting and large roots are more likely to split than small ones.
- Irrigation and herbicides do not significantly affect the splitting of roots.

Cavity spot

This disorder appears as a cavity in the cortex, in most cases the subtending epidermis collapses to form a pitted lesion. This is mainly due to deficiency of calcium and increased accumulation of potassium.

Control measures

- Increase in calcium level in the growing medium by liming.

Root forking

Roots get forked in poorly prepared soil having clods. Roots in touch of the lumps of undecomposed organic matter also get forked.

Control measure

- Land should be prepared finly so that the primary root may grow freely.

Disease

Alterneria blight (*Alternaria radicina, Alternaria dauci*)

This seed borne fungal disease first appears as dark brown lesion on the foliage with yellow edge and under favourable weather condition the entire foliage dried up and withers.

Control measures

- Seed treatment with thiram 3g per kg of seeds.
- Clean cultivation as infected debris provide the further source of infection.
- Spraying of mancozeb (0.25%) or chlorothalonil (0.2%) or propiconazole (0.1%) at regular intervals.

Leaf spot (*Cercospora carotae*)

The symptom of this fungal disease first appears as elongated lesions along the edge of the leaf segment, resulting in lateral curling. In dry weather, the spots are light tan in colour, whereas in humid weather the spots are darker in colour.

Control measures

- Seed treatment with thiram 3g per kg of seeds.
- Clean cultivation as infected debris provide the further source of infection.
- Spraying of mancozeb (0.25%) or chlorothalonil (0.2%) or propiconazole (0.1%) at regular intervals.

Insect pests

Carrot rust fly (*Psila rosae*)

The slender, yellowish-white maggot burrows into the roots, often causing it to become mis-shapen and subject to decay and render them unfit for consumption and marketing.

Control measures

- Spraying the crop with 0.1% dimethoate (Rogor) 2-3 times.

Yield

- Asiatic type : 25–30 tonnes/ha
- European type: 10–15 tonnes/ha.

Bulb Crops

18. Onion

Scientific Name: Allium cepa
Family: Alliaceae (Amarylidaceae)
Origin: Central Asia
Chromosome No. 2n = 16

Introduction

Alliums are among the oldest cultivated plant species. References to edible onion can be found in the Bible, Koran, and in the inscriptions of ancient civilizations of Egypt, Rome, Greece and China. Onion is an important vegetable crop grown in almost all parts of the world. The four major onion growing countries in the world are China, India, USA and Turkey and other important countries for onion are Italy, Egypt, Netherlands, Russia, Thailand, Indonesia, Korea, Japan and Brazil. It is used in almost all kinds of Indian cuisine. The important onion growing states in India are Maharashtra, Karnataka, Tamil Nadu, Andhra Pradesh, Gujarat, Punjab, Haryana, Rajasthan, Uttar Pradesh, Bihar and Madhya Pradesh.

Importance and use

- Onion is the most common and indispensable items in every kitchen as vegetable, spice and condiment.
- It is used fresh as salad or cooked as vegetable.
- It is a universal flavouring agent for different vegetarian and non-vegetarian dishes all over the world.
- Green leaves and mature and immature bulbs are eaten raw as salad or used in preparation of various types of curry.
- It is used in soups, sauces and for seasoning different foods.
- Dehydrated onion powder is in great demand which reduces transport cost and storage losses.
- It is also used for making different pickles.
- White onions are generally used for dehydration.
- Dried onion flakes can be reconstituted by cooking in water.
- Intact bulb is odourless but if it is cut it produces a distinctive odour.
- A colourless, odourless, water-soluble amino compound known as alliin is present in uninjured bulbs. On injury of the cells, an enzyme, allinase comes in contact with alliin and causes its breakdown into sulphur containing product allicin (diallyl thiosulfinate) which gives typical odour of fresh onion. It is unstable and undergoes non-enzymatic degradation leading to formation of mono, di and trisulfides present in the volatile oil.

- The pungency of onion is due to a volatile oil known as allyl propyl disulphide and its concentration in the bulb is highest just before the fall of top in the field.
- The outer skin colour is due to the presence of a flavanoid called quercetin.
- The Indians prefer red and pungent types while in the markets of Japan, Europe and America mild, juicy and brown or yellow varieties are mostly preferred.

Nutritional value

- Onion bulbs are rich in minerals like calcium (180 mg/100 g fresh) and phosphorus (50 mg/100g fresh).
- The bulbs are fairly rich in carbohydrate (11g/100 g fresh).
- It also contains good amount of protein (1.2 g/100 g fresh) and vitamine A (35 IU/100 g fresh).
- Dry matter content varies from 10.6 to 14.8% and on average white cultivars contains more dry matter percentage than red ones.
- Steam distillation of 100 g fresh bulb yields 0.002 to 0.03 g oil containing methyl and propyl compounds.

Medicinal value

- Onion bulbs contain different flavonoids, the plant secondary metabolites like, quercetin (28.4-48.6 mg), kaempferol (0.2 mg), myricetin (0.1 mg), luteolin (0.1 mg), etc. per 100 g fresh which are potential antioxidants.
- The highly biologically reactive organosulfur compounds responsible for the characteristic odour and taste have potential influence on phase-I (carcinogen activation) and phase-II (carcinogen detoxification) enzymes.
- Consumption of onion result in fewer tendencies to form blood clots and lower the levels of cholesterol and lipoproteins associated with heart disease.
- Onions are diuretic.
- It is applied on bruises, boils and wounds.
- It relieves heat sensation.
- Bulb juice is used as smelling on hysterial fits in faintness.
- It also relieves insect bites and soar throat.

Adaptation in India

Onion is an outstanding example of adaptation carried through from the very early times before the Christian era. Onion has got a Sanskrit equivalent 'Palandu' mentioned in "Apastamba Dharma Sutra-I" during the Sutra period dated back 800 B.C. to 300 B.C. which signifies its very early introduction in India. It is also mentioned in "Charaka-Samhita", a famous early medical treatise of India, around 6th century A.D. The popularity of onion in India however, has been promoted by Mughals. Onion is originally a native of Central Asia, the temperate region, with perennial/biennial habit and long day bulbing character. In North India, rabi (winter) onion production has been developed with such biennial types. Huge genetic diversity available in the rabi onion in northern and eastern India have been utilized mainly by mass selection and other population improvement schemes.

The concentration of onion cultivation in peninsular and western India resulted in the evolution of tropical and short day cultivars. In western Maharashtra and Gujarat two crops, one in kharif and other in rabi season are regularly raised. The adaptations to hardy conditions of high temperature, high rainfall and short photoperiod condition have resulted in the development of wide diversity of tropicalized cultivars in western and peninsular India. These wide arrays of germplasm have been improved mainly by selection to develop varieties for kharif season only or for both rabi and kharif seasons. Vegetatively propagated multiplier onion has been established mainly in Tamil Nadu.

Commercial varieties of onion generally fall under *Allium cepa* (common onion) and *Allium cepa* var. *aggregatum* (multiplier onion or potato onion). Large number of land races and local cultivars has been developed in India through selection of genotypes from the wide germplasm base developed due to wide arrays of cultivation commensurate with out-crossing breeding behaviour.

Almost all varieties grown in India are short day or intermediate type cultivars. Varieties belonging to common onion in India are grouped according to growing season viz., winter or rabi onion and kharif onion. Rabi onions are basically short day or intermediate type cultivars and generally require 12-13 hours day length for bulb formation and development. Rabi onion can broadly be classified according to the skin colour viz., red, white and yellow. Kharif onions require comparatively shorter day length of 10-11 hours for bulb formation and development. True long day varieties like, Brown Spanish require day length of more than 14 hours and lower temperature for proper development of bulbs.

Multiplier onions are hardy and early maturing types grown for closely packed clusters of small sized bulblets. This group of onion generally does not produce seeds and thus are propagated vegetatively by bulblets.

A good number of local cultivars are grown in India in different names like, Sukhsagar, Chhanchi, etc. in West Bengal; Niphad, Nasik Red, Nasik White Globe, Poona Red, Bombay Red, Bombay White, Mathewad, Shirwal, Kagar, Peth, Dhulia, Fangira, Lohru, etc. in Maharashtra; Patna Red, Patna White in Bihar; Bellary Red, Bellary White, Telagi Red, Bangalore Rose, Chikballpur Local, Krishnapuram Local, Kempgende in Karnataka; Cuddalorae, Podusa, Dindigal Red, Mutlore, Natu, Pallu Vengayam in Tamil Nadu; Rampuram Red in Andhra Pradesh; Saurastra Local in Gujarat; Kalyanpur Red, Kalyanpur Lalgol in Uttar Pradesh; Panipat Local, Rajpur Local, Hisar Local, Bahadurgarh Local in Haryana; Ludhiana Local in Punjab; etc.

Breeding behaviour

- The flowers are borne in simple umbels at the apex of a floral stem which is hollow when mature and the number of seed stems varies from 1-20 or more.
- The number of flowers per umbel may vary between 50 and 2000 depending on the varity, time of planting, size of bulb, etc.
- Onion is basically a cross pollinated crop due to protandry and wide span of flower opening in the umbel.
- The flowers are protandrous in nature and stigma becomes receptive one or two days after anthers have all dehisced.
- The delay in receptivity of stigma reduces self fertilization of single flower.
- The anthers of inner whorl dehisce first and it occurs between 9.00 a.m and 5.00 p.m.
- An inflorescence may continue opening for 2 weeks or more and a plant may be in bloom for more than 30 days.
- It is pollinated chiefly by honey bees (*Apis* sp.) through transferring the pollens among the flowers within an umbel and between different plants.
- Blow-flies are also excellent pollinators as they go from flower to flower.
- Male sterility in onion, cytoplasmic-genic in nature is determined by an interaction between a cytoplasmic factor (*N* for normal cytoplasm and *S* for sterile cytoplasm) and a nuclear gene (*ms* for sterility and *Ms* for ferlility).
- The male sterile line has the genetic constitution *Smsms* and the male fertiles are *SMsMs, SMsms, NMsMs, NMsms, Nmsms.*
- Genetic constitution of the male sterility maintainer is *Nmsms*.

Origin and taxonomy

Onion has been cultivated for 5000 years or more but does not exist as a wild species. Onion is originally a native of Central Asia. It is thought to have domesticated in the mountainous regions of central Asia comprising Turkmenia, Uzbekistan, Tadzhikistan, Afganistan, Pakistan and north Iran. Its secondary centre of origin was in the Mediterranean region. Its cultivation had spread in large areas of Western Asia, extending from Palestine to India. While place of origin of onion is still a mystery, there was a reference to onion as food, medicine or religious objects date back to 3200 B.C. in Egypt. It was mentioned by the Greek and Roman and Indian authors in 600 B.C to 79 A.D. It was being cultivated in India about 600 B.C.

Onion entered northern Europe through Italy and North Africa in 500 A.D. It was introduced in the USA by the Spanish explorers around 1625.

Onion is classified under the class Monocotyledon, Super order Liliflorae, Order Asparagales, Family Alliaceae, Tribe Allae and Genus *Allium*. The genus *Allium* has more than 500 species and the eight cultivated species used as vegetable are *Allium cepa* (common onion), *Allium cepa* var. *aggregatum* (multiplier onion or potato onion), *A. cepa var viviparum* (top onion or tree onion) , *A. cepa* var. *ascalonicum* (shallot), *A. sativum* (garlic), *A. schoenoprasum* (chieve), *A. fistulosum* (welsh onion or Japanese bunching onion), *A.ampeloprasam* var. *porrum* (leek), *A. tuberosum* (Chinese chieve) and *A. chinensis* (rakkyo). There are several closely related wild *Alliums* crossable with onion are *A. vavilovii*, *A. oschaninnii*, *A. pskemense*, *A. galanthum*, etc.

Open pollinated improved varieties

- Winter or rabi onion: Pusa Red, Pusa Ratnar, Pusa Madhabi, Punjab Selection, Arka Niketan, Arka Bindu, Agrifound Light Red, Udaipur 101, Udaipur 103, Sukhsagar, Punjab Red Round, Agrifound Rose, Punjab Naroya, Punjab Selection, Kalyanpur Red Round, Patna Red, VL Piaz-3, Agrifound Rose, Arka Bindu, Hisar-2, etc.
- Kharif onion: Niphad-53, Arka Kalyan, Agrifound Dark Red, Baswant-780.
- Red skinned varieties: Pusa Red, Pusa Ratnar, Pusa Madhabi, Punjab Selection, Arka Niketan, Arka Bindu, Agrifound Light Red, Udaipur 101, Udaipur 103, Sukhsagar, Punjab Red Round, Agrifound Rose, Punjab Naroya, Punjab Selection, Kalyanpur Red Round, Patna Red, VL Piaz-3, Hisar-2, Agrifound Rose, Arka Bindu, Niphad-53, Arka Kalyan, Agrifound Dark Red, Baswant-780, etc.
- White or silver skinned varieties: Pusa White Flat, Pusa White Round, Punjab-48, Udaipur – 102, N-257-9-1, Phule Safed, etc.
- Yellow skinned varieties: Early Grano, Brown Spanish, Bermuda Yellow, Arka Pitambar,Phule Suvarna, etc.
- Varieties of multiplier onion: CO-1, CO-2, CO-3, CO-4, MDU-1, Agrifound Red, etc.
- Varieties with small bulb: Agrifound Rose, Arka Bindu
- Varieties suitable for export: Agrifound Rose, Arka Bindu, Niphad-53, Nasik Red, Agrifound Light red, Agrifound Dark Red, Arka Niketan, Pusa Madhbi, Pusa Red, Patna Red, etc.
- Varieties suitable for dehydration: Pusa White Flat, Pusa White Round, Punjab-48, Udaipur-102, N257-9-1, etc.
- Public sector hybrids: Arka Kirtiman, Arka Lalima, MOH-2.

Cultural requirements

- Onion is basically a cool season crop but can be grown in a wide range of climatic conditions under both tropical and subtropical conditions.
- Photoperiod and temperature both influence bulb development and flowering however, photoperiod is comparatively more important for bulb formation and temperature is more important for flowering.

- Almost all varieties grown in India are either short day or intermediate type varieties.
- The rabi onion varieties require relatively high temperature and 12-13 hours day length for proper bulb development.
- The kharif onion varieties require relatively low temperature and 10-11 hours day length for proper bulb development.
- The long day onion varieties require more than 14 hours day length and lower temperature for proper development of bulbs hence, can only be cultivated in the hill region of India.
- Long day varieties do not bulb under short day conditions where as short day varieties if planted under long day condition will develop early bulbs.
- It grows well in mild climate without extreme of high or low temperatures.
- The optimum temperature for good vegetative growth before bulbing is between 13-24°C with 70% relative humidity.
- Ideal temperature for bulb development is 15-25°C.
- The plants are hardy and can withstand freezing temperature.
- In the rabi season a sudden rise in temperature may result in early bulb maturity and smaller sized bulbs.
- Very low temperature at the vegetative phase results early bolting.
- Generally low temperature influences the development of seed stalk.
- It does not thrive well in places where average rainfall exceeds 75-100 cm in the monsoon periods.
- Onion grows well on light loam, sandy loam or clay-loam, deep friable and fertile soil rich in organic matter.
- In heavy soil bulb development is restricted and crops mature late compared to light soils.
- Onion is sensitive to high acidity and alkalinity and ideal soil pH ranges between 5.8 and 6.5.

Nursery management

- Seed beds are prepared finely, well drained, 15-20 cm raised, 1.0m wide and of convenient length.
- Fine and fully decomposed farmyard manure or compost @ 3-4 kg/m^2 should be well mixed to the beds.
- Drench the beds with formaldehyde (4.0%) or captan @ 2-3g/litre and cover with polythene sheet for 7 to 10 days to avoid damping off disease.
- 8-10 kg seeds/ha is required for open pollinated varieties of common onion.

- About 1000 square metre area is required to produce seedlings for one hectare.
- Before sowing treat the seeds with thirum or captan @ 2-3g/kg of seeds to avoid damping off.
- Sowing of seeds in lines spaced at 5-7cm distance is recommended.
- The seeds after sowing should be covered with finely sieved farm yard manure or compost or vermicompost followed by light watering by rose can.
- Cover the seed beds with straw or dry grass till seed germination and sprinkle water regularly.
- Seedlings should be drenched with 0.2-0.3% thiram or captan twice, 15 days and 30 days after sowing.
- Side dressing with single super phosphate and muriate of potash and spraying with 0.5% urea produce healthy seedlings.
- Time to time hoeing, weeding and irrigation are required to raise healthy seedlings.
- Seedlings may be raised under low poly tunnels having side open to protect them from heavy rains.
- The seedlings are ready within 6-7 weeks after sowing for kharif crop and 8-9 weeks for rabi crop.
- Under-aged seedlings do not stand well in the field and over-aged seedlings may bolt early.

Sowing time

Onion is grown in two distinct seasons viz., kharif and rabi in the plains of India although sowing and transplanting season vary with the region.

- Early kharif crop: April-May (Tamil Nadu, Karnataka, Andhra Pradesh)
- Kharif crop: May-June (Tamil Nadu, Karnataka, Andhra Pradesh, Maharashtra, part of Gujarat, Rajasthan, Punjab and other parts).
- Late kharif crop: August – September (Maharashtra and part of Gujarat).
- Early rabi crop: September (West Bengal, Orissa).
- Rabi crop: September-October (Tamil Nadu, Karnataka, Andhra Pradesh).
- Rabi crop: October-December (West Bengal, Orissa and North and north-eastern part).
- Rabi crop: November-December (Maharashtra and part of Gujarat).

Sowing for rabi crop in the hills is done in September-October and for summer crop with long day varieties sowing is done in October end to early November.

Land preparation

- Prepare the land in well advance with repeated ploughings (at least 4-5 ploughing) to a fine tilth.
- Remove all stubbles, weeds, etc. from the land with proper leveling.
- Add well rotten organic manure preferably farm yard manure or compost @ 20-25 t/ha during land preparation and level the land properly.
- Divide the field into beds and channels.

Transplanting / direct planting

- Transplant the seedlings in the main field when they are 6-7 weeks old for kharif crop and 8-9 weeks old for rabi crop.
- The variety with small bulb like Agrifound Rose can be directly seeded by broadcasting or in rows in the main field with 20-25 kg seeds per hectare.
- Multiplier onions are directly planted in the main field with 1.0-1.2 tonnes of bulblets of 1.5 to 2.0 cm size per hectare.
- Irrigate the seed beds before lifting of seedlings to facilitate easy pulling of seedlings.
- Late afternoon hours are best for transplanting the seedlings in the main field for better establishment.
- Irrigate the field lightly just after transplanting or direct seeding.

Spacing

- Common big size onion :15cm × 10cm
- Small pickling onion :8 cm × 5cm
- Multiplier onion :30 cm × 15cm

Nutrient management

- Soil should be liberally manured and fertilized for getting good yield.
- Organic manure should be applied atleast one month before transplanting the seedlings.
- Apply a general fertilizer dose of 100 kg N, 60kg P_2O_5 and 100kg K_2O per hectare for open pollinated improved varieties.1/2 N along with entire phosphate and potash fertilizers should be given as basal and rest N should be top dressed in two split doses at 30 and 45 days after transplanting.
- Complete the top dressing before bulb initiation as delayed application results in thick neck or doubles.
- To get better results apply 3 tonnes of vermicompost/ha along with 50% of recommended dose of NPK ferlilizers.

- Application of micronutrients particularly copper sulfate 4 g/l and boron (borax 3 g/l) enhance bulb yield.
- Apply vermicompost @ 3.0 t/ha as broadcasting during final land preparation along with 50% of recommended dose of NPK ferlilizers.
- Apply *Azotobactor* or *Azospirillium* @ 800 g/ha as seedling root dip.
- Prepare a suspension of bacterial inoculants in sufficient water.
- Dip the seedlings in the bacterial suspension in shade for 2-3 hours in the afternoon hours.
- Transplant the treated seedlings immediately.

Irrigation

- Onion being a shallow rooted crop requires light irrigations.
- Irrigate the field immediately after transplanting.
- Soil moisture condition should be kept optimum.
- Frequent and light irrigation at weekly interval is best (2 acre-inch per irrigation).
- Depending on rains 8-10 irrigations are enough for kharif onion.
- Late kharif or early rabi onion generally require 12-15 irrigations.
- Rabi onion generally requires 15-20 irrigations.
- Irrigatin must be given at bulb formation stage.
- For rabi onion, stop irrigation when tops mature and start falling.
- For kharif onion, stop irrigation 10 days before harvesting.

Interculture and weed control

- Onion is a poor competitor of weeds hence, weeds must be kept down to get good yield.
- Onion is a shallow rooted crop and planted closely and there is a chance of root pruning by cultivation.
- Two light hoeings at the early stages of growth give good weed control.
- Chemical weed control gives better results.
- Apply Basaline @ 1 litre/ha or Stomp @ 3.35 litre/ha immediately after transplanting and before first irrigation along with one hand weeding is the best.

Harvesting the crop

- Harvest the rabi crop one week after 50-70% neck fall stage.
- In kharif crop, since tops do not fall harvest the bulbs soon after the color of leaves changes to slightly yellow and tops starts drying.

- For green onion harvest the tender plants 75-80 days after transplanting.
- If soil is light harvest the crop by hand pulling or use hand implements or modified potato digger.

Post harvest handling and storage

- After harvest cure the bulbs in the field for big, small and multiplier onions.
- Purpose of curing or drying is to remove excess moisture from the outer skin and neck of the bulb which helps in minimizing shrinkage of bulb, developing skin colour and reducing the disease infection during storage.
- The rabi onions are cured in the field for 3-5 days after tops are cut leaving 2.0-2.2.5 cm above the bulb and then the bulbs are again cured in shade for 7-10 days to develop the skin and remove field heat.
- The kharif onions are generally cured for 3-4 days along with the tops.
- Cold storage temperature of 0°C with 65-70% relative humidity is ideal for 6-8 months long storage in dormant condition and reasonably free from decay.
- Onion bulbs can be stored at 24.0 to 30.0°C for 5-6 months without excessive loss in weight and sprouting.
- Dormancy of onion is longer both at lower (0°C) and higher (30°C) than at moderate temperature of 10-15°C.
- For low cost storing, make a hut like structure under the shade with thatched roof and sides covered by bamboo sticks with provision of good air circulation.
- Store onion in these sheds by spreading them on racks or tiers having 2-3 layers of bulbs.
- Inspect periodically and remove damaged, diseased and sprouted bulbs immediately from the stock.
- To check sprouting apply pre-harvest spray of maleic hydrazide 2500ppm, 75-90 days after transplanting in kharif onion only.
- Irradiation of bulbs with very low dose of 4-9 krad gamma rays effectively controls sprouting of onion.
- Jute mesh bags are common for packaging onion in India however, in foreign countries plastic netted bags are mostly used.

Plant protection measures

Disease

Damping off (*Pythium*, *Rhizoctonia* ,*Phytophthora*, etc.)

It is very destructive fungal disease especially in nursery beds of kharif and early rabi crop. Seed rotting and both pre-emergence and post-emergence damping off of seedlings causing thin or irregular stand frequently occur.

Control measures

- Treat the seeds with thiram/ captan / mancozeb @ 3.0g /kg of seeds.
- Drenching the nursery beds 7 days before sowing with thiram/ captan or any copper fungicide @ 3g / litre of water.
- Cover the beds with transparent polythene sheets before sowing and left to open sun for at least 20 days, known as soil solarization.
- Spraying the young seedlings with 0.25% mancozeb or metalaxyl-mancozeb (0.25%) or cymoxanil-mancozeb (0.25%) or fosetyl-Al (0.25 %) with a sticker at weekly intervals.
- Provision of proper drainage to the beds.
- Remove the affected seedlings from beds as soon as the symptoms are visible.
- Avoid flooding the beds to check spread of the disease.

Purple blotch (*Alternaria porri*)

The common symptoms of this important fungal disease occur on leaves and flower stalks as small, shunken, whitish flecks with purple coloured centre. Finally the leaves and stems fall down from point of attack. The most favourable temperature for disease development is 28-30°C temperature with 80-90% relative humidity.The disease appears mostly in the kharif and seed crop.

Control measures

i. Give atleast three summer ploughing to reduce disease severity.
ii. Spray the crop with mancozeb @ 2.5g/l or chlorothalonil @ 2g/l or Difenconazole@ 0.5g/l with a sticker 4-5 times starting from 15 days after transplanting.

Stemphylium blight (*Stemphylium vesicarium*)

This foliar fungal disease is particularly serious in the seed crop. Infection occurs in the form of small yellow to pale orange spots or streaks on one side of leaves or flower stalks causing fall over of seed stalks or leaves.

Control measures

- Spray the crop with mancozeb @ 2.5g/l or chlorothalonil @ 2g/l or difenconazole @ 0.5g/l or propiconazole (0.1%) with a sticker 4-5 times starting from 15 days after transplanting.

Basal rot (*Fusarium oxysporum f. sp. cepae*)

Common symptoms of this fungal disease are wilting and rapid dying-back of leaves from tip with the roots turn pinkish as the plants approach maturity. The bulbs become soft and a semi watery decay is found advancing from base of the scales leaves.

The infection may also be carried over to the storage where bulb rotting takes place in the same manner.

Control measuresd

- Adopt long crop rotation (at least 5 years) excluding onion and related crops like, garlic, leek, etc.
- Treat the seeds with thirum @ 3g/kg of seeds.
- Incorporation of green manuring is also effective.
- Application of lime to raise the soil pH to 7.2
- Soil application of *Trichoderma viride* @ 3 kg/ ha along with bio-manure is effective.
- Adopt mixed cropping with sugarbeet, maize or wheat in the infected field.

Downy mildew (*Peronospora destructor*)

This fungal disease is prevalent in cool and humid weather condition. On the surface of the leaves or flower stalks, violet growth of this fungus is noticed which later becomes pale green yellow and finally leaves or seed stalks collapse. In storage, bulbs of infected plants become soft, shriveled and amber in color.

Control measures

i. Adopt crop rotation with 4- year break in onion or garlic cultivation.
ii. Control weeds and keep the field clean.
iii. Remove the primary infected onion plants.
iv. Post harvest heating of mother bulb at 41°C to destroy the pathogen.
v. Spraying the young seedlings with 0.25% mancozeb or metalaxyl-mancozeb (0.25%) or cymoxanil-mancozeb (0.25%) or fosetyl-Al (0.25 %) with a sticker 4-5 times at an interval of 10 days.

Insect pests

Thrips (*Thrips tabaci*)

It is the most injurious pest of onion. The nymphs and adults are slender, fragile and blackish or yellowish in color. Both of them are found lacerating the leaf epidermis and lapping the exuding sap, producing silver white patches known as 'silver top'. Severe infestation resulted drying of leaves from tip to the base gradually.

Control measures

- Adopt crop rotation.
- Destroy and burn the weeds and plant debris on which adults hibernate.
- Apply insecticides like acetamiprid @ 0.5g/l observing the infestation of thrips after transplanting.

- Spray the crop with malation @ 1ml/l or endosulphan @ 1.5ml/l or Nuvacron @ 1ml/l of water 4 times at 15 days interval starting after transplanting.

Garlic mite (*Aceria tulipae*)

The mite colonize along the mid rib vein of dorsal surface of leaf where they feed and multiply. All the active stages viz., adults, nymphs and eggs are found on the affected site. The leaf of infected plant get curled and twisted resulting stunted growth of the plant. Severely infested crop produce very low yield.

Control measures

- Soaking of seed material into 0.05% dicofol solution for ten minutes before planting is effective.
- Spraying of infected standing crop with wettable sulphur @ 2 g/l of water or 0.05% dicofol is effcetive.

Onion fly (*Hylemia antigua*)

Adult appears like house fly. Maggots are small, white and devoid of legs. Maggots enter the bulbs through roots and attack the tender portions. Infested plants turn yellowish-brown and finally dry up.

Control measures

- Follow crop rotation.
- Apply granular insecticide like Phorate 10g @ 25kg/ha.
- Spray the crop with malathion 1ml/l or Thiodan 1.5ml/l.

Yield

- Big sized common onion in rabi season: 25-30 t/ha.
- Big sized common onion in kharif season: 20-25 t/ha.
- Small sized common onion in rabi season: 16-20 t/ha.
- Multiplier onion: 15-18 t/ha.

19. Garlic

Scientific Name: Allium sativum
Family: Alliaceae (Amarylidaceae)
Origin: Central Asia
Chromosome No. 2n = 2x =16

Introduction

Alliums are among the oldest cultivated plant species. The most widely cultivated *Alliums* next to common onion is garlic. Wild ancestor of garlic, *Allium longicuspis*

Regel was known in Egypt as early as 3000 B.C. and also to the ancient Greeks and Romans. The five major garlic growing countries in the world are USA, Egypt, China, Korea and India and other important countries for garlic are Thailand, France, Spain and Yugoslavia. Garlic is used as a spice and condiment throughout India. The important garlic growing states in India are Madhya Pradesh, Gujarat, Orissa, Maharashtra and Uttar Pradesh.

Importance and use

- Garlic is a common spice and condiment crop.
- It is a flavouring agent particularly for different non-vegetarian dishes.
- The flavour of garlic cloves is more powerful than that of the other bulbs crops.
- Green leaves are used in chutney and pickles.
- It is used in soups, sauces and for seasoning different foods.
- Dehydrated garlic is in great demand particularly for export.
- A colourless, odourless, water-soluble amino compound known as alliin is present in uninjured garlic cloves. On injury of the cells, an enzyme, allinase comes in contact with alliin and causes its breakdown into sulphur containing product allicin (diallyl thiosulfinate) which gives typical odour of fresh garlic. It is unstable and undergoes non-enzymatic degradation leading to formation of mono, di and trisulfides present in the volatile oil.
- Garlic oil is an item of commerce.
- Fresh garlic cloves on steam distillation yield 0.1-0.5% essential oils.
- The chief constituents of the oil are diallyl disulfide, diallel trisulfide, allyl-propyl disulfide and a small quantity of diethyl disulfide and diallyl disulfide possess the true garlic odour.
- Garlic of more than 40mm size is exported to Quatar, Saudi Arabia, UAE, Baharin, Kuwait, Bangladesh and Sri Lanka.
- Demand now from all over the world is only for bigger cloved (40-60 cm diameter) with 10-15 cloves in each bulb and such big cloved garlic are the long day cultivars suitable for growing in the hill condition.

Nutritional value

- Garlic has higher nutritive value than other *Allium* bulbs crop.
- Garlic is the rich source of carbohydrates (29 g/100 g fresh), protein (6.3 g/ 100 g fresh) and phosphorous (310 mg/100 g fresh).
- It is a rich source of thiamine (0.16 mg/100g fresh) and riboflavin (0.23 mg/ 100 g fresh).
- It contains fair amount of calcium (30 mg/100g fresh).
- Ascorbic acid content is quite high in green garlic.

Medicinal value

- The highly biologically reactive organosulfur compounds responsible for the characteristic odour and taste have potential influence on both phase-I (carcinogen activation) and phase-II (carcinogen detoxification) enzymes.
- Garlic cloves contain different flavonoids, the plant secondary metabolites like, quercetin, kaempferol, myricetin , luteolin , etc. which are potential antioxidants.
- Allicin (diallyl thiosulfinate), which has hypocholesterolemic action, reduces the cholesterol concentration in human blood.
- Garlic consumption also reduces serum lipids perhaps by decreasing their synthesis and/or increasing their excretion through the intestinal tract.
- Garlic oil also reduces blood sugar and this hypoglycemic effect may be due to increased glycogenesis in the liver and better utilization of glucose in the peripheral tissue.
- Garlic is described as hot stimulant, carminative and anti-rheumatic.
- Garlic oil is a powerful anticeptic mainly due to presence of allicin.It is used as vermifuge for expelling round worms and has long been recommended for cure of a number of ailments viz., wounds, ulcers, pneumonia, bronchitis, atopic dyspepsia and gastro-intestinal disorders.
- Inhalation of garlic oil or garlic juice has generally been recommended by doctors in cases of pulmonary tuberculosis, rheumatism, sterility, impotency, cough and red eyes.
- Garlic is a gastric stimulant and helps in the digestion of food and absorption of nutrients from it.

Adaptation in India

Garlic from its original home in Central Asia reached India long ago. Under domestication, it has become exclusively vegetatively propagated crop, mostly through cloves and bulbils, in some cases. During long period of garlic cultivation in India many cultivars have been developed mainly due to vegetative mutation and human selection. However, vegetative propagation has considerably restricted the variability needed for the development of large number of cultivars as seen in onion.

The classification of garlic is mainly based on bolting and non-bolting habit although most of the garlic types do not bolt naturally. In most of the garlic types grown in India bulbs are formed under short day condition and critical day length being 12.00 hour. Long day cultivars are suitable for high altitude condition. Bulbs are generally silvery-white in colour although pink coloured bulbs are also found. Bulbs of all the commercial varieties contain multiple cloves ranging from 20-40 of 3.5-5.0 cm size with creamy flesh colour. However, single cloved bulb is also found in some local cultivars.

A number of local cultivars are grown in India in different names like, Fawari, Regalle Gaddi, Madrasi, Tabiti, Creole, Jamnagar, Katki, etc.

Breeding behaviour

- Most of the garlic types have non-bolting habit, hence are propagated vegetatively through cloves and bulbils.
- Chances of getting genetic variation through genetic exchange by normal cross-fertilization are most of the times not possible in garlic.
- Vegetative mutation during long period of garlic cultivation is mostly responsible of developing many cultivars.
- *In vitro* culture of somaclonal variants opens up the possibility genetic improvement of garlie.
- Clones with bolting habit with fertile seeds have been found in Japan.
- In bolting types of clones, the usual sexual breeding methods are followed as in other vegetatively propagated vegetables.

Origin and taxonomy

Garlic had its original home in Central Asia particularly the high mountain plateaus of Pamir and Tian Shan. Its wild ancestor, *Allium longicuspis* Regel is native to Central Asia. From the extended regions of Turkmenia in the south to the north east-ward in Pamir and Tien Shan regions of Central Asia, it spread to Mediterranean region in Prehistoric times, where its virtues are still cherished perhaps more than in any other region of the world. It was known in Egypt as early as 3000 B.C. and also to the ancient Greeks and Romans. It spread to the eastern Mediterranean region, Caucasus and North Africa which are considered to be the secondary centre of origin. It reached China and is grown in China and India since a long time. However, its introduction in Europe and the rest of Asia was much later. It was used in England as early as first half of the 16th century. The Spanish and Portuguese settlers introduced garlic in North and South America at the beginning of the 15th century. The Anglo-Saxon settlers in Australia and Central and Southern Africa started its cultivation in the beginning of the 19th century.

Garlic *Allium sativum* L. is classified under the class Monocotyledon and belongs to the section *Porrum* of the family Alliaceae. The classification of garlic is mainly based on bolting and non-bolting habit however, controversy remains on the name of the subspecies for bolting and non-bolting type. In one classification, the non-bolting species is called *Allium sativum* var. *sativum* and the two subspecies *Allium sativum* var. *ophioscorodon* and *Allium sativum* var. *scordoprasum* come under bolting type. In another classification the non-bolting subspecies is *Allium sativum* var *vulgare* and the bolting subspecies is *Allium sativum* var. *sagitatum* and each subspecies is divided into different ecotypes. Three ecotypes namely, Continental, South Russian and East Mediterranean come under non-bolting subspecies and another three ecotypes viz., Asian, Caucasian and Seaside belong to the bolting subspecies.

Clonally selected improved varieties

- Varieties with silvery-white bulb: Agrifound White (G-41), Yamuna Safed (G-1), Yamuna Safed-2(G-50)), G-282, Agrifound Parvati, Fawari, Rajalle Gaddi, Madrasi, Tahiti, Jamnagar, Prinkindel, Sweta, HG.1, HG-6, Pusa Sel-10, LCC1, ARU-52, Almora, etc.
- Varieties with pink bulb: Godavari
- Varieties suitable for export: Agrifound Parvati, Yamuna Safed, Agrifound White, Jamnagar Ladwa, Malek, etc.
- Varieties suitable for hill: Agrifound Parvati

Cultural requirements

- Garlic is basically a frost-hardy plant, requiring cool and moist period and relatively dry period during bulb maturity.
- Garlic generally grows in mild climate without extremes of high or low temperatures even though it can be grown under a wide range of climatic condition.
- Photoperiod and temperature both influence bulb formation development.
- Bulb development is better under cool temperature condition.
- Exposure of dormant cloves or young plants around 13-20° C depending on the varieties for 1-2 months hastens bulbing.
- Garlic is grown as a rabi crop in the plains of India.
- Both long day and short day cultivars are present in garlic.
- Almost all varieties grown in Indian plains are short day type varieties.
- Some long day type varieties are grown in hill condition.
- The short day garlic varieties grown in Indian plains require about 12.00 hours day length for proper bulb development.
- The long day garlic varieties grown in the hills require about 14 hours day length and lower temperature for proper development of bulbs.
- Garlic are grown on wide ranges of soil such as sandy loam, silt loam and heavy clay soils.
- Well drained loam or medium-loam soil rich in organic matter is ideal for garlic.
- Development of bulbs is not normal when garlic is grown in sandy or heavy clay soil.
- In heavy soil bulb development is restricted and crops mature late compared to light soils.

- Garlic is sensitive to high soil acidity and alkalinity and ideal soil pH ranges between 6.0 and 7.0.
- High alkaline and saline soils are not suitable for garlic cultivation.

Propagation

- Garlic is vegetatively propagated by single cloves.
- Bulbils produced in some varieties are also used as planting material.
- Micropropagation through *in vitro* culture is employed to produce disease free clones.

Planting time

The time of planting differs from region to region.

- August-October: Madhya Pradesh, Karnataka, Maharashtra and Andhra Pradesh.
- September- November: Northern parts of India.
- October-November: Gujarat, West Bengal and Orissa
- March-April and October-November: Hills

Land preparation

- The field is ploughed to fine tilth by giving four to five ploughing with sufficient intervals between two ploughing.
- Ploughing should be shallow, not more than 8-10 cm depth.
- Planking should be done for leveling.
- The width of plots should be such that manual intercultural operations are possible.
- The normal width of a bed should be about 1.8 m and length may vary accordingly to the level of land.

Direct planting

- Cloves of 8-10 mm diameter are the best for planting.
- Bigger cloves from outer side of the bulb should be selected.
- Long and slender cloves present in the centre of the bulb should not be selected because such cloves give poorly developed bulbs.
- About 500 kg cloves are required for planting in one hectare land.
- A population of 3, 33,000 plants generally remain in one hectare land.
- Irrigate the field lightly just after direct seeding.

Methods of planting

- Dibbling: It is one of the commonly adopted planting method .Fields are divided into small plots convenient for irrigation. Cloves are dibbled 5-7.5cm deep keeping their growing ends upward.
- Furrow planting: It is also a common method of planting garlic where shallow furrows are opened with hand hoe. The cloves are dropped in the furrows and then are covered lightly with loose soil.
- Broad casting: Cloves are scattered evenly in well leveled field and then are covered by harrowing and field is divided into small plots for irrigation. This method is mostly followed in Gujarat.
- Planting of bulbils: In Lahaul and Spiti district of Himachal Pradesh bulbils are first sown in nursery and then transplanted after 45 days in main field at proper distance.

Spacing

The planting distance depends on varieties and different cultivation region of India. Wider spacing generally results in thicker necks of the bulbs.

- Big bulb and clove :15 cm × 10 cm
- Small bulb and clove :10 cm × 8 cm

Nutrient management

- Soil should be liberally manured and fertilized for getting good yield.
- Apply 15-20 tonnes of farmyard manure per hectare before planting the cloves.
- Apply a general fertilizer dose of 100 kg N, 65 kg P_2O_5 and 100 kg K_2O per hectare for the improved varieties.1/2 N along with entire phosphate and potash fertilizers should be given as basal and rest N should be top dressed one month after planting the cloves.
- Complete the top dressing before bulb initiation as delayed application results in thick neck of the bulb.
- Excessive nitrogen application results in thick neck of the bulb and sprouting before harvest.
- Soil application of borax upto 10 kg / ha increase bulb size and yield.
- Foliar application of micronutrients particularly manganese (0.1% $MnSO_4$), boron (0.2% boric acid), copper (0.02% $CuSO_4$) and zinc (0.02% $ZnSO_4$) stimulate dry matter accumulation in the cloves.

Irrigation

- Garlic being a shallow rooted crop requires light irrigations.
- Irrigate the field immediately after planting the cloves.
- In general garlic needs irrigation at an interval of 8 days during vegetative growth and of 10-15 days during bulb maturation.
- Frequent and light irrigation at weekly interval is the best to keep optimum soil moisture.
- Over watering should be avoided.
- Irrigation should be stopped when the crop reaches maturity at least 15 days before harvesting when the crops first begin to dry.

Interculture and weed control

- Garlic is a poor competitor of weeds hence, weeds must be kept down to get good yield.
- Garlic is shallow rooted crops and planted closely and there is a chance of root pruning by cultivation.
- First weeding should be done after one month of planting and second after two months of planting.
- Chemical weed control gives better results.
- Apply pendimethalin @ 3.35 l/ha after planting and before first irrigation plus one hand weeding at 45 days after planting is the best.

Harvesting the crop

- The crop is ready for harvest when tops turn yellowish or brownish and show signs of drying up and bend over.
- It becomes ready for harvesting 4-5 months after planting depending on variety, climate and soil.
- It is generally harvested in March-April in the plains and July-September in the hills.
- Harvesting is done by pulling out the plants along with the tops.
- The bulbs are dried for a week or so under shade.

Post harvest handling and storage

- After harvest the bulbs should be cured in the field in the field about a week.
- Purpose of curing or drying is to remove excess moisture from the outer skin and neck of the bulb which helps in minimizing shrinkage of bulb, developing skin colour and reducing the disease infection during storage.

- Curing also allow the bulbs to become compact and go into dormant stage.
- The bulbs are cured for 7-8 days in shade either with tops or after cutting top leaving 2.5 cm of the stalk.
- During curing the bulbs are covered along with the tops of each other to avoid damage from the sun.
- Curing in shade may be done on a floor having ventilation from bottom or on wire racks.
- After curing before the bulbs are put for storage or marketing, the thick necked, split, injured, diseased or bulbs with hallow cloves are sorted out.
- Thoroughly cured garlic bulbs keep fairly well in ordinary well ventilated rooms.
- For low cost storing, make a hut like structure under the shade with thatched roof and sides covered by bamboo sticks with provision of good air circulation.
- Store garlic in these sheds by spreading them on racks or tiers having 2-3 layers of bulbs.
- Garlic with dried leaves can be stored by hanging in well-ventilated room.
- Bulbs are generally stored for 6-8 months at relative humidity higher than 70% at any temperature however, after this period moulds are developed and root growth is started.
- Cold storage temperature of 0?C-2.2?C with 60-70% relative humidity is ideal for 6-8 months long storage in dormant condition and reasonably free from decay.
- Best cold storage condition is provided by packaging the bulbs singly in polyethylene bags, and stored at 0-1.6?C temperature with 32-35 % relative humidity.
- To prolong the storage life of bulbs and check sprouting in storage upto 300 days apply pre-harvest foliar spray of maleic hydrazide 2500- 3000 ppm, 3 weeks before harvest.
- Irradiation of bulbs with very low dose of 6 krad cobalt 60 gamma rays effectively controls sprouting and prolongs storage life of the bulbs.

Plant protection measures

Diseases

Purple Blotch (*Alternaria porri*)

The common symptoms of this important fungal disease occur on leaves as small, water soaked, shunken spots gradually extends and colour of the spot changes from olive to purple. Centre of the spot is always concentric and surrounding margins is deep purple. Further, large purple areas are developed forming dead patches resulting into fall over of the leaves. The most favourable temperature for disease development is 28-30°C temperature with 80-90% relative humidity.

Control measures

- Management schedules same as mentioned in onion.

Basal rot (Fusarium oxysporum f. sp. cepae)

Common symptoms of this fungal disease are progressive yellowing, wilting and rapid dying-back of leaves from tip with the roots turns pinkish as the plants approach maturity. Later, most of the roots rot. The bulbs become soft and a semi- watery decay is found advancing from base of the scales leaves. The infection may also be carried over to the storage where bulb rotting takes place in the same manner.

Control measures

- Management schedules same as mentioned in onion.

Downy mildew (Peronospora destructor)

This fungal disease is prevalent in cool and humid weather condition. Yellowish spot first appear on the upper surface of the leaves. Later, violate growth of the fungus appears on the infected leaves. The leaves turn yellow and in severe cases, show die back. Bulbs of infected plants become soft, spongy, shriveled and of poor keeping quality.

Control measures

- Management schedules same as mentioned in onion.

White rot (Sclerotium cepivorum)

This soil borne fungus invades roots and leaf sheaths resulting in killing of stem plate, roots and leaf sheath. Leaves become yellow and flaccid and finally whole plant is killed.

Control measures

- Treat the cloves by dipping 30 minuts in 0.3% solution of carbendazim and dried in shade before planting.
- Incorporation of green manuring is also effective.
- Soil application of *Trichoderma viride* @ 3 kg/ ha along with bio-manure is effective.
- Spraying the crop with 0.1% carbendazim / carbendazim (1g) + mancozeb (2g)/l or captan @ 2.5 g/ l along with sticker checks the rapidity of the infection.

Black mould (Aspergillus niger)

This fungus cause the storage disease causing rotting of the bulb and development of black mould occur when the bulbs are stored in high temperature and high humidity condition.

Control measures

- Storage of bulbs in cool and well ventilated place.
- Spraying the bulbs with 0.2% captan or carbendazim 0.1% at the time of storage checks the disease.

Soft rot (*Erwinia carotovora* pv. *carotovora*)

This common bacterial disease is caused in storage.Bulbs with green necks and which are not properly cured is attacked by these bacteria. The rot begins from the neck of the bulb. Cloves of the bulbs are water- soaked and soft. Foul odour emits and bacterial oozes come out upon squeezing the neck of the bulb.

Control measures

- Avoidance of mechanical injury to the bulb during harvesting.
- Proper curing till the neck is tight and scale leaves are dry.
- Bulbs with green neck should not be kept in storage.
- Spraying the bulbs before storage with 0.5% each of Bavistin + Plantomycin is effective in keeping the bulb disease-free.

Insect pests

Thrips (*Thrips tabaci*)

It is the most injurious pest of onion. The nymphs and adults are slender, fragile and blackish or yellowish in color. Both of them are found lacerating the leaf epidermis and lapping the exuding sap, producing silver white patches known as 'silver top'. Severe infestation resulted drying of leaves from tip to the base gradually.

Control measures

- Adopt crop rotation.
- Destroy and burn the weeds and plant debris on which adults hibernate.
- Apply insecticides like acetamiprid @ 0.5g/l observing the infestation of thrips after transplanting.
- Spray the crop with malation @ 1ml/l or endosulphan @ 1.5ml/l or Nuvacron @ 1ml/l of water 4 times at 15 days interval starting after transplanting.

Cut warm (*Agrotis ipsilon*)

It is a highly polyphagus and become active with the onset of cool weather. The nocturnal caterpillar remains in the soil crevices during day time and cut the seedlings at ground level at night.

Control measures

- Clean cultivation and breaking the big soil cloths.

- Mechanical destruction of hiding larvae at base of affected plant in the early infested crops.
- Dusting the soil around the plant with 4% endosulphan @ 2kg a.i./ha.

Yield

- Average yield: 10-20 t/ha depending on the variety and climate.
- Recovery of clove in bulbs ranges from 86-96% depending on the variety.

Greens and Pot herbs

20. Amaranthus

Scientific Name: Amaranthus spp.
Family: Amaranthaceae
Origin: Asia, Europe, Central America, Latin America
Chromosome No.: 2n = 2x = 32, 34

Introduction

There are two types of amaranthus: leaf and grain type. The leaf amaranths are used as pot herb vegetable. It is the most common leafy vegetable grown during summer and rainy season in India. Among the leafy types, *Amaranthus tricolor* L. occupies the predominant position in India with different morphological forms in colour and shape of leaves and other prominent cultivated species are *A. blitum* and *A. tristis*.

Importance and use

- Leaf amaranthus types are generally used as pot herb vegetable.
- Young plants with tender leaves of grain amaranthus are also used as leafy vegetable.
- It fits well in crop rotations because of short duration and considerable biomass yield per unit area.
- It is easily cultivated both in kitchen garden and on commercial scale.
- The fresh tender leaves and stems of *Amaranthus* are delicious when cooked like other fresh leafy vegetables.
- The tiny seeds of grain amaranthus are popped or parched and milled for flour or gruel. It is also used to make "ogi" (fermented product).
- The leaves and tender stems of amaranthus are rich in protein, minerals, Vitamin A and C.

Nutritional value

- Amaranthus leaves are good source of protein, mineral, calcium, iron and vitamins particularly provitamin A and C.
- 100 g of edible part provide 4 g protein, 6.3 g carbohydrates, 1g fibre, 83 mg phosphorus, 397 mg calcium, 25.5 mg iron, 341 mg potassium and 247 mg magnesium.
- 100 g of edible part provide 9200 I.U vitamin A, 99 mg vitamin C and 0.1 mg riboflavin.
- Dry matter content of amaranthus plant showed positive correlation with crude fibre and phosphorus content at all the stages of harvest.
- There is variation in chemical constituent in the leaves of different cultivars.

Medicinal value

- Consumption of leaf amaranthus protects against cancers of digestive tract, stomach, colon and rectum because it contains considerable quantity of antioxidant compounds such as β carotene and ascorbic acid.
- It prevents vitamin A deficiency related night blindness and xeropthalmia in the childrens.
- Amaranthus leaves and tender stems contain considerably high dietary fibre which help in reducing the problem of constipation.
- Leaf amaranthus contains some anti-nutrient compounds like oxalates and nitrates which may cause kidney stone and methaemoglobin in blood, respectively if it is consumed in high quantity regularly.

Adaptation in India

Leaf amaranth, said to be the native of India, is the most common leafy vegetable crop grown in India. Several kinds of vegetable amaranth under the genus *Amaranthus* and belonging to the different species namely, *tricolour*, *tristis*, *hypochondriacus*, *polygonoid*, *viridis*, *lividus* and *caudatus* are grown in India. The cultivars vary considerably in plant height, size and colour of leaves, branching habit, flowering time and response to cutting. Interspecific and interval hybridization in nature have caused such wide variation. Mostly green and red types are cultivated and both these types are nutritionally similar. In India there are many local cultivars of *Amaranthus* belonging to different species grown by the farmers under different names.Some of the very popular indigenous cultivars of West Bengal are Katoa Local Red, Katoa Local Green, Jaba Kusum, Sureswar, etc. Improvement of *Amaranthus* in India has been done by selection of superior type from indigenous germplasm. The varieties may be grouped accordingly to the species under which they belong.

Breeding behaviour

- Amaranth is a monoecious crop and generally both self and cross-pollinated by wind (anemophilous).
- Arrangement and sequence of anthesis of the unisexual flowers favour a combination of self and cross-pollinations.
- Abundant pollen production in the *Amaranth* species leads to a high degree of self-pollination.
- Dominant green perianth colour can be employed as marker character to direct the extent of cross-pollination in grain types, which ranges from 16–40 per cent.
- *Amaranthus* species are characteristically wind pollinated but the grain species with colourful inflorescence are occasionally visited by bees.
- Grain types of amaranthus favour more to cross-pollination, while in vegetable types comparatively high propensity of self-pollination occurs.
- The vegetable amaranths are predominantly self-pollinating due to presence of high percentage of (10-25%) of staminate flowers per glomerule and more number of axillary glomerules.
- From the breeding point of view, the small closely grouped flowers in glomerules make emasculation extremely difficult.
- The most satisfactory method of effecting crosses is to pollinate heavily as soon as the stigma is receptive and to remove the staminate flowers by hand.
- The stigma of the pistillate flowers become receptive several days prior to the opening of the staminate flower and it is more pronounced in grain types.

Origin and taxonomy

The origin of *Amaranthus* species is still uncertain. The centers of diversity for *Amaranthus* are Central and South America, India, South East Asia and the secondary diversity is in West and East Africa.The predominant leafy amaranthus under *A. tricolor* has been originated somewhere in south Asia or southeast Asia. Most probably it is native to India and from where it spread to the neighbouring countries. *A. dubius* has probably originated in Central America. It has much diversity in Central America, Indonesia and India. All grain Amaranthus have originated in Central and South America particularly, Guatemala (*A. cruentus*), Argentina and Peru (*A. hypochondriacus*), high Andes of South America (*A. caudatus*) with Hisdustani subcontinent as an important secondary centre of diversity.

Amaranthus belongs to the genus *Amaranthus* and family Amaranthaceae. This family comprises of 65 genera and 850 species. The genus includes 50-60 species, both cultivated and wild, and used for grains and leaves. The genus *Amaranthus* is classified into two sections: *Amaranthus* and *Blitopsis*. All leaf amaranthus species come under the section *Blitopsis* and all grain type species under the section *Amaranthus* .The important leafy type cultivated *Amaranthus* species are: *A. tricolour, A. tristis, A. hypochondriacus, A. polygonoid, A. viridis, A. lividus* (*A. blitum*) and *A. caudatus.* The most popular grain species are *A. hypochondriacus, A. cruentus* and *A. caudatus* and the most important weed species are *A.spinosus, A.viridis* and *A.graecizans.*

Improved varieties

A. tricolor: CO-2, CO-5, Pusa Badi Chaulai, Pusa Kiran, Pusa Lal Chaulai, Arka Arunima, Arka Suguna, Amt-105, Amt-237

A. lividus (A. blitum): Chhoti Chaulai, Pusa Kirti

A. dubius: CO-1

A. tristis: CO-3

A. hypochondriacus: CO-4

A. polygonoid: Sirukeerai

Cultural requirements

- Amaranthus can be grown in a wide range of soils, however, sandy-loam slightly acidic in nature is preferred.
- Heavy clay soils with very poor drainage or sandy soil with very poor water holding capacity are unsuitable for its cultivation.
- It requires a very good tilth and thorough field preparation for getting quick growth.
- Amaranthus is widely distributed in both temperate and tropical regions of the world with optimum temperature requirement of 20°-30°C.
- It is basically a crop of humid tropics but grown in temperate climate during summer season.
- Amaranthus species that grow under varying climatic condition differ in day length requirements and respond differently to change in photo and thermoperiodism.
- Well-drained fertile loam soils are ideal for its cultivation.
- The ideal pH range is between 5.5 and 7.5, but some strains of *Amaranthus* can be gown in soil with pH as high as 10.
- Since the crop is directly sown, and the seed size is small, the field should be free from weeds and clods.
- Amaranthus is susceptible to water logging.

Land preparation

- Prepare the land in well advance with repeated ploughings (at least 4-5 ploughings) to a fine tilth.
- Remove all stubbles and weeds etc.
- After three to four harrowing, level the land to bring the field into a fine tilth.
- Apply well rotten organic manure (preferably Farmyard manure) @ 25/ha at the time of last ploughing.

Sowing time

Amaranthus is warm season crop and it can be grown throughout the year in tropical and subtropical regions. In south and east Indian condition amaranthus is grown almost throughout the year. In north Indian condition, generally two sowings are practiced.

- Summer crop: February – March
- Rainy season crop: June – July

Sowing of seeds

- Seeds are sown by broadcast method or by drilling in lines 20-30 cm apart.
- A convenient bed size in the field is 2.0 × 1.2 m.
- In South India, there is also a practice of transplanting amaranthus especially the variety Badi chaulai, either a pure crop or along with the border of beds of other vegetables.
- In direct growing, the seeds are mixed either with fine soil or sand for its even distribution.
- Because of small seed size, shallow sowing about 1-1.5 cm deep is recommended.
- Seed rate is about 2 kg/ha for direct sowing and 1kg/ha for transplanting crop.
- Seed treatment with IBA at lower concentration of 0.1-1 ppm increase germination.

Spacing

- The plant spacing within the row is maintained at about 20-25 cm at the time of thinning.
- Sometime transplanted after raising the seedling during rainy season with a spacing of 45 cm from row to row and 30 cm from plant to plant.

Nutrient management

- Apply 25 tonnes of FYM per hectare at the time of land preparation.
- Amaranthus is short duration crop and is good feeder of macronutrients.
- Apply a general fertilizer dose of 90 kg N and 50 kg each of K_2O and P_2O_5/ha

- Whole of $P_2O_5$and K_2O and 1/3 of N are applied before sowing or transplanting at the time of field preparation.
- The balance of N is applied 3 weeks after sowing or transplanting and next dressing after first cutting.
- In non-clipping type where whole plant is uprooted at harvest, full dose of NPK is given as basal.
- Leaf number and stem diameter is not generally affected by N application but plant height, leaf area and fresh weight increase with increasing N from 0-110 kg/ha.
- Application of N also increases the plant crude portion, carotene and chlorophyll contents in the leaves and stem.

Irrigation

- In the absence of sufficient soil-moisture at the time of sowing, apply first irrigation immediately after sowing and further irrigation are to be given once a week with gentle flow of water.
- During summer, subsequent irrigation are given at 3-5 days intervals.
- In rainy season, irrigation is given as and when required.

Interculture and weed control

- Practice shallow inter cultivation to remove the weeds from field, 3-4 weedings are needed.
- Amaranthus is mostly sown directly to the field, so it needs to be thinned properly to get remuneration.
- Generally 10-15 days after sowing, it is thinned to a spacing of about 20-25 cm between plants, leaving one or two plants per hole.

Harvesting the crop

- In general, the plants are pulled as a whole, washed and send to the market as tender greens
- Optimum stage of harvest in amaranthus is about 25-35 days after sowing. At this stage, the performance of this types are better with increase in leaf weight, stem weight, leaf length, leaf breath, stem diameter, plant height as well nutritious and palatable.
- In *A. tristis*, a total of 10 clippings can be obtained with the first clipping starting on the 25 days after sowing and thereafter at weekly interval.
- In *A. blitum*, 6 clippings are done in a duration of 60 days
- In case of intermediate and late flowering cultivars of *A. cruentas* continuously cutting back is more profitable method of harvesting than uprooting at the optimum commercial stage.
- The highest vegetable yield is obtained from black seeded cultivars and highest seed yield from white seeded cultivars.

Post harvest handling and storage

- Amaranthus does not stand storage for more than few hours under ordinary conditions.
- About 3-4 days storage life can be obtained when green leaves are stored in cool storage condition.

Plant protection measures

Diseases

Leaf spot (*Cercospora sp.*)

This disease is characterized by the presence of numerous small, brown, circular spots on the leaves. In the beginning, the spots are small, roundish with concentric rings but later these spots increase in size and some times they coalesce.

Control measures

- Spraying with Blitox (0.3%) or difenconazole (0.05%) or chlorothalonil (0.2%) 2-3 times at 15 days interval

White rust (*Albugo bliti*)

This disease is characterized by white, blister like circular or irregular pustules on the lower surface of the leaf and opposite each pustule on the upper surface, a yellow patch developes. Severe infection causes death of the leaves giving the field a blighted look.

Control measures

- Spraying the crop with 0.2-0.3% mancozeb or 0.25% metalaxyl-mancozeb

Insect pests

Leaf eating caterpillar (*Hymenia recurvalis*)

It is a serious pest of Amaranthus and many other vegetable crops. Larvae bind the leaves together by their webbing and eat away by scraping the leaf tissues. In severe cases the leaves dry up and the plant die.

Control measure

- The crop should be sprayed with 3-4% neem seed karnel extract at regular interval as prophylactic measures as soon as the pest is observed.
- Conserve and restoration of natural enemies, *Apantalis* sp.
- Spraying of malathion 0.05% or DDVP 0.05% is effective.

Red spider mite (Tetranychus urticae)

The mites attack all type of Amaranthus during summer months which colonize at the ventral surface of leaf and produce numerous whitish specks. In case of severe attack the leaf fall off and dry out.

Control measure

- Botanical pesticides like neem or karanja may be applied as prophylactic measures.
- Mass release of predatory mite, *Amblyseius longispinosus* is very effective against the red spider mite complex.
- Spraying of propergite @ 1.5 ml/l or dicofol @ 2.5 ml/l is effective.

Yield

10-15 tonnes per hectare depending on the variety, number of cuttings, location, season and cultural practices.

21. Spinach beet *or* Palak

Scientific Name: Beta vulgaris var. bengalensis
Origin : Indo Chinese region
Family : Chenopodiaceae.
Chromosome No.: 2n = 2x = 18

Introduction

Palak is one of the most common leafy vegetables of the tropical and subtropical regions and is grown widely in India particularly in northern plain of Uttar Pradesh, Bihar, West Bengal, Rajasthan, Punjab and Haryana. It is also grown in Maharashtra, Gujarat and Madhya Pradesh. This vegetable crop is not so popular in the southern states. Leaves of this crop might have been first used in Bengal, the var. *bengalensis* stands for this reason.

Palak is different than spinach (*Spinacia oleracea*) not only cytologically (2n = 12) but also with respect to many characters. Palak produces only bisexual flowers while spinach produces staminate, pistillate and /or bisexual flowers.

Importance and use

- Palak or spinach beet or beet leaf is generally used as pot herb vegetable.
- Young plants with tender leaves are used as leafy vegetable.
- It can tolerate high temperature unlike spinach which is purely a cool season crop.
- It fits well in different crop rotations because of its adaptation to high temperature condition, short cropping duration and appreciable biomass yield per unit area.
- It is easily cultivated both in kitchen garden and on commercial scale.

- The fresh tender leaves of palak are delicious and can easily be cooked with other vegetables like, potato, onion, and brinjal or fried alone.
- Delicious vegetarian dishes are also prepared with cheese and potato.

Nutritional value

- Palak leaves are good source of protein, mineral, calcium, iron, phosphorus and vitamins particularly provitamin A and C.
- 100 g of edible part provide 3.4 g protein, 6.5 g carbohydrates, 0.7 g fibre, 30mg phosphorus, 380 mg calcium and 16.2 mg iron.
- 100 g of edible part provide 9770 I.U vitamin A, 70 mg vitamin C and 0.56 mg riboflavin.
- There is variation in chemical constituent in the leaves of different cultivars.

Medicinal value

- Palak is the rich and cheap source of vitamin A as compared to carrot.
- It prevents vitamin A deficiency related night blindness and xeropthalmia in the childrens.
- Palak is mildly laxative besides other medicinal values.
- Palak provide enough roughage which is important for balanced diet.

Adaptation in India

Palak as a leafy vegetable is known from Charaka Samhita days, c. 600 B.C. which is the testimony of its long history of cultivation in India. In India, it was perhaps first cultivated in Bengal. Although, despite such a long history of cultivation the genetic variability observed in this species is not wide enough, commensurate with outcrossing breeding behaviour. However, all the improved varieties have been developed utilizing indigenous variability.

Breeding behaviour

- Palak is a long day crop and produces seeds with the advent of spring after passing the winter.
- Palak does not require vernalization for production of seed stalks.
- All cultivars of palak are annual type and hence, their seeds can be produced freely in the plains of India.
- Palak bears hermaphrodite flowers but varying levels of cross-fertilization takes place in this crop due to long anthesis and anther dehiscence time, open inflorescence like a large spike which sways in wind, prolonged stigma receptivity and conspicuous production of very fine pollens which drift in air for long time.
- It is anemophilous crop and cross-pollination is accomplished by flowing air although considerable self-pollination also takes place.

- To maintain the purity of the varieties, isolation distance of at least 1000 m is safe for seed production.
- Anthesis starts in the morning at about 7.00 a.m. and continues up to 5.00 p.m. depending on the prevailing weather conditions.
- Anther deshiscence occurs at about 8.00 a.m. and continues upto 6.30 p.m.
- The stigma becomes receptive 6 hours before and 8 hours after anthesis.
- Pollen remains viable on the day of anthesis.
- For experimental hybridization purpose matured buds of suitable size on a spike are selected one day prior to anthesis. Emasculation is done with forceps and the whole spike is bagged. Pollination is carried out next day by dusting the pollens from the spike of the intended male parent on the emasculated buds. Spike from the male parent may also be bagged along with the emasculated spike for effecting pollination.
- Palak crosses with beet root, sugar beet and swiss chard.

Origin and taxonomy

Palak has most probably been originated in the Indo-Chinese region.It was known in China in 647 A.D. It was first recorded in India in the medical treatise, Carak Samhita, c. 600 B.C. The palak, *Beta vulgaris* var. *bengalensis* Hort. belongs to the family Chenopodiaceae. It is closely related to Swiss chard, beet root and sugar beet, all having the same chromosome number. The sea beet, *Beta vulgaris* ssp. *maritime* is the ancestor of beet root and palak.

Improved varieties

- Punjab Green, Punjab Selection, Pusa Jyoti, HS-23, All Green, Pusa Harit, Jobner Green, Pusa Bharati, Banerjee's Giant, Pusa Palak, Ooty-1, Pant Composite, Arka Anupama.

Cultural requirements

- Palak grows well both under subtropical and tropical climate.
- It grows well in the winter season of the tropical elimate.
- In mild temperature condition it can be grown throughout the year.
- It can tolerate frost better than other vegetables and also tolerate warm weather but high temperature (35°-40°C) leads to early bolting.
- Sandy loam soil is considered best for palak cultivation.
- It can tolerate slightly alkaline soil, but best quality and high yield can be obtained in the soil with pH 7.0.
- Palak is a highly salt tolerant vegetable and can be grown successfully in saline sodic soils.

Land preparation

- Soil is prepared by 3-4 ploughings followed by planking so that soil is well pulverized and leveled.
- After preparation of soil thoroughly, beds of convenient size and irrigation channels are made.

Sowing time

- Early spring: January-February
- Rainy season: June-July
- Main crop: August-November
- Hill condition: April - June

Sowing of seeds

- Seeds are sown by broadcast method or by drilling in lines 20-25 cm apart and 5-8 cm plant to plant (after thinning).
- Seeds are broadcast in well leveled beds and mixed with soil thoroughly.
- Seed rate is 25-30 kg/ha for summer crop and 10-15 kg/ha for winter crop.
- Seeds should be soaked overnight to hasten the germination.
- Seeds should be sown to a depth of 2-3 cm.
- Seeds generally germinate after 7-8 days of sowing.

Spacing

- 20-25 × 5-8 cm.

Nutrient management

- Palak is short duration crop and is good feeder of macronutrients.
- Apply a basal dose of 25-30 tonnes of farm yard manure.
- Apply a general fertilizer dose of 100 kg N and 60 kg each of K_2O and P_2O_5/ha.
- Whole of P_2O_5 and K_2O and 1/3 of N are applied at the time of field preparation.
- The balance of N is applied 4 weeks after sowing and next dressing after first cutting.
- Foliar application of 1.0-1.5% area gives better yield.

Irrigation

- It is advisable to give a good pre-sowing irrigation and a light irrigation a few days after sowing for better germination of seeds.
- In spring-summer season, irrigation is required at 6-7 days interval.
- In autumn-winter season, irrigation is required at 10-15 days interval

Interculture and weed control

- 2 to 3 hoeings-cum-weedings are required to keep away the weeds from the field and to loosen the soil for proper aeration.
- Hand weeding is advocated in the early stage.
- After every cutting, weeding and hoeing should be done.

Harvesting the crop

- The crop is ready for harvesting in about 3-4 weeks after sowing.
- The crop is generally harvested by uprooting the whole plant at tender stage.
- During seemer and rainy season only one harvesting by uprooting the crop is done.
- Its first flush of leaves become ready for picning 3-4 weeks after sowing.
- Subsequent cutting should be taken at an interval of 20-25 days depending upon the variety and season.
- About 6-8 cuttings can be taken depending on the variety.

Post harvest handling and storage

- After harvesting, it is trimmed, graded and tied in bunch before marketing.
- Palak has a large leaf surface to weight ratio and has very high respiration rate.
- It is essential to cool the product immediately and rapidly to prevent wilting and weight loss.
- Best cooling temperature is 2.0- 2.8 °C for 10 minutes by vacuum cooling.
- The leaves cannot be stored for longer period, so it should be sent to the market immediately after harvesting.
- Small bundles are made to give a better look.
- Damaged, diseased and yellow leaves are discarded before bunching.
- Washing of leaves is not advocated because it results in rotting in the centre of the bundle.
- Under low temperature (0°C) and high relative-humidity (90-95%), leaves can be stored for 10-14 days.

Plant protection measures

Diseases

Damping off (Pythium sp.)

Seedlings are attacked and killed by this fungus. The affected seedlings look pale green and brownish lesions are found at the collar region at the advanced stage. Collar rotting occurs subsequently and the seedlings topple over the ground.

Control measures

- Seed treatment with thiram or captan @ 3 g/kg of seed or with Bavistin 2 g/kg of seed.
- Soil drenching with mancozeb @ 3g/l of water.
- Spraying the crop with mancozeb (0.25%) or metalaxyl-mancozeb (0.25%).

Leaf Spot (*Cercospora beticola*)

Brown dead circular spots are found on the diseased leaves.

Control measures

- Spraying the crop with 0.2 % chlorothalonil or 0.1% propiconazole at 15 days interval.

Insect pests

Aphid (*Aphis* sp.)

Small, oval, soft bodied nymphs and adults are found in large number on the tender portions of the plant. They suck the cell sap from the tissue which causes curling and ultimately drying up of leaves.

Control measures

- Spraying the crop with 0.05% dimethoate (Roger 2 ml/l), 0.08% malathian (Cythion 2 ml/l) at 10 days interval

Leaf eating caterpillar (*Laphygma exigua*)

Young caterpillars feed on leaves by scraping gregariously. Older caterpillars feed voraciously the entire leaf. It causes economic defoliation ranging from 16-80%.

Control measure

- Spraying the crop with malathian (Cythion 2 ml/l) at 10 days interval dissolved in 625 litres of water.

Yield

- 15 – 30 tonnes/ha depending on the variety, number of cuttings, location, season and cultural practices.

Peas and Beans

22. Garden pea

Scientfic Name: Pisum sativum ssp. hortense
Family: Leguminosae
Origin: Central Asia, Ethiopia and Mediterranean basin
Chromosome No.:2n=2x=14

Introduction

Pea is a commonly grown leguminous vegetable in the world. It is a cool seaon crop of the tropical and subtropical regions. The important pea producing countries are USA, China, India, France, U.K. Netherlands, Hungery, Russia, Egypt and Australia. Green peas are all-time favourite vegetables. It is grown in Uttar Pradesh, Madhya Pradesh, Himachal Pradesh, Punjab, Haryana, Rajasthan, Maharashtra, Bihar and Karnataka, contributing to 67% of the total production.

Importance and use

- Peas are cultivated for the fresh green seeds, tender green pods, dried seeds and foliage as herb.
- Green peas are eaten cooked as a vegetable, and are marketed fresh, canned, or frozen while ripe dried peas are used whole, split, or made into flour.
- In some parts of the world, dried peas are consumed split as pulse, roasted, parched or boiled.
- Green peas are the number one processed vegetable specifically in UK and USA.
- Leaves are used as a pot herb in India, Burma and parts of Africa.
- Some cultivars of snowpea are grown for their tender green pods, which are eaten cooked or raw as pod wall of this pea contain less fibre.
- Pea is being used in a growing snack market. One snack item is prepared by soaking the peas overnight and frying them in palm oil or coating them with other food items such as rice flour before frying for the purpose of imparting different flavours. Another product is prepared by finely grinding the peas and extruding them under pressure to create different shapes.
- Cultivars such as Alaska, Super Alaska, and Supergreen have long been the standard type of canning pea. Canning cultivars usually have a tough skin that holds its shape during canning.
- Wrinkled-seeded garden peas are sweeter than smooth seeded types.

Nutritional value

- Pea is a rich source of protein (25%), amino acids and sugars (12%) along with carbohydrates and vitamins.

Medicinal value

- Oil from ripened seed has antisex hormonic effects; produces sterility and antagonizes effect of male hormone.
- In Spain, flour is considered emollient and resolvent, applied as a cataplasm.
- It has been reported that seeds contain trypsin and chymotrypsin which could be used for contraceptive, ecbolic, fungistatic and spermicide.

Adaptation in India

Archaeological evidences suggest the existence of pea (*Pisum sativum* L.) in India before the advent of the Aryans. It was used in pre-historic period during Indus valley civilization (C.2500-300 B.C) as evidenced by the carbonized pea seeds discovered in Mohan-ga-Daro and Harappa excavations. This pea must be round and non-sweet pulse type (*P. sativum var. arvense*). The sweet garden pea (*P.sativum* var. *hortense*) varieties developed in temperate regions of Europe might have been introduced in India during the British regime. Some of the hardy and round seeded garden pea varieties like Bonia, Kaip, Asauji, etc. may possibly be the early introductions of the Britishers. These were less sensitive to wilt even under warmer conditions and have been used in different breeding programmes. So, introduction of exotic varieties form the mainstay of breeding garden pea in India. They are either used as varieties as such or utilized in combination breeding programmes to develop new varieties.

Breeding behaviour

- It is highly self-pollinated crop due to hermaphrodite sex form, homogamy i.e. maturation of stigma and anthers at the same time.
- Stamens and pistil in opened flower remain enveloped together inside the tube like structure of jointed petals called "keel" leading to cleistogamous like self-pollination.
- Flowers open only once between 5.00 a.m. and 11.00 a.m.
- Though the flowers open in the morning, the dehiscence of anthers starts much earlier even 24 hours before anthesis however, dehiscence of anthers is maximum at anthesis.
- Stigma becomes receptive and pollen fertility at its maximum on the day of anthesis.
- Stigma remains receptive for 3 days after emasculation at 16° C and for one day at 20° to 24°C.
- Pollens are viable from the time of anther dehiscence until few days thereafter.
- Since the dehiscence of anthers is much in advance of the blooming, the emasculation needs to be carried out in mature flower buds in the preceding afternoon hours.
- Emasculation of matured buds in the afternoon hours followed by hand pollination in the next day morning between 8.00 and 10.00 a.m. is the best method of hybridization.
- The flower buds which are to open in one or two days are selected for emasculation. After giving a slit in the keel with a pair of forceps, the stamens are removed. The pollination is done in the next morning by rubbing the matured anthers of the male parent against the stigma of the female parent.

- Pollens are collected next morning from a freshly opened flower. The standard and wings of the flower from the intended male parent are removed. By slight depression of the keel, stigma laden with the pollen grains protrudes out which it self can be used as a brush for pollination.

Origin and taxonomy

Central Asia (Kazakhistan, Uzbekistan, Tazikistan, Turkmenistan, Kyrgyzstan, Afghanistan), Abyssinia (Ethiopia) and the Mediterranean basin are the primary centres of origin of pea with the near-east (Syria, Lebanon, Israel, Jordon) as the secondary centre of origin. In another view, *Pisum* is considered to have been originated in Ethiopia from where it spread during pre-historic times to Mediterranean region, Central Europe, the Near East, and subsequently to rest of the world. The wild progenitor of *Pisum sativum* is unknown. Archaeological evidence shows that smooth seeded form of field pea was cultivated in the Near East and Europe as early as 7000 BC. The somatic chromosome number of pea is 2n=2x=14. Based on crossability and cytogenetic evidence, the number of species within the genus *Pisum sativum* includes the following sub-species.

Garden peas : *P. sativum* var. *hortense* Asch. & Graebn

Field peas: *P. sativum* var. *arvense* (L.) Poi

Edible podded peas: *P. sativum* var. *macrocarpon*

Early dwarf pea : *P. sativum* var. *humile*

Wild form: *P. stivum elatius*

Wild form: *P. stivum sysiacum*

Improved varieties

- Early types: Arkel, Harbhajan, Alaska, Hisar Harit, Early Badger, Meteor, Lincoln, Asauji, Alderman, Azad P-1, Jawahar Matar-3, Punjab-88, Matar Ageta-6, Azad P-3, PM-2, Pant Sabji Matar-3, VL Ageti Matar-7.
- Medium/Late types: Bonneville, Arka Ajit, New Line Perfection, Jawahar Matar-1, Jawahar Matar-4, Jawahar peas-83, Kashi Nandini, NDVP-8, NDVP-10, Pant Uphar, VL Matar-6 and Vl Matar-8.
- Powdery mildew resistant: Jawahar Matar-5, Jawahar peas-83, Jawahar Matar-6, Jawahar Matar-15, Jawahar Matar-54.
- Edible podded type: JP-19, UN-53-6, Sylvia.

Cultural requirements

- Peas require a cool, relatively humid climate and are grown as cool season crop.

- As a winter annual, pea tolerates frost to –2°C temperature in the seedling stage, although top growth may be affected at –6°C. Winter hardy peas can even withstand –10°C.
- The optimum temperature levels for the vegetative and reproductive periods of peas were reported to be between 21 and 16°C, and 16 and 10°C (day and night), respectively. Temperature above 27°C shorten the growing period and adversely affect pollination.
- A hot spell is more damaging to peas than a light frost. Peas can be grown successfully during mid-summer and early fall in temperate countries.
- Seeds can germinate even at 5°C temperature however, optimum temperature is about 22°C
- For very early crops, a sandy loam is preferred; for large yields where earliness is not a factor, a well-drained clay loam or silt loam is preferred.
- It can be grown on many types of soils—light sandy loam to clay soil however, well-drained, loose, friable and heavy soils are ideal.
- Pea does not thrive on acidic soils. It is very sensitive to saline and alkaline conditions. The pH of 6.0–7.5 is ideal. If it is lower than 6.0, liming improves the soil condition.

Sowing time

- In central and eastern parts of India: October–November
- In hilly areas: May
- In Mandla, and Shahdol, (Amarkantak hills) of Madhya Pradesh: February–March.
- Peninsular India : June-July

Land preparation

- Prepare the land in well advance with repeated ploughings (at least 4-5 ploughing) to a good tilth.
- Remove all stubbles, weeds, etc. from the land.
- Thorough preparation of land is necessary for better growth and spread of root system which fix appreciable atmospheric nitrogen.
- Add well rotten organic manure preferably farm yard manure or compost @ 15-20 t/ha during land preparation and level the land properly.

Sowing of seeds

- Pea is a direct seeded crop.
- Viable, healthy, well-mature and pure seeds should be used for sowing. About 100-120 kg/ha seed is enough for early crop and 80-90 kg/ha for mid-season/late crop.

- The seeds must be treated before sowing to avoid losses due to fungal diseases. Mixture of thiram + Bavistin (2g + 1g/kg seed) is recommended for seed treatment. The seeds should be mixed thoroughly with the required quantity of fungicide before sowing.
- Pea seeds can be drilled by local implements or by tractor hauled seed drills. The seeds should be sown 2–3 cm deep in the soil. Soaking seeds of wrinkled seeded type for overnight is essential.
- Generally broadcasting of seeds is done but line sowing in rows on both sides of the raised bed is the best.

Spacing

- The drilling or dibbling can be done in rows placed 30 cm apart for early crop and 45 cm apart for mid-season/late crop. The spacing of 10 cm from plant-to-plant is ideal.

Nutrient management

- Pea being a leguminous crop, fixes atmospheric nitrogen through root nodules and the comparative nitrogen requirement is less. Soaking of seeds with 30 g Rhizobium bacteria (*Rhizobium leguminosarum*) in 10% molasses solution per kg seed is effective for increased nodulation.
- Apply 15-20 tonnes/ha of thoroughly decomposed FYM or compost well in advance.
- A basal dose of 50 kg N/ha at the time of sowing stimulate early growth.
- Application of 75 kg P_2O_5 for early varicties and 100 kg P_2O_5 for mid and late varicties along with 40 kg K_2O as bassal is optimum.
- Foliar application of 0.1% ammonium molybdate at flowering increase yield.

Irrigation

- Water requirement of pea is comparatively low.
- For proper germination, pre-sowing irrigation is a must if the soils are dry. Irrigation frequency depends on the nature of soils and winter showers.
- Irrigation after sowing results poor germination of seeds.
- Normally 2–3 irrigations are needed for mid-season peas or late sown peas. If the soils are lighter and shallow, the number can be increased.
- Excess of moisture results in yellowing of crop reducing the yield. Soil moisture deficit reduces growth and also affects nodulation.
- Furrow irrigation is normally adopted for irrigating peas but sprinkler system of irrigation is much better.
- Moisture stress at flowering and subsequent pod filling stage is most undesirable affecting the yield and quality of pods.

Interculture and weed control

- There is a heavy weed growth in pea right from sowing till harvest. This is due to wide spacing given for hand picking of green pods as well as due to slow growth of pea during earlier stages.
- Usually 2 hand-weedings are sufficient. Weeds can also be controlled by early mechanical means but deep harrowing damages the roots. Late interculture will also damage the crop due to trumping and mechanical breakage of tender and succulent stems and branches.
- The weeds can also be controlled by application of pre-sowing herbicide (fluchloralin @ 0.75 a.i./ha) or (pendimethalin @ 1.0 kg a.i./ha).
- Tall varicties need staking.
- At the spreading stage of the vine, the plants are supported on bamboo sticks or any twiggy sticks.

Harvesting the crop

- Harvesting green pods must be done at proper maturity stage.
- The pods start losing their quality rapidly after reaching the edible stage.
- Green pods of the early, mid and late varicties are ready for harvest 50-55, 60-65 and 70-75 days after sowing, respectively.
- The green pod pickings should be done during the morning or evening. The pea plants are very tender with soft stem and therefore pickings should be done gently. A small jerk damages the vines thereby injuring the plants. Moreover, repeated picking should not be done.
- Only 2-3 pickings in early crop and 3-4 in mid-season/late crop should be done.

Post harvest handling and storage

- The green pods may be handled carefully.
- Proper storage at low temperature (0^0C) and relative humidity (88-92%) may prolong the availability duration of green peas for 2-3 weeks and reduce the losses due to fungal diseases by arresting the metabolic breakdown.
- The pods will freeze at –10°C.
- Adequate packaging may be done in gunny bags, baskets lined with jute cloth, bamboo baskets, corrugated fibre board boxes and plastic containers.

Plant protection measures

Diseases

Powdery mildew (*Erysiphi polygoni*)

White, powdery fungal growth develops on leaf surfaces, petioles, and stems. The growth is primarily asexual spores called conidia. It usually develops first on shaded lower leaves, and on leaf undersurfaces. Older plants are affected first. Infected leaves usually wither and die. The disease becomes more severe in late sown crop.

Control measures

- Spray the crop with Bavistin or Benlate or Topsin M @ 1g/l of water 2 to 3 times at 15 days interval.
- Two sprays of 0.1% Bavistin followed by two sprays of 0.1% karathane or tridemorph effectively control the disease.

Rust of Pea (*Uromyces pisi*)

Orange colour pustules developed on the foliage of pea and all the above ground parts of the plant. Infected leaves usually wither and die.

Control measures

- Spray the crop with Bavistin or Benlate or Topsin M or tebuconazole @ 1g/l of water 2 to 3 times at 15 days interval.

Fusarium will (*Fusarium oxysporum f. pisi*)

This seed and soil borne fungal disease is more damaging in warm conditioin. Early infection cause death of the plant and late infection cause poorly filled pods.

Control measures

- Seed treatment with thiram of Bavistin @ 3.0 g/kg of seed.
- Spraying the crop ith carbendazim (1g) + maneozeb (2.0g) per litre at weekly interval.

Bacterial blight (*Pseudomonas pisi*)

In this bacterial disease, the affected plants develop watery, olive-green blisters on stems and leaf bases and water soaked oily spots appear on pods and leaves.

Control measures

- Use of disease free healthy seeds.
- Seed treatment with 0.01% streptocycline solution for 30 minutes before sowing.
- Spraying of 0.01% streptocycline + 0.2% copper oxychloride.

Insect pest

Pod borers (*Helicoverpa armigera*)

It is the most serious pest of the pods in some parts of the country. The caterpillar after hatching bore into the developing pods and feed on the seeds therein. Severely infested crops are subjected to havoc economic yield loss leaving behind most of the affected fruits.

Control measures

- . Remove and destroy the infected pods at the initial stage of attack.
- Spray the crop with Nuvacron @ 1 ml/l, or Thiodan @ 1.4 ml/l or malathion @ 2 ml/l or Sevin W. P. @ 4 g/l of water.
- Biological control measure by applying a nuclear polyhedrosis virus (NVP) @ 250 LE/ha/week starting from first appearance of young leaves.
- Spray the crop with 5% neem seed kernel extract reduces the incidence.

Aphids (Macrosipum pisi)

The gregarious aphids suck sap from growing tips and later from the whole plant which causes curling of leaves, reduces vigour, yield and quality. They also transmit mosaic viruses.

Control measures

- Spray the crop with 5% neem seed kernel extract along with 2% starch reduces the incidence.
- Encourage lady bird beetles in the crop field to predate upon the aphid species.
- Spray the crop with Rogor @ 1.5ml/l of water.
- Apply carbofuran granules in the field (Furadon 3G @ 1.0 Kg a.i. /ha) before sowing of the seeds.

Yield

- Early variety : 2.5-4.0 t/ha.
- Mid-season variety : 6.0-7.5t/ha.
- Late variety : 8.0-10.0 t/ha
- Shelling percentage : 35-50

23. French bean

Scientific Name: Phaseolus vulgaris
Family: Leguminosae
Origin: Guatemala
Chromosome No.: 2n = 22

Introduction

French bean is an important leguminous vegetable crop throughout the world. It is consumed as tender pods, shelled green beans and dry beans. It is largely grown in hilly areas of Himachal Pradesh, Jammu and Kashmir and north-eastern states during summer and as winter and autumn crop in parts of Utter Pradesh, Maharashtra, Karnataka, Andhra Pradesh and other parts of the country. In northern plains, it is cultivated on a limited scale as autumn or spring crop, because of susceptibility to low as well as high temperatures. French bean is an important vegetable crop of USA, Latin America, Canada, all European countries, Nepal, Bangladesh and Pakistan.

Importance and use

- There are 3 types of French bean : “Bush tuype” with short internode, “semi-pole type” with long internods and “Pole type” with longer internode.
- The green pods are commonly used as a vegetable; they have a mild flavour and should only be cooked for a short time.
- The immature seeds are boiled or steamed and used as a vegetable.
- The mature seeds are dried and stored for future use. They must be thoroughly cooked before being eaten and are best soaked in water for about 12 hours prior to this.
- They can be boiled, baked, pureed, ground into a powder or fermented into ‘tempeh’ etc. The powdered seed makes a protein-enriching additive to flour, it can also be used in soups etc. The seed can also be sprouted and used in salads or cooked. The roasted seeds have been used as a coffee substitute.
- In USA large quantity of tender pods are processed to produce canned and frozen products.
- Young leaves - raw or cooked can be used as a potherb. The very young leaves are sometimes eaten as a salad while, the older leaves are cooked.

Nutritional value

- French bean is highly nutritions vegetable.
- It contains proteins (1.7 g), calcium (50 mg), phosphorus (28 mg), iron (1.7 mg), carotene (132 mg), Thiamine (0.08 mg), Riboflavin (0.06 mg) and vitamin C (24.0 mg) per 100g of edible pods.
- French beans are filled with healthy dietary fibers.
- French beans also help in improving the immune system of the body as they are rich source of Vitamin C and A.

Medicinal value

- The green pods are mildly diuretic.
- Dietary fibers are beneficial for the prevention of cholesterol and sugar levels after having a meal.

- French beans are also helpful in energizing the body as they are rich in iron, the nutrients found in haemoglobin which helps in giving energy to the body.
- The presence of copper in French beans, in turn, helps in the proper synthesis of haemoglobin.
- The dried mature seeds are used in the treatment of diabetes.
- People suffering from severe migraine attacks can benefit from eating French beans as they are a very good source of riboflavin, the source that helps in mitigating migraine attacks.
- French beans have many anti-inflammatory nutrients like beta-carotene and, of course, Vitamin C. These nutrients help in preventing many diseases and ailments where there are chances of inflammation, such as asthma, rheumatoid arthritis and osteoarthritis. They also contribute in preventing fatal diseases like colon cancer.
- The seed is diuretic, hypoglycemic and hypertensive. Ground into a flour, it is used externally in the treatment of ulcers.
- When bruised and boiled with garlic they cure intractable coughs.

Adaptation in India

French bean (*Phaseolus vulgaris* L.), which is utilized both for the tender green pods as vegetable and dry seeds as pulse has been introduced in India comparatively in recent times from Europe during the 17th century. Most of the local types native to North-eastern region are pole types. So, introduction plays very important role in breeding French bean in India. In fact, most of the popular varieties grown in India are exotic varieties.

Breeding behaviour

- It is highly self-pollinated crop due to hermaphrodite sex form, homogamy i.e., maturation of stigma and anthers at the same time, however, dehiscence of anthers may take place much earlier than anthesis.
- Beans are primarily self-pollinated and self-fertilized, but natural crossing often occurs as a result of pollinator's pressure, principally by large bees.
- Stamens and pistil in opened flower remain enveloped together inside the tube like structure of jointed petals called "keel" leading to cleistogamous like self-pollination.
- Flowers open only once between 7.00 and 9.00 a.m. depending on the weather condition and never closes upon fertilization, though the corolla is shed within few days of blooming.
- Though the flowers open in the morning, the dehiscence of anthers is much earlier by evening hours of the previous day.

- The stigma is receptive two days before flowers open but receptivity is maximum on the day of anthesis.
- Even though the pollen is shed from the anthers in the evening before the flower opens, the pollen is not transferred to the stigma at this time unless the flower is agitated or tripped by a bee.
- A short time after opening, the stigma retracts into the keel and comes in contact with pollen from the anthers and self-pollination occurs.
- After pollination, the pollen tube grows toward the ovary and fertilization takes place about 8 to 9 hours later. Pollen tubes from foreign pollen may grow more quickly than self-pollen, leading to a higher incidence of cross-fertilization.
- Pollen viability is at maximum between 5.00 and 9.00 a.m. on the day of anthesis.
- Since the dehiscence of anthers is much in advance of the blooming, the emasculation needs to be carried out in mature flower buds in the preceding afternoon hours.
- French bean flowers are highly sensitive and drop off easily with slight mechanical disturbance or injury.
- To prevent drying out emasculated bud, leaflet may be folded and pinned around the bud. A tissue paper can be used to cover and protect the bud.
- Being a large flowered leguminous vegetable crop, it is sufficient to open only one side of the flower bud by careful removal of the keel which remains spirally around the stamen and style. Then emasculation of the anthers is done carefully.
- Emasculation of matured buds in the afternoon hours followed by hand pollination in the next day morning between 6.00 and 8.00 a.m is the best method of hybridization.
- Pollens are collected next morning from a freshly opened flower. The standard and wings of the flower from the intended male parent are removed. By careful removal of the spiral keel, stigma and hairs beneath covered with the pollen grains itself can be used as a brush for pollination.

Origin and taxonomy

French beans are thought to have been originated in warm temperate regions of South and Central America. Southern Mexico and Guatemala are considered to be the primary centre of origin, while secondary centres lies in Peru-Bolivia-Ecuador region of American continent. The cultivation of French beans was started 7000 years ago, by the Indian tribes settled in Tehuacan Valley of Mexico and in Callejon de Huaylas, Peru. When Christopher Columbus returned from his second voyage to the New World in the year 1493, he brought French beans with him in the Mediterranean region. French beans were earlier considered to be rare to find and expensive but soon became one of the commonly used beans in the 19th century. In France, French beans were introduced in the year 1597 by the Conquistadors. In Europe, the French bean spread rapidly in the 16th and 17th centuries and reached England by 1594. It was brought to India from Europe during 17th century. *Phaseolus aborigineus* is known as the progenitor of *P. vulgaris*. During the process of domestication in common beans, several morphological changes like growth habit, testa colours, pod structure, etc. have occurred.

Improved varieties

The French beans are classified into string and stringless beans based on the extent of fibre in pods. Wax-podded varieties are not common and mostly confined to home garden. The bean varieties are grouped according to two basic growth habit—pole or climbing beans and bush beans, which are dwarf in nature. Important varieties grown in India are:

- Bush type: Pusa Parvati, Arka Komal, Arka Subidha, Contender, Bountiful, Jampa, Giant Stringless, Pant Anupama, VL Boni-1, YED-1
- Pole type: Kentucky Wonder, Pusa Himlata, TKD-1, KKL-1.
- Resistant to angular leaf spot: Pant Anupama, Lakshmi, SVM-1.
- Resistant to powdery mildew: Contender, Pusa Parvati.
- Resistant to wilt: Jampa.
- Resistant to mosaic disease: Pant Anupama, Contender, Pusa Parvati.

Cultural requirements

- French bean is a day nentral and cool weather crop but thrives well in wide range between cool to moderate temperature.
- A mean air temperature of 20°–25°C is optimum for its growth and high pod yield. Extreme high temperature interfere with pod filling, while low temperatures are unfavourable for vegetative growth.
- It cannot tolerate frost.
- Its seeds do not germinate below 15°C and plants drop blossoms in hot or rainy weather.
- A favourable soil temperature is 18°–24°C. French bean is grown over a wide range of well-drained, alluvial friable soils, but it cannot withstand extreme acidic and alkaline soils.
- It requires a warm sunny situation in a rich well-drained preferably light soil with plenty of moisture in the growing season.
- Clay soils impede the emergence of seedling leading to uneven or poor stand.
- The optimum soil pH is 5.5–6.8. Liming is needed if soil pH is less than 5.5.
- For optimum nitrogen fixation, good soil aeration is required.

Sowing time

- August to September and January to February in plains.
- March to beginning of May in hills.
- September–November in southern and western India.
- October in both north-east and central zone.

Land preparation

- Prepare the land in well advance with repeated ploughing (at least 4-5 ploughing) to a good tilth.
- Remove all stubbles, weeds, etc. from the land.
- Better growth and spread of roots is necessary for adequate fixation of atmospheric nitrogen by root nodules.
- Add well rotten organic manure preferably farm yard manure or compost @ 15-20 t/ha during land preparation and level the land properly.

Sowing the seeds

- French bean is a direct seeded corp.
- Seeds are dibbled in the hills in ridge and furrow method of planting.
- Irrigation is given through the furrows.
- Viable, healthy, well-matured and pure seeds should be used for sowing. About 25-30 kg/ha seed is enough for pole type and 50-75 kg/ha for bush type.
- The seeds must be treated before sowing to avoid losses due to fungal diseases. Mixture of thiram + Bavistin (2g + 1g/kg seed) is recommended for seed treatment. The seeds should be mixed thoroughly with the required quantity of fungicide before sowing.
- Seeds can be drilled by local implements or by tractor hauled seed drills. The seeds should be sown 2–3 cm deep in the soil.

Spacing

Pole type : 90-100 cm × 15-20 cm

Bush type : 30-45 cm × 10-15 cm

Nutrient management

- French bean, being a leguminous crop fixes atmospheric nitrogen through root nodules and the comparative nitrogen requirement is less. Soaking of seeds with 30 g Rhizobium bacteria (*Rhizobium phaseoli*) in 10% molasses solution per kg of seed is effective for increased nodulation.
- Apply 15-20 tonnes/ha of well decomposed FYM or compost well in advance.

- Apply a general fertilizer dose of 50 kg N, 75 kg P_2O_5 and 60 kg K_2O/ha.
- Full of P_2O_5 and K_2O and half of N is given as basal and rest half N during earthing up after 3rd week of sowing.
- French bean also responds to micronutrients. The foliar application of B, Cu, Mo, Zn, Mn and Mg each of 0.1% is effective in enhancing quality and pod yield.

Irrigation

- Pre-sowing irrigation must be given for better seed germination.
- French bean is shallow-rooted. It is sensitive to both water excess and water stress conditions.
- About 6–7 irrigations are needed during the growing season.
- The plants are susceptible to water stress at critical periods of growth—pre-blooming, flowering and pod-filling stages.
- The frequency of irrigation depends upon season, soil type and organic-matter content. Thus, the moisture level should be near to field capacity particularly during flowering and pod-formation stage.
- Deformed pods can result from water stress due to low moisture or excessive evapo-transpiration losses. Flooding in French bean causes anoxia leading to root rot. Therefore, moisture should be evenly distributed throughout the growth period.
- About 300–350 mm of water is usually sufficient for French bean.
- The crop should be irrigated as soon as the soil moisture tension reaches 0.5 bar in the top 10 cm of the soil.

Interculture and weed control

- Beans are susceptible to injury during cultural operations after plants begin to flower. Cultivation or working among plants, where foliage is wet should be avoided, because spores of anthracnose and angular leaf-spot diseases are easily spread under wet conditions.
- Bush beans do not require staking, while pole varieties are staked for obtaining high yield. The vine grows clockwise around the support. Two wooden or iron poles are used at the two ends of the rows and lines of the wires are stretched. Fine rope or jute string is fixed in zigzag fashion between the lines of the wires.
- In temperate climate, pole beans are grown in mixture with corn as the stalk is used for support of the vines.

- Annual grasses cause main problem in the French bean planted early in the growing season. The weed crop competition starts 2–3 weeks after planting. Inadequate weed management practices reduce the yield drastically.
- In most of the beans hoeing is done by hand. The narrow spacing reduces crop-weed competition and hoeing increases yield. About 2 weedings may be needed till the plant can smother the weeds. A pre-emergence application of Alachlor @ 2–2.5 kg a.i./ha is recommended for an effective control of weeds.
- In hills, French beans are intercropped with maize, amaranth and foxtail millets. The maize + French bean + wheat cropping sequence is commonly followed. In plains, French bean is intercropped with early potato. In mid-hills of Himachal Pradesh, pole beans are sometimes grown as relay after tomato crop to use the same stakes.
- Application of plant growth regulators like, 2 ppm, PCPA, 20-25 ppm NAA or 50 ppm GA increase pod set in unfavourale condition.

Harvesting the crop

- French beans are harvested when pods are tender (approximately 7–12 days after flowering).
- Bush beans mature relatively in a short period of 45-50 days requiring 2–3 pickings, while pole beans take 60–75 days for first picking and require 3–5 pickings.
- The beans are generally harvested by hand. Delayed harvesting reduces the quality of pods as they become fibrous.
- Mechanical pickers are also devised for harvesting of beans especially for processing type varieties. These employ 'Once over' destructive harvest, which strips leaves and remove pods from the plants.
- Shelled beans are harvested when their seeds achieve full size and become relatively firm. The seeds are separated from pods and empty pods discarded.

Post harvest handling and storage

- French beans are highly perishable and rapid cooling after harvesting is important to maintain quality.
- After harvesting, beans are washed and culled and diseased, inferior pods are destroyed. Before these are placed in storage or transit, the beans should be cooled as rapidly as possible.
- Optimum storage and transit temperatures are 5°–7°C. Under these conditions storage life increases up to 20–25 days. At temperatures below 1°–2.5°C, chilling injury occurs 10–12 days after storage.

- In India, bulk of French bean is consumed as fresh, while in advanced countries like USA more than 80% crop is processed. Processors have strict requirements for specific varieties such as deep seed coat/white colour and uniform crop maturity though white colour is preferred. The varieties with fleshy pods are widely used for commercial freezing.

Plant protection measures

Physiological disorder

- Flower initiation and development are greatly delayed under sub-optimal temperatures, especially below 10°C, where fertilization may not occur, producing small and mis-shapened pods.
- Blossoms drop and ovule abortion are the common problems at high temperature (35°C). Therefore, planting crop at suitable time and place are of utmost importance.
- Transverse cotyledon cracking takes place when dry seeds of beans are sown in wet soils. This occurs since susceptible cultivars imbibe water rapidly. Seed moisture content below 12 % also enhances cracking of cotyledons. White-seeded cultivars are more prone to this malady. The resistant cultivars show a germination of 60–80%, while susceptible ones 20% or less. Therefore, resistant varieties with hard seed coat, optimum seed moisture content and planting of crop at suitable time are essential to avoid this disorder.
- The necrosis of hypocotyl is associated with low Ca content in seed after germination. Soils rich in Ca and Mg offset this problem.

Diseases

Bean anthracnose (*Colletotrichum lindemuthianum*)

It is most serious seed borne fungal disease of bean. The most characteristic symptoms of the disease are black, sunken, crater like cankers on the pods. The lesions remain isolated by yellow-orange margins and under high humidity, dull salmon coloured ooze is formed in the centre.

Control measures

- Seed treatment with 0.2% Bavistin or 0.3% thiram.
- Spraying the seedlings with 0.3% captan or 0.1% Bavistin or chlorothalonil 0.2% or hexaconazole 0.1%.
- Spray the crop with hexaconazole 0.1% followed by 0.3% Blitox after removal of the infected twigs at 10 days interval.

White mold (Sclerotinia sclerotiorum)

Symptoms are usually first visible about one week after full bloom because blossoms generally are the first part of the plant to be colonized by the fungus. Leaves, stems, and pods in contact with the colonized blossoms can then become infected, provided moisture is present. Initial symptoms on these tissues are pale colored, water-soaked lesions. These lesions enlarge and within a few days, become covered with a white, cottony fungal growth. Leaves of severely diseased plants become yellow and eventually turn brown and fall off. As the disease progresses, the infected plants wilt and the leaf canopy opens. The fungus may eventually invade and kill all above ground parts of the plants.

Control measures

- Successful control of white mold with fungicides depends upon spraying at the proper time and completely covering the plants with the chemical, especially inside the plant canopy.
- Application of carbendazim / benomyl @1.5 g/l or carbendazim (1g) + mancozeb (2g)/l at 10 days intervals for 2-3 times is very much effective.

Bean rust (Uromyces fabae)

In this fungal disease, dark brown coloured rust pustules are developed on the foliage of bean with chlorotic hallo. Infected leaves usually wither and die.

Control measures

- Spray the crop with Bavistin or Benlate or Topsin M or tebuconazole @ 1g/litre of water 2 to 3 times at 15 days interval.

Bean common mosaic virus (BCMV)

It is still an important disease of beans worldwide. Typical symptoms in bean consist of green mosaic and downward cupping along the main vein of each leaflet. Green vein-banding, blistering, and malformation are common in leaves of the same plant. Plants are reduced in size, and pods may be mottled and malformed.

Control Measures

- BCMV is most effectively controlled by growing resistant cultivars and using virus-free seed.

Insect pests

Stem fly (Ophiomyia phaseoli)

It is a polyphagous pest. The adult flies are metallic black with hyaline wings and the newly hatched maggots about a millimeter long are white but become yellow at later stages. The maggots mine the leaf, bore inside the petiole and tender stem tunneling inside which cause girdling effect and death of the plant. The adults puncture the leaf and the affected leaves become yellow and dry up.

Control measures

- Remove and destroy all affected plants.
- Spray with Nuvacron 1.5 ml/l of water twice at weekly interval from the time of complete germination of seeds
- Spray the crop with 3-4% neem seed extract reduces the incidences.

Hairy caterpillers (*Ascotis imparata* and *Spilosoma obliqua*)

The larvae cause characteristic skeletonization of the leaves during the early gregarious stage and later they completely denude the plants.

Control measures

- Remove and destroy the leaves with gregarious larvae during the early stage.
- Spray the crop with Nuvacron @ 1 ml/l, or Thiodan @ 1.4 ml/l or malathion @ 2 ml/l or Sevin W. P. @ 4 g/l of water
- Biological control measure by applying a nuclear polyhedrosis virus (NVP) @ 250 LE/ha/week starting from first appearance of young leaves.
- Spray the crop with 5% neem seed kernel extract reduces the incidence

Aphids (*Aphis craccivora*)

The gregarious aphids suck sap from growing tips and later from the whole plant which causes curling of leaves, reduces vigour, yield and quality. They also transmit mosaic viruses.

Control measures

- Spray the crop with 5% neem seed kernel extract along with 2% starch reduces the incidence.
- Encourage lady bird beetles in the crop field to predate upon the aphid species.
- Spray the crop with Rogor @ 1.5 ml/l of water.
- Apply carbofuran granules in the field (Furadon 3G @ 1.0 Kg a.i. /ha) before sowing of the seeds.

Yield

- Bush type: 6.0-8.0 t/ha.
- Pole type: 12.0-15.0 t/ha.
- Dry seed: 1.5-2.0 t/ha.

24. Vegetable cowpea

Scientific Name: Vigna unguiculata
Family: Fabaceae (Leguminasae)
Origin: Africa
Chromosome no: 2n = 2x = 22

Introduction

Cowpea is in cultivation from very ancient times in the tropics of old world. The cowpea, whether utilized for green pods as vegetable or dry seed as pulse, forms an important component of farming systems from the arid to the humid tropics. In fact, it probably has the greatest potential among all food legumes in the semi-arid to sub tropical areas. Cowpea cultivars grown for the immature green pods as vegetable are variously known as asparagus bean, snake bean and yard long bean and when grown for dry or immature seeds, they are known as black-eye pea, kaffir pea, China pea and southern bean. Cowpea is well adapted to stress and has excellent nutritional qualities. It is a key dietary staple for the poorest sector of many developing countries of Africa and Latin America and greatly improves an otherwise bland and unbalanced diet based on sorghum and millets. Cowpea is grown extensively for dry seed in the African countries particularly, Nigeria, Niger, Burkina Faso, Ghana, Kenya, Uganda, Malawi, Tanzania, Togo and Senegal. The major cowpea producers in Latin America are Brazil, Venezuela, Peru, Panama, El Salvador and Haiti. Asian production includes that of yard long bean as vegetable and major cowpea producing countries are India, SriLanka, Bangladesh, Myanmar, China, Korea, Indonesia, Nepal, Pakistan, Phillippines, Thailand and Malayasia. In the USA cowpea is grown both for immature pod and seed and dry seed.

In India, cowpea is known since Vedic times. It is grown widely throughout the year for all forms – tender pod, dry seed, fodder, green manure and cover crops both as sole and mixed crop.

Importance and use

- It is the key dietary staple for the poorest sector of many under developed and developing countries of Africa, Latin America and Asia.
- Cowpea is the crop of all round utilization, grown for tender pods and seeds, dry seeds as pulses, green leaves and even roots.
- The typical vegetable cowpea has higher monosaccharide: polysaccharides ratio.
- The immature pods and seeds as well as dry seeds besides being widely used as fresh vegetable and pulse are also frozen and canned.
- Some unusual utilizations of cowpea include baking powder for biscuits, vegetable milk and legume starch.
- It is tolerant to drought and most of the soil stress, thus can be grown over a wide range of environmental conditions.
- Effective cowpea – *Rhizobium* symbiosis can fix about 150 kg nitrogen per hectare.
- It is also grown for hay, silage, pasture for all types of live stock.
- It is also used as fodder and green manure crop.

Nutritional value

- Cowpea pods are good source of protein, fiber, minerals, calcium and vitamins particularly vitamin A and vitamin C.
- It contains 8.0 g carbohydrates, 4.3 g proteins and 0.6 g fat, 2.0g fiber per 100 g of edible portion.
- Amino acid profile particularly lycine, leucine and phenylalanine contents are relatively high in cowpea which greatly improves the protein quality of cereals.
- It provides high of 0.9 g minerals per 100 g of edible part.
- Tender fruits contain 80 mg calcium, 74 mg phosphorus and 2.5 mg iron per 100 g fresh.
- It provides 941 I.U of vitamin A per 100g edible portion.
- It is a good source of Vitamin B like, 0.09 mg Riboflavin and 0.07 mg Thiamine per 100 g fresh.
- It contains 13.0 mg of vitamin C per 100 g fresh.
- Antinutritional factors like, trypsin inhibitor, chymotripsin etc. present in dry seeds can easily be removed by boiling in fresh water before cooking.

Medicinal value

- Among the food legumes it is particularly important for the supply of antioxidants and especially β-carotene in the foods.
- Cowpea is particularly important for the supply of vitamin B in the foods.
- Cowpea contains high amount of dietary fibre which gives protection against certain types of cancer particularly colon cancer, regulates transit and lowers blood cholesterol.

Adaptation in India

Cowpea is grown throughout India for its tender pods different length and colour (green, dull green, whitish, purple) to be cooked as vegetable or for dry seeds used as pulses and foliage as fodder or green manure. In India, cowpea cultivars grown for tender pods as vegetable are either true Sequipedalis forms (sub species *sesquipedalis*) originally introduced from southeast Asia and integrades (intermediate types) of Unguiculata (sub species *unguiculata*) and true Sequipedalis forms, while the cultivars grown for dry seed are either true Unguiculata forms or integrades of Unguiculata and Biflora (sub species *cylindrica*) forms and the cultivars grown for fodder are either Biflora (sub species *cylindrica*) or integrades of Unguiculata and Biflora forms.

Breeding behavior

- It is highly self-pollinated crop due to hermaphrodite sex form, homogamy i.e. maturation of stigma and anthers at the same time and dehiscence of anthers much earlier than anthesis.
- Stamens and pistil in opened flower remain enveloped together inside the tube like structure of jointed petals called "keel" leading to cleistogamous like self pollination.
- Infestation of some beetles which bore through the keel tube leads to some cross pollination in hot and humid environment.
- Flowers open only once between 7.00 and 9.00 a.m. On cloudy days the flowers may open even in the afternoon.
- Though the flowers open late in the morning, the dehiscence of anthers is much earlier which may vary from 10.00 p.m at night to 1.00 a.m.
- Anther dehiscence is influenced by environmental factors like presence of moonlight, a clear sky and a dry warm atmosphere and during dark nights the dehiscence tends to be delayed.
- Stigma becomes receptive and pollens are fertile on the day of anthesis.
- Pollen remains viable for 42 hr at room temperature of 28.5°C and relative humidity 91%.
- Since the dehiscence of anthers is much in advance of the blooming, the emasculation needs to be carried out in mature flower buds in the preceding evening.
- Cowpea flowers are highly sensitive and drop off easily with slight mechanical disturbance or injury.
- To prevent drying out emasculated bud, leaflet may be folded and pinned around the bud. A tissue paper can be used to cover and protect the bud.
- Emasculation of matured buds in the afternoon hours followed by hand pollination in the next day morning between 8.00 and 10.00 a.m is the best method of hybridization.
- Pollens are collected next day morning from a freshly opened flower. The standard and wings of the flower from the intended male parent are removed. By slight depression of the keel, stigma covered with the pollen grains protrudes out which it self can be used as a brush for pollination.

Origin and taxonomy

The cowpea is one of the most ancient human food sources and has probably been used as a crop plant since Neolithic times. All evidence points to its originating in

Africa, although where the crop was first domesticated is uncertain. Among the African countries Ethiopia, central and south Africa and West Africa are considered to be the probable centres of domestication of cowpea. In India, cowpea is known since Vedic times. However, it is generally agreed that the cowpea is of African origin as conspecific wild forms are found in Africa but are absent in Asia. However, India is considered the modern centre of diversity of cowpea cultivars. The earlier view was that the progenitor of cowpea was the wild var. *dekindtiana* of sub-species *dekindtiana* but later it was suggested that var. *mensensis* of sub-species *dekindtiana* was the most likely progenitor of cowpea.

In the African context, the role of cowpea is predominantly that of pulse and African use of pulse continues, but after the introduction of Unguiculata forms (sub-species *unguiculata*) to southeast Asia and India, two other subspecies were evolved: sub-species *sesquipedalis* (vegetable type) mainly in southeast Asia and sub-species *cylindrica* (fodder type) in India and southeast Asia under the predominant influence of human selection. The other sub-species *textilis* was probably selected from *unguiculata* forms in West Africa for fibre from its long peduncle.

Cowpea belongs to the sub-family Faboideae under the family Fabaceae (Leguminoceae) with in the order Fabales. Classification of cowpeas was a matter of debate as they were ranked as species, varieties or sub-species. In the general classification, cowpeas belong to the botanical species *Vigna unguiculata L.* with four cultivated sub-species, *unguiculata, sesquipedalis, cylindrica* and *textilis* and two wild sub-species *dekindtiana* and *mensensis.* Some authorities however, do not consider the cultivated sub-species as distinct and lumped them under one sub-species *V. unguiculata,* sub-species *unguiculata* and differentiate them by intraspecific category "cultigroup". The sub-species *unguiculata, sesquipedalis, cylindrica* and *textilis* are renamed as cultigroup Unguiculata, Sesquipedalis, Biflora and Textilis. Similarly, the wild sub-species *dekindtiana* and *mensensis* were put under single sub-species *dekindtiana* and distinguished them by varietal category.

Improved varieties of vegetable cowpea

- Indeterminate, viny type: Pusa Barsati, Arka Garima, Yard Long Bean.
- Determinate and bushy type: Pusa Phalguni, Pusa Dofasli, S-408, Pusa Komal, Asseem, Pusa Rituraj, Narendra Lobia-1, Bidhan Barbati-1, Bidhan Barbati-2,CO-2, Arka Garmia, Arka Samrudhi, Sel. 263, Arka Suman.
- Compact, bushy and dual type: Pusa Rituraj, Narendra Lobia-1, Asseem.

Cultural requirements

- Cultivars of the viny and vegetable type under *V. unguiculata,* sub-species *sesquipedalis* (cultigroup Sesquipedalis) are photosensitive in nature and basically short day crop.

- Integrades or intermediate types of the sub-species *unguiculata* (cultigroup Unguiculata) and *sesquipedalis* (cultigroup Sesquipedalis) mainly cultivated as vegetable cowpea show both photo sensitive and photo non-sensitive characters.
- Vegetable cowpea cultivars developed through combination breeding involving the cultivars of sub-species *unguiculata* (cultigroup Unguiculata) and *sesquipedalis* (cultigroup Sesquipedalis) generally show photo non-sensitive character.
- Cultivars under sub-species *unguiculata* (cultigroup Unguiculata) and sub-species *cylindrica* (cultigroup Biflora) generally show photo non-sensitive character.
- It grows best in long and warm growing season.
- It grows well under wide temperature range of 21°C-35°C.
- The best temperature for its seed germination is 18°C to 25°C.
- Better growth and development takes place at 21°C to 27°C
- It is susceptible to low temperature and frost.
- It does not grow well at high temperature.
- Cowpea is a drought hardy crop and comes well under rain fed condition tolerating moderate dry spells.
- It can not withstand heavy rainfall and water logging condition.
- It can grow on almost all types of soil of average fertility with optimum pH range of 5.5 to 6.5.
- Well drained, fertile, sandy to sandy loam soils are ideal.

Land preparation

- Prepare the field with 4-5 deep ploughing to a depth of 25 cm.
- Remove all the weeds, stubbles and level the land by laddering.
- Manures should be mixed with the soil at the time of last ploughing.
- The plant is susceptible to water logging so land should be well drained.
- Cultivars of indeterminate and bushy types are generally grown on flat beds of convenient size.
- Cultivars of the indeterminate, viny types are generally sown in the mounds and trained along the stakes.
- Beds and mounds are prepared with the mixture of 15-20 tonnes of FYM per hectare and basal dose of fertilizers are mixed with the soil.

Sowing time

Cowpea can be grown in spring-summer, rainy seasons, rainy-autumn and early autumn season. In places having mild climate, where summer and winter are moderate, it can be grown round the year.

- January- February : Spring-summer season
- May-June : Summer-rainy season
- July-August : Rainy-autumn season
- August-September : Early autumn

Seed sowing

- Cowpea seeds are directly sown in the field.
- Seeds are sown by hand dibbling in the bed.
- The seed can be sown by fertilizer cum seed drill or by dibbling in line which is operated by a tractor, bullock or manual labour.
- Sowing is also done in a well-prepared and leveled field by broadcasting method.
- As long as possible emphasis should be given on line sowing method.
- Inoculate the seed with *Rhizobium* culture before sowing for quick nodulation on the roots which fix atmospheric nitrogen.
- About 400 - 500 gram *Rhizobium* culture is sufficient for treating the seeds required for one hectare.
- Determinate compact types are sown in rows by drilling behind the plough.
- 20-25 kg seed/ha is sufficient for bush type cultivars grown in spring-summer season.
- 12-15 kg seed /ha is sufficient for bush type cultivars grown in rainy season.
- 8-10 kg seeds/ha is sufficient for indeterminate, viny cultivars grown in rainy-autumn and early autumn season.
- Seeds require sufficient moisture for germination.
- It is better to do seed sowing in pre-irrigated moist soil than irrigate the field after sowing.

Spacing

- Indeterminate, viny type cultivars as sole crop: 75 × 60 cm
- Bushy and vigorous cultivars : 50-60 × 50-60 cm
- Bushy and semi-vigorous cultivars : 45-50 cm × 25-30 cm

Nutrient management

- Soaking the seeds with Rhizobium bacteria (*Rhizobium vignae*), 30 g in 10% molasses solution per kg seed is effective for increased nodulation.
- Although cowpea is a legume crop, it responds well to the application of fertilizers particularly phosphatic fertilizers.
- Phosphorus helps in better nodulation in the roots through *Rhizobium* symbiosis.
- Its requirement of nitrogen fertilizer is comparatively much less because of being the legume crop.
- Well rotten farm yard manure at the rate of 15-20 t/ha should be applied at the time of land preparation.
- Apply a general fertilizer dose of 25 kg N, 75 kg P_2O_5 and 60 kg K_2O per hectare. 1/2 N along with full P and K fertilizers and farm yard manure should be given as basal and rest N should be top dressed 25-30 days after sowing during earthing up.
- The N, P, K fertilizers as basal should be applied at the time of sowing in bands about 7-8 cm to the side and slightly deeper than the seeds.
- Nitrogen fertilizers after 25-30 days of sowing should be top dressed by broadcasting around the plant 6-7 cm away from the plant stem.

Irrigation

- Cowpea is shallow rooted crop and requires comparatively less moisture for its growth and development.
- Pre-sowing light irrigation helps in better germination of seeds.
- Irrigate the crop after germination if moisture is insufficient particularly during spring-summer season.
- It is a hardy crop and can tolerate low soil moisture.
- The crop must be irrigated during its most critical stages i.e. flowering and pod development stages.
- Irrigation prior to the flowering helps in good pod setting and irrigation should be given after the pods have set.
- It is essential to maintain available soil moisture atleast above 50% during flowering and pod development.
- Irrigate regularly at 7-10 days interval during dry periods to get high yields particularly during spring-summer and early autumn season.
- It is sensitive to heavy irrigation and water logging.
- Light irrigation is most suitable for successful luxurious growth.

Interculture and weed control

- Indeterminate, viny cultivars must be given support for better growth, development and pod yield.
- The crop needs to be trained along the stakes made up of bamboo or jute sticks upto about 1.0 m high above the ground.
- In mixed cropping, the pole type cultivars get support on the stalks of maize and sorghum after their earheads are harvested.
- At the initial stages of weed growth, shallow hand weeding and hoeing are sufficient to check the weed growth.
- At the later stages of crop growth, the weeds are kept under check due to the thick canopy of the crop.
- Earthing up about four weeks after sowing helps in root aeration as well as weed control.
- Spraying of growth regulator malic hydrazide at 50-200 ppm concentration just before flowering increase the yield of pod.
- Application of herbicides effectively controls weeds.
- Pre-sowing application of fluchloralin @ 2litres/ha checks the weed growth for 20-25days.

Harvesting the crop

- The crop duration of the indeterminate, viny types is about 125-160 days and of the indeterminate, bush types is 80-100 days depending on the varieties.
- Harvest the pods periodically by hand picking when they are fully grown and succulent and have not become more fibrous.
- Normally harvesting starts, 60-65 days after sowing in bushy cultivar and 75-80 days after sowing in case of indeterminate, viny cultivars.
- In pole types about 5-9 pickings of tender pods can be done depending on the variety.
- In bush types generally harvesting is completed by 2-3 pickings of tender pods.
- As harvest is delayed, the total yield increase but the quality falls rapidly.

Post harvest handling and storage

- Remove fibrous and damaged pods from the harvested lots.
- Wash the selected pods in running water before sending to market.
- Quality of the tender pods deteriorates very quickly under ambient condition because of their very high respiration and transpiration rate.

- High transpiration rate makes the pods leathery rendering them unfit for marketing and consumption.
- Fresh green cowpea pods can be kept for two weeks at O°C and 90-95% relative humidity.
- Fresh pods can be kept for about 15-16 days in cold storage.
- For dry seed purpose, the pods are allowed to full maturity on plants and then the crop is harvested and threshed after proper drying.
- The harvested pods are dried for a few days before threshing.
- About 8-9% seed moisture is best for safe storage of dry seeds.

Plant protection measures

Diseases

Powdery mildew (Erysiphe polygoni)

This common fungal disease generally appears late in the season during dry summer. Faint, slightly discolored specks are first formed on leaves. From these specks grayish white powdery growth of the mycelium and spores spread over the leaves, stem and even fruits. In severely attacked plants, fruits either do not set or remain very small.

Control measures

- Growing resistant varieties like Pusa Komal.
- Spray the crop with 0.5% sulfex or other formulation of wettable sulphur as soon as disease appears.
- Spray the crop thrice with systemic fungicide like Bavistin or tridemorph @ 0.1% concentration at weekly interval.

Anthracnose (*Colletotrichum lindemuthianum*)

This seed borne fungal becomes most severe in high rainfall areas of subtropical climatic condition. Tissues of necrotic spots on leaves collapse become thin and papery and produce pinkish slimy mass in high humidity condition. Similar lesions appear on the cotyledons and stems of the young seedlings.

Most characteristics symptoms appear on the pods where dark brown, sunken lesions with raised, reddish or yellowish margins are found. Under high humidity condition, dull salmon or pinkish coloured oozes or slimy mass is formed at the centre of the lesion.

Control measures

- Use healthy and disease free seeds.
- Clean cultivation and crop rotation.

- Treat the seeds with carbendazim @ 2.5g/kg of seeds or thiram @ 2.5 g/kg of seeds.
- Spray the crop with 0.1% Bavistin or hexaconazole or 0.2% chlorothalonil at 10 days interval.

Ashy stem blight (*Macrophomina phaseolina*)

This soil borne fungal disease becomes severe in warm and humid climate. In early stages, it causes damping off and killing of young seedlings where black cankerous lesions appear on the cotyledons. Collar portion of the plant becomes brown and vascular bundle of the roots also turns brown resulting in dry root rot.

Control measures

- Use disease free and healthy seeds.
- Dry seed treatment with carbendazim/captan/thirum @ 2-3g/kg of seed.
- Apply adequate organic matter in soil.
- Soil application of *Trichoderma viride* @ 3 kg/ ha along with manure is effective.
- Spraying the crop with 0.1% carbendazim / carbendazim (1g) + mancozeb (2g)/l or captan @ 2.5 g/l along with sticker checks the rapidity of the infection.

Bacterial blight (*Xanthomonas vignicola*)

It is both soil and seed borne bacterial disease. The primary damage is caused through high mortality of the seedlings, especially when the seeds used for sowing are from a severely diseased crop. Infected leaves show light yellow, irregular to circular spots with necrotic brown centre, later changing to straw colour. Dark green water-soaked spots of variable shape and size appear on pods which later becomes yellow and dry.

Control measures

- Growing resistant varieties like, Pusa Komal.
- Use of disease free healthy seeds.
- Seed treatment with 0.01% streptocycline solution for 30 minutes before sowing.
- Spraying of 0.01% streptocycline + 0.2% copper oxychloride.

Mosaic (Cowpea mosaic virus)

Mosaic virus of cowpea is seed borne as well as vector transmitted (aphid). The affected leaves develop a typical mosaic of broad and raised dark-green patch along with chlorotic spots. The disease plant remains stunted with reduced and malformed leaves.

Control measures

- Use the seeds that have been harvested from mosaic free plants.
- Remove and destroy the affected plants as early as possible.
- Spray the crop with systemic insecticides like Rogor or Metasystox @ 2 ml / l of water to control the insect vector

Insect pests

Stem fly (Ophiomyia phaseoli)

It is a polyphagous pest. The adult flies are metallic black with hyaline wings and the newly hatched maggots about a millimeter long are white but become yellow at later stages. The maggots mine the leaf, bore inside the petiole and tender stem tunneling inside which cause girdling effect and death of the plant. The adults puncture the leaf and the affected leaves become yellow and dry up.

Control measures

- Remove and destroy all affected plants.
- Spray with Nuvacron 1.5 ml/l of water twice at weekly interval from the time of complete germination of seeds.
- Spray the crop with 3-4% neem seed extract reduces the incidences.

Pod borer (Adisura atkinsoni, Exelastis atomosa, Helicoverpa armigera and spotted borer *Maruca testulalis*)

Among the different borer complex attack cowpea, the spotted borer is found to be the most destructive under Bengal basin. The caterpillar after hatching bore into the developing pods and feed on the seeds therein. Severely infested crops are subjected to havoc economic yield loss leaving behind most of the affected fruits.

Control measures

- Remove and destroy the infected pods at the initial stage of attack.
- Spray the crop with Nuvacron @ 1 ml/l, or Thiodan @ 1.4 ml/l or malathion @ 2 ml/l or Sevin W. P. @ 4 g/l of water
- Biological control measure by applying a nuclear polyhedrosis virus (NVP) @ 250 LE/ha/week starting from first appearance of young leaves.
- Spray the crop with 3-4% neem seed kernel extract reduces the incidence

Aphids (Macrosiphum pisi, Aphis craccivora, Myzus persicae)

Aphis craccivora is the most damaging to cowpea and other leguminous crop grown in the Gangetic plains of West Bengal. The gregarious aphids suck sap from growing tips and later from the whole plant which causes curling of leaves, reduces vigour, yield and quality. They also transmit mosaic viruses.

Control measures

- Spray the crop with 3-4% neem seed kernel extract along with 2% starch reduces the incidence.
- Encourage lady bird beetles in the crop field to predate upon the aphid species.
- Spray the crop with Rogor @ 1.5ml/l of water.
- Apply carbofuran granules in the field (Furadon 3G @ 1.0 kg a.i. /ha) before sowing of the seeds.

Flower thrips (Haplothrips verononiae, Frankliniella sulphurea)

Flower thrips are very destructive to cause havoc bud and flower drops resulting poor yield. The adult and nymphs suck sap from the petals, calyx and developing ovary.

Control measures

- Spray the crop with acetamiprid @ 0.5 g/l at 10 days interval at the early stage.
- Application of 3-4% neem seed kernel extract at later stage of the crop is useful.

Root knot nematode (Meloidogyne spp)

Nematodes attack the roots and produce galls there which cause stunting of the plants and severely reduce yield.

Control measures

- Plough the field in summer months.
- Apply oil cakes, more specifically neem cake in the field during land preparation.
- Apply granular insecticides like Thimet 10G @ 25 kg/ha or Furadon 3G @ 35 kg/ha in soil during land preparation.

Yield

Yield of cowpea varies according to region, cultivar and duration of the crop.

Tender pods

- Indeterminate, viny cultivars: 12.0 to 18.0 t/ha
- Bush type cultivars: 7.0 to 12.0 t/ha

Dry seeds: 1.5- 2.0 t/ha.

25. Hyacinth bean

Scientific Name: Lablab purpureus (L).Sweet.
(Syn. Dolichos lablab Roxb.)
Family: Fabaceae (Leguminaceae)
Origin: India
Chromosome No.: 2n = 2x = 20, 22

Introduction

Hyacinth bean is one of the most ancient among the cultivated plants. India is probably its place of origin and wild forms are found here. It is a tropical crop and mainly grown South Asia (India, Bangladesh), South-east Asia and Africa. It is a common bean in Madhya Pradesh, Maharashtra, some ares of Gujarat, Tamil Nadu, Andhra Pradesh, Haryana, eastern Uttar Pradesh and West Bengal. In India, it is grown as a monocrop, intercrop or mixed crop with ragi, sorghum, bajra, maize or castor. Indeterminate pole types as monocrop are grown over the bower. In mixed cropping, the pole type hyacinth bean plants get support on the stalks of ragi and sorghum after their earheads are harvested.

Importance and use

- It is one of the most ancient among cultivated plants.
- It is also called field bean, Indian bean, Indian butter bean,lablab bean, dolichos bean,Egyptian bean, Australian bean, bonavist bean, lobia bean, sem and waby salad bean.
- It is an excellent pod vegetable and mostly grown for fresh whole tender pod.
- In India it is grown as a monocrop, intercrop or mixed crop with ragi, sorghum, bajra, maize or castor.
- Generally cultivars belonging to *Lablab purpureus* var. *typicus* are grown for soft and edible pods as vegetable.
- Edible podded, bush type cultivars have been developed through combination breeding approach involving the viny cultivars of *Lablab purpureus* var. *typicus* and bushy cultivars of *Lablab purpureus* var. *lignosus.*
- Fresh seeds as vegetable and dry seeds as pulse grain are also widely used.
- Generally cultivars belonging to *Lablab purpureus* var. *lignosus* are grown for dry seed as pulse.
- Dry seeds of the cultivars belonging to *Lablab purpureus* var. *typicus* are also used as pulse particularly at the end of the season.
- This crop starts fruiting in winter and continues intermittently in spring and pods are available regularly for 3-5 months
- Cultivation cost of this crop is very less.
- Although the crop is indigenous and grown all over the country, compact large areas of commercial production are uncommon.
- Dolichos bean is also grown as green manure crop and cover crop.
- It is also grown as ornamental plant in the USA.

Nutritional value

- The pods are one of the rich sources of protein.
- It contains 6.7 g carbohydrates, 3.8 g proteins and 0.7 g fat per 100g of edible portion.
- It provides high of 0.9 g minerals per 100 g of edible part.
- Tender fruits contain 210 mg calcium, 68mg phosphorus 55 mg sodium, 34mg magnesium and 1.7 mg iron per 100 g fresh.
- It provides 312 I.U of vitamin A per 100 g edible portion.
- It is a good source of Vitamin B like, 0.06 mg Riboflavin, 0.1mg Thiamine and 0.7 mg Nicotinic acid per100 g fresh.
- It contains 9 mg of vitamin C per 100 g fresh.
- Antinutritional factors like trypsin and chymotrypsin inhibitor etc. present in dry seeds can easily be removed by boiling in fresh water before cooking.

Medicinal value

- Hyacinth bean is particularly important for the supply of vitamin B in the foods.
- Hyacinth bean contains high amount of dietary fibre which gives protection against certain types of cancer particularly colon cancer, regulates transit and lowers blood cholesterol.

Adaptation in India

Hyacinth bean has been originated in India and later it was introduced to China, Western Asia and Egypt. Various forms and cultivars of both the cultivated types *viz.*, viny, soft and edible podded vegetable type (*Lablab purpureus* var. *typicus*) and fibrous podded pulse type (*Lablab purpureus* var. *lignosus*) are available in India. Huge genetic diversity in hyacinth bean exists in India because of its probable primary gene center in Indian subcontinent (Hindustani centre). Large variation in petiole color, pod colour and shapes and plant types are available in both *Lablab purpureus* var. *typicus* and *Lablab purpureus* var. *lignosus*. A number of improved varieties of particularly vegetable type (*Lablab purpureus* var. *typicus*) have also been developed through purification and subsequent improvement over the indigenous cultivars. However, edible podded, bush type cultivars have also been developed in India through combination breeding approach involving the viny cultivars of *Lablab purpureus* var. *typicus* and bushy cultivars of *Lablab purpureus* var. *lignosus*.

Breeding behaviour

- It is highly self-pollinated crop due to hermaphrodite sex form and homogamy i.e. maturation of stigma and anthers at the same time.

- Stamens and pistil in opened flower remain enveloped together inside the tube like structure of jointed petals called “keel” leading to cleistogamous like self pollination.
- The extrovert stigma through the keel tube in a few varieties leads to cross pollination by insects upto 6-10%
- Anthesis starts in the morning by 8.00 a.m. and continues up to 5.00 p.m.
- Anther dehiscence generally occurs before anthesis and continues from 5.00 a.m. to 2.00 p.m.
- Stigma becomes receptive and pollens are fertile on the day of anthesis.
- The inflorescence produces flowers for about 10-18 days, giving an adequate period for hybridization.
- Emasculation of matured buds in the afternoon hours followed by hand pollination in the next day morning between 8.00 and 10.00 a.m is the best method of hybridization.
- It needs good care for emasculation because the flowers, though large are prone to injury.

Origin and taxonomy

Hyacinth bean has been originated in India as wild forms of this bean are found in this country. From India, it was introduced to China, Western Asia, Egypt and other tropical countries of South and South-east Asia and Africa.

It belongs to the small genus *Lablab* under the family Fabaceae (Leguminaceae) and the order Fabales. Vegetable and pulse type hyacinth bean belong to two distinct botanical varieties under *Lablab purpureus* viz., *Lablab purpureus* var. *typicus* (soft podded vegetable type) and *Lablab purpureus* var. *lignosus* (fibrous podded pulse type).

The hyacinth bean has chromosome number 2n = 22. However, two groups were identified in *Lablab purpureus,* one having 20 small chromosomes and the other with 22 large chromosomes. General characteristics of the vegetable (*Lablab purpureus* var. *typicus*) and pulse (*Lablab purpureus* var. *lignosus*) types are given below.

Lablab purpureus var. typicus (syn. Dolichos lablab var. typicus)

- A viny and indeterminate type with soft edible pods as the pod walls has less fibre.
- Perennial and photosensitive type.
- Grown for tender pods as vegetable.
- The long axis of the seed is parallel to the suture.

Lablab purpureus var. lignosus (syn. Dolichos lablab var. lignosus)

- Generally bushy and determinate type with pods having high fibre content.
- Pod walls have high fibre content and they have also characteristics aroma.

- Annual and photo non-sensitive type.
- Grown for fresh dry seeds as pulse.
- Seeds are at right angle to the suture.

Improved varieties

- Viny, photo sensitive and vegetable type: Pusa Early Prolific, Deepaliwal, JDL-37, JDL-79, JDL-53, T- 1, K-6802, HD-18, HD- 60, CO-1, CO-2, CO-3, CO-4, CO-5, CO-10, Rajni, Dasarwal, KDB-403, KDB-405, Pusa Sem-2, Pusa Sem-3, Swarna Utkrisht, BCDBM-1.
- Compact, photo non-sensitive and vegetable type: Arka Joy, ArkaVijay, CO-6, CO-7,CO-8,CO-11, CO-13.
- Compact, photo non-sensitive and dual type: Konkan Bhusan,, Vijay, Vani, Wal Konkan-1, Hebbal Avare-1, Hebbal Avare-3, Hebbal Avare-4.

Cultural requirements

- Cultivars of the viny and vegetable type under *Lablab purpureus* var. *typicus* are photosensitive in nature and basically short day crop.
- Cultivars under *Lablab purpureus* var. *lignosus* and and those which have been developed through combination breeding involving the cultivars belonging to *Lablab purpureus* var. *typicus* and *Lablab purpureus* var. *lignosus* are photo non-sensitive in nature and can be cultivated through out the year in favourable temperature condition.
- It is a cool season and drought tolerant crops.
- The best temperature for its seed germination is 18°C to 27°C.
- Better growth and development takes place at 21°C to 27°C temperature.
- Temperature range of 12°C to 18°C is considered most suitable for better fruit set.
- It is susceptible to frost.
- It does not grow well at very high temperature.
- It can grow on almost all types of soil of average fertility with optimum pH range of 6.5 to 7.8.
- Sandy loam, silt loam and clay loam soils are best for its cultivation
- Very rich soil leads to more vegetative growth and less pod yield.
- It can not stand water-logging.

Land preparation

- Prepare the field with 4-5 deep ploughing to a depth of 25 cm.
- Remove all the weeds, stubbles and level the land by laddering.

- Manures should be mixed with the soil at the time of last ploughing.
- The plant is susceptible to water logging so land should be well drained.
- As a monocrop, it is generally grown on flat or raised beds, about 1.0 to 1.5 m wide and seed are sown on both sides of the bed.
- Mounds in the bed method of planting are prepared with the mixture of 15-20 tonnes of FYM per hectare and basal dose of fertilizers are mixed with the soil.

Sowing time

Soft podded indeterminate cultivars belonging to *Lablab purpureus* var. *typicus* are photosensitive in nature and flowers under short day condition with the advent of cool temperature hence, are sown during late rainy season or early autumn season. Determinate and compact cultivars though photo non-sensitive in nature yet are also cultivated during this season because of high propensity of flowering and fruiting under cool season.

- June-July
- July-August

Seed sowing

- Hyacinth bean seeds are directly sown in the field.
- Seeds are sown by hand dibbling in the bed.
- Inoculate the seed with *Rhizobium* culture @ 30 g/kg of seeds before sowing for quick nodulation on the roots which fix atmospheric nitrogen.
- Determinate compact types are sown in rows by drilling behind the plough.
- 20-30 kg seed/ha is sufficient for pole type cultivars.
- 35-40 kg seed /ha is sufficient for bush type cultivars.
- Seeds require sufficient moisture for germination.
- It is better to do seed sowing in pre-irrigated moist soil than irrigate the field after sowing.

Spacing

- Pole type cultivars as sole crop: 1.5 m × 75 cm; 1.5 m × 1.0 m
- Pole type cultivars as mixed crop: 1.0 m × 75 cm
- Compact and bush type cultivars: 75 cm × 75 cm

Nutrient management

- Soaking the seeds with *Rhizobium* bacteria, 30 g in 10% molasses solution per kg seed is the best for symbiotic nitrogen fixation.

- Hyacinth bean responds well to manuring and application of particularly phosphatic fertilizers.
- Phosphorus helps in better nodulation in the roots through *Rhizobium* symbiosis.
- Its requirement of nitrogen fertilizer is comparatively much less because of being the legume crop.
- Well rotten farm yard manure at the rate of 15-20 t/ha should be applied at the time of land preparation.
- Apply a general fertilizer dose of 30 kg N, 75 kg P_2O_5 and 60 kg K_2O per hectare.1/2 N along with full P and K fertilizers and farm yard manure should be given as basal and rest N should be top dressed 30 days after sowing.
- The N,P,K fertilizers as basal should be applied at the time of sowing in bands about 7-8 cm to the side and slightly deeper than the seeds.
- Nitrogen fertilizers after 30 days of sowing should be top dressed by broadcasting around the plant 6-7 cm away from the plant stem.

Irrigation

- Pre-sowing light irrigation helps in better germination of seeds.
- Irrigate the crop after germination if moisture is insufficient.
- It is a hardy crop and can tolerate low soil moisture.
- The crop must be irrigated during its most critical stages i.e. flowering and pod development stages.
- Moisture stress during flowering cause huge flower drop.
- Irrigate regularly at 7-10 days interval during dry periods to get high yields.

Interculture and weed control

- Indeterminate, pole type cultivars grown on trellis give better crop stand and pod yield mainly because of better use of sunlight by maximum number of leaves.
- The crop needs to be trained over low trellis of 1.5 m high above the ground.
- The indeterminate types are also given support by bamboo poles.
- In mixed cropping, the pole type cultivars get support on the stalks of ragi and sorghum after their earheads are harvested.
- At the initial stages of weed growth, shallow hand weeding and hoeing are sufficient to check the weed growth.
- At the later stages of crop growth, the weeds are kept under check due to the thick canopy of the crop.

- Application of herbicides effectively controls weeds.
- Pre-sowing application of fluchloralin @ 2litres/ha checks the weed growth for 20-25 days.

Harvesting the crop

- The crop duration of the pole types is about 210-240 days and of the bush types is 90-160 days depending on the varieties.
- Harvest the pods periodically by hand picking when they are fully grown and succulent and have not become fibrous.
- Normally harvesting starts, two months after sowing in bushy cultivars and three months after sowing in case of pole type cultivars.
- In pole types about 9-12 pickings of tender pods can be done.
- In bush types generally harvesting is completed by 2-3 pickings of tender pods.

Post harvest handling and storage

- Remove fibrous and damaged pods from the harvested lots.
- Wash the selected pods in running water before sending to market.
- Quality of the tender pods deteriorates very quickly under ambient condition because of their very high respiration and transpiration rate.
- High transpiration rate makes the pods leathery rendering them unfit for marketing and consumption.
- The best storing temperature is 4.5° to 7.0°C with 65-70% relative humidity.
- Fresh pods can be kept for about 15-20days in cold storage.
- Dried seeds can be kept for 2-3 years at room temperature under dry condition.

Plant protection measures

Diseases

Powdery mildew (*Laveillula taurica var macrospora*)

This fungal disease generally appears late in the season during dry summer.Faint, slightly discolored specks are first formed on leaves. From these specks grayish white powdery growth of the mycelium and spores spread over the leaves, stem and even fruits. Defoliation occurs in severe cases.

Control measures

- Management schedules same as mentioned in cowpea.

Downy mildew (*Peronospora spp*).

This fungal disease appears as white cottony, patches that are formed on the lower surface of leaflets, while upper surface turn yellow, brown and eventually dries up.

Control measures

- Spray the crop with 0.25% Dithane M-45 or metalaxyl-mancozeb (0.25%) at 15 days interval during growth period.

Anthracnose (*Colletotrichum lindemuthianum*)

This seed borne fungal becomes most severe in high rainfall areas of subtropical climatic condition. Tissues of the necrotic spots on leaves collapse, become thin and papery and produce pinkish slimy mass in high humidity condition. Similar lesions appear on the cotyledons and stems of the young seedlings. Most characteristics symptoms appear on the pods where dark brown, sunken lesions with raised, reddish or yellowish margins are found.

Control measures

- Management schedules same as mentioned in cowpea.

Ashy stem blight (*Macrophomina phaseolina*)

This soil borne fungal disease becomes severe in warm and humid climate. In early stages, it causes damping off and killing of young seedlings where black cankerous lesions appear on the cotyledons. Collar region of the plant becomes brown and vascular bundle of the roots also turns brown resulting in dry root rot.

Control measures

- Management schedules same as mentioned in cowpea.

Bacterial blight (*Xanthomonas* spp.)

This seed and soil borne bacteria causes watery, olive-green blisters on stems and leaf bases. Water-soaked spots also appear on leaf and pods. Infected leaves show irregular, sunken, yellow to brown necrotic spots with narrow yellowish halo. In severe infection, defoliation occurs. Dark green, water-soaked spots may appear on the pods which later become yellow-brown and dry.

Control measures

- Use disease-free and healthy seeds.
- Treat seeds with 0.01% streptocycline solution for 30 minutes before sowing.

Mosaic virus

Yellow mosaic of hyacinth bean is transmitted by white fly while mosaic virus is seed borne and aphid transmitted. Characteristic mosaic pattern on the affected leaf occurs due to the infection of these viruses.

Control measure

- Use the seeds that have been harvested from mosaic free plants.
- Remove and destroy the affected plants as early as possible.
- Spray the crop with systemic insecticides like Rogor or Metasystox @ 2 ml/1 of water to control the insect vector.

Insect pests

Stem fly (Ophiomyia phaseoli)

It is a polyphagous pest where the maggots mine the leaf, bore inside the petiole and tender stem tunneling inside which cause girdling effect and death of the plant. The adults puncture the leaf and the affected leaves become yellow and dry up.

Control measures

- Remove and destroy all affected plants
- Spray with Rogor or Nuvacron 1.5 ml/l of water twice at weekly interval.
- Spray the crop with 5% neem seed extract reduces the incidences.

Pod borers (*Adisura atkinsoni, Exelastis atomosa, Helicoverpa armigera*)

The caterpillars after hatching bore into the developing pods and feed on the seeds therein.

Control measures

- Remove and destroy the infected pods at the initial stage of attack.
- Spray the crop with Nuvacron @ 1 ml/l, or Thiodan @ 1.4 ml/l or malathion @ 2 ml/l or Sevin W. P. @ 4 g/l of water.
- Biological control measure by applying a nuclear polyhedrosis virus (NVP) @ 250 LE/ha/week starting from first appearance of young leaves.
- Spray the crop with 5% neem seed kernel extract reduces the incidence.

Aphids (*Macrosiphum pisi, Aphis craccivora, Myzus persicae*)

The gregarious aphids suck sap from growing tips and later from the whole plant which causes curling of leaves, reduces vigour, yield and quality. They also transmit mosaic viruses.

Control measures

- Spray the crop with Rogor @ 1.5ml/l of water
- Apply carbofuran granules in the field (Furadon 3G @ 1.0 kg a.i. /ha) before sowing of the seeds.

Root knot nematode (*Meloidogyne* spp).

Nematodes attack the roots and produce galls there which cause stunting of the plants and severely reduce yield.

Control measures

- Plough the field in summer months.
- Apply oil cakes, more specifically neem cake in the field during land preparation
- Apply granular insecticides like Thimet 10G @ 25 kg/ha or Furadon 3G @ 35 kg/ha in soil during land preparation.

Yield

Tender pods

- Pole type cultivars: 20.0 to 40.0 t/ha depending on the cultivar.
- Bush type cultivars: 4.0 to 6.0 t/ha

Dry seeds: 1.0-1.5 t/ha.

Tropical Tuber Crops

26. Elephant foot yam

Scientific Name: (Amorphophallus paeoniifolius (Dennst) Nicholson)
Family: Araceae
Origin: South east Asia
Chromosome No.: 2n=26

Introduction

Among tropical tuber crops elephant foot yam has become a popular vegetable due to high productivity in a short growing season and high net returns. It contains vitamins, minerals, energy and also has medicinal and therapeutic values. In India, this crop is traditionally cultivated in West Bengal, Andhra Pradesh, Tamil Nadu, Bihar, Gujarat, Kerala and Jharkhand.

Importance and use

- The edible plant part is corm, the modified stem.
- The crop has gained tremendous prospects in the Gangetic plains of eastern India.
- The corm is primarily used as vegetable after boiling for curries and baking.
- Petioles of unopened leaves can also used as vegetable.
- The corm can be also used for making various value added products like flour, dried cubes, pickles, medicines, etc.

- Corms have considerable potential to be an alternative food source, but it has not been realized its full potential because of the palatability problems associated with its content of certain anti-nutritional factors like oxalate, trypsin inhibitors and acridity.
- The acridity is believed to be due to the combination of needle shaped crystals (raphides) of calcium oxalate and a chemical irritant, the nature of which has not been resolved till date, although it is reported to be either a diglucoside of 3,4-dihydroxy benzaldehyde or an unidentified proteinase or hormone or sapotoxin.
- Being vegetatively propagated, the rate of multiplication is very low, so it takes long time to reach the elite varieties for the end users.
- Huge quantity of seed corms is required to plant per unit area.
- Like other crops, there is no agency or company available in the country to supply bulk quantity of optimum size (500-600 g) seed corms of elite varieties at a reasonable rate.
- In intensive cropping zone, the farmers are not keen to keep the crop in their field up to maturity and go for early winter vegetables, instead. In this situation, the crisis of seed corm during planting is of common occurrence.
- The vast mono-cropped areas of eastern part of India can be exploited through seed corm production programme.

Nutritional value

- It is a good source of carbohydrates, rich in minerals particularly potassium and calcium and vitamin A and B.
- 100g fresh corm supply 18.4g carbohydrates, 2.0g protein, 0.1g fat, 434 IU vitamin A, 0.06 mg thiamine, 0.02 mg riboflavin, 6.0 mg vitamin C, 387 mg Ca, 2.4 mg iron and 38 mg phosphorus.
- In addition, glucose, galactose, rhamnose and xylose are also present as free sugars.
- The soluble oxalate content of smooth cormed non-irritant cultivars is found to be lower and safe from the view point of accumulation of urinary oxalate leading to kidney stones.

Medicinal value

- Corms are stomachic, carminative, used in piles, dysentery and having blood purifier property.

- Fresh corms have stimulant and expectorant properties.
- Macerated corms are externally applied to relieve pain of rheumatic swellings.
- Its root are used in boils and opthalmia and also as emmenagogue.

Adaptation in India

The Old World genus *Amorphophallus* has long been dispersed with the migration of the sea-faring people of Ind-Malaysia-Pacific region and believed to predate the culture of rice. The cultivation of *A. paeoniifolius* is rather restricted to Malaysia, and the drier parts of India and Sri Lanka. It is, however, reported to grow wild in Southeast Asia extending to Java, the Philippines and the Pacific. Initially, local irritant types are primarily grown in homestead areas in the tropical parts of the country. After the introduction of smooth cormed non-irritant types from east Godavari region of Andhra Pradesh during early 80's , the crop is gaining tremendous popularity in the Gangetic plains of eastern India. Now the crop is regarded as the most important cash crop in this region.

Breeding behaviour

- It is highly heterozygous and vegetablively propagated crop with out-breeding habit.
- The plants are characterized by protogyny and predominant cross pollination in nature.
- Self-fertilization is prevented mainly by strong self-incompatibility mechanisms.
- The crop cycle of one to three years makes the occurrence of inflorescence very rarely.
- In the tropics, close to equator, the inflorescence will open fully and release pollen after 16-25 days, depending on the genotype, environmental condition and the size of the inflorescence.
- The inflorescence is produced without a leaf.
- They appear on the thick peduncles which may be short or long, depending on the genotypes and species.
- A single spadix is subtended by a large blotched spathe.
- Since the flowers are naked, the opening of the spathe is considered as anthesis.
- Flowers are unisexual.
- The female part is separated from the male by a sterile band of flowers.
- Ovules are anatropous or semi- anatropous.
- The spadix terminated in a large appendix which can be elongated or short or wide.
- Stigmas become receptive before pollen is released.

- The period of stigma receptivity and shedding of pollen usually overlap within the same inflorescence.
- The main pollinators are insects attracted by the bad odour released by the inflorescence.
- The chemical responsible for the bad odour is dimethyl oligo sulphide.
- The release of odour is associated with the beginning of stigma receptivity, occurring in the afternoon before the spathe opens.
- The odour attracts insects mainly flies from neighbouring areas, and also from other inflorescence of the same species. The latter are the most important because they bring pollen and distribute it on the stigma of female flowers which are already receptive.
- The odour remains intense for several days, sometimes for more than a week. It is usually so strong that it can attract flies from several hundreds of meters away.
- In Java, the presence of small beetles (Nitidulidae) gathered around the base of the spadix at the time of receptivity indicated that they may be the primary pollinators.
- Presence of of extreme protogyny coupled with the delay in the opening of the spathe is responsible for the lack of seed set in *A. paeoniifolius* under natural conditions.
- There is profuse seed set in the wild species under natural conditions.
- Fruits are numerous, ovoid, red, containing one or three seeds.
- The seeds have high germination capacity (about 75 %) although they have dormancy which can be removed by washing in running water for six days.
- The number of sprouts varied from one to six per seed indicates the presence of polyembryony.
- Since the seed set itself has been achieved only in recent times, the genetic improvement of this crop is just started.

Origin and taxonomy

Elephant foot yam is originated in Asia and Asia-Pacific region. It belongs to the family Araceae and sub-families Lasioideae. The genus *Amorphophallus* includes some 100 species from Africa, India, and Malaysia to Australia. The species that are utilized are *A. konjac,* C. Koch *(syn. A rivieri* Durien var. *konjac* Engler*), A. oncophyllus* Prain, *A. variabilis* Blume and *A. campanulatus* (Roxb.) Blume (syn. *A. paeoniifolius* (Dennst.) Nicolson). A. *konjac* is cultivated and utilized in Japan and warmer parts of China. *A. oncophyllus* and *A. variabilis* are utilized in Indonesia. The most spectacular and largest inflorescences are produced by *A. titanium* Becc. The tallest inflorescence was ever recorded was 2.5m. It is considered to be an exotic species because of its inflorescence. It rarely flowers out of its original environment (Indonesia). The most important species is *A. campanulatus* which occur in wild and cultivated forms. Cultivated genotypes of *A. campanulatus* are classified according to corm shape, corm surface shape and colour, corm flesh pigmentation, petiole size, petiole pigmentation, leaf structure, inflorescence shape, size and colour, yield and quality characteristics.

The somatic chromosome number is 2n=2x=26. The induced tetraploids do not show any superiority over normal diploids.

Improved varieties

- Smooth cormed non-irritant type: Gajendra, Sree Padma, Bidhan Kusum, NDA-9.
- Irritant type: Santragachi, local cultivars.

Cultural requirements

- Elephant foot yam grows well in hot climate at a temperature of 25-35°C.
- Humid climate helps in the initial stages of the crop growth whereas, dry climate facilitates tuber bulking.
- Well-distributed rainfall of 1000-1500 mm is helpful in good crop growth and tuber yield.
- Well-drained, fertile, sandy loam soil is ideal for elephant foot yam cultivation.
- The crop can be grown in other soil types including laterite soils also by raising the crop in pits filled with well-decomposed farmyard manure and good sandy loam soil.
- Water stagnation at any stage adversely affects crop growth, corm bulking and corm quality.

Planting time

- February-March as irrigated crop in the Gangetic plains.
- June-July as rainfed crop.

Land preparation

- Prepare the land well in advance with repeated ploughing (at least 4-5 ploughing) to a good tilth.
- Remove all stubbles, weeds, etc. from the land.
- Add well rotten organic manure preferably farm yard manure or compost @ 20-25 t/ha in the pits.

Planting the seed tuber

- Availability of quality seed material is the major constraints in elephant foot yam cultivation.

- Whole corm size of 500-750 g is recommended for commercial cultivation.
- However, due to non availability of the recommended size, farmers usually cut the corm into small pieces from big tubers.
- The big size corm of 2-3 kg is cut into 4-6 pieces by vertically retaining part of the apical bud.
- Dipping of the cut tubers in thick cow-dung slurry (1:1) mixed with captan (0.3%) reduces the corm rot in the field and also increases sprouting.
- After removing the whole seed corm or corm pieces from cow-dung slurry, they should be dried in shade for 6-8 hours before planting.
- Small and whole tubers should be preferred over cut tubers for commercial cultivation.
- If nursery facility is available, cut tubers after treatment can be planted on raised beds during the first week of June for sprouting.
- The cut tubers should be closely kept in lines 20 cm apart in the nursery beds.
- The bed should be lightly covered with soil and paddy straw and light irrigation should be regularly provided.
- The tubers sprout in about 3 weeks.
- The sprouted tubers should be transferred to the main field with the onset of monsoon.
- For obtaining a good commercial crop, the planting should be done in pits (50 cm × 50 cm × 30 cm) filled with Farm Yard Manure, good soil and fine sand in the ratio of 1:1:1.
- The distance of the pits is decided depending on the size of planting material being used.
- For seed materials ranging from 500-600 g, a distance of 75 cm × 75 cm is recommended.
- If the seed materials are more than one kg, a spacing of 90 cm x 90 cm should be provided.
- If the seed materials are below 500 g, a spacing of 60 cm × 60 cm should be provided.
- The cut tubers are planted 4-6″ below the soil surface with the apical portion upward and the pits should be covered with soil followed by light earthing up after planting. Paddy straw is usually covered over the soil as mulch material.

Nutrient management

- Nutrient requirement of elephant foot yam is rather high.
- Well decomposed farmyard manure @ 20-25 t/ha should be given before 15 days of final land preparation.
- If the planting is done in pits, the FYM mixed with soil should be filled in pits only.

- The fertilizer dose should be decided depending on the soil type and nutrient status.
- A fertilizer dose of 150:100:150 kg/ha of nitrogen, phosphorus and potash has been found to be optimum.
- Full dose of phosphorus and one-third quantity of nitrogen and potash is applied at the time of planting.
- The remaining dose of nitrogen and potash is applied in two split doses at 60 and 90 days after planting, immediately after weeding and followed by earthing-up.
- Use of *Azotobactor* as corm dipping (2% solution) and soil application (9 kg/ha) at 35 days after planting increases the corm yield.

Irrigation

- If the planting is done in February-March, frequent irrigation should be provided immediately after planting to initiate sprouting of corms.
- Mulching with paddy straw after planting the corms helps in conservation of moisture, early sprouting, checks early weed growth and reduces the number of irrigation.
- Depending on the soil moisture availability, irrigation should be given at regular intervals till the arrival of monsoon for the irrigated crops.
- Care should be taken to prevent water stagnation at every stage of crop growth.
- Atleast 8-10 irrigations are required throughout the growth period.

Interculture and weed control

- After 30-35 days of planting, when the sprouts start emerging, weeding is done followed by first earthing-up.
- Second earthing up should be followed after first top dressing of N and K at 60 days of planting.
- Light earthing up if required, may be done after second top dressing at 90 days of planting.
- To keep the crop healthy, plant protection measures should be strictly followed.
- During the initial period of 2 months after planting, crops like leafy vegetables, cowpea etc. can be taken up as inter-crop for additional benefits.

Crop rotation and Intercropping

- Crop rotation of elephant foot yam (mid-March to mid-November) - wheat/ mustard/ Bengal gram/peas (mid-November to Mid-March) or Elephant foot yam (mid-June to mid-December)-okra/colocasia/maize/mustard (January to mid-June) has been found suitable.

- Inter-cropping of elephant foot yam is very successful in coconut, banana and other plantations for getting additional income.

Harvesting the crop

- Yellowing of the leaves and drying up of the plants indicate that the crop is ready for harvest.
- Tubers are harvested by digging using crowbars or spades after 6-7 month.
- While harvesting, care should be taken to avoid injury to the tubers.

Post harvest handling and storage

- After harvesting, soil and roots should be removed from the tubers.
- Damaged tubers should immediately be disposed-off for local sale.
- The tubers should be graded and dried in shade for 4-5 days before marketing.
- Use of baskets made of bamboo/palm leaves has been found to be ideal. The tubers can be kept in layers sandwiched with paddy straw or dry banana leaves.
- Elephant foot yam corms are always in great demand due to their acceptability as a popular vegetable and usefulness as major component of several indigenous medicinal preparations.
- The tubers can be also used for making various value added products like flour, dried cubes, pickles, medicines etc.

Plant protection measures

Diseases

Collar Rot (*Sclerotium rolfsii*)

The collar region of the plant is attacked by the fungus and water soaked lesions appear on them. The leaves become yellow. The stem shrinks and collapses due to rotting to the collar region. A thick white spreading mat of mycelia with globular slerotia can be seen on the affected tissue. Heavy rains, high humidity, heavy soils, poor drainage are some of the pre-disposing factors.

Control measures

- Since the pathogen is soil-borne, cultural practices such as crop rotation, removal of plant debris, and improvement of drainage system will minimize the severity of the disease.
- Corm should be treated with carbendazim (0.2%) before planting.
- Application of *Trichoderma viridae* @10 kg/ha as soil treatment is found effective.

- Drenching of collar region with carbendazim (1g) + mancozeb (2g) or Thiram (2g) + carbendazim (1g) / litre of water twice at monthly interval commencing from the first appearance of the symptom would effectively control the disease.

Mosaic

The characteristic symptom is mosaic mottling on the leaves. The disease causes more proliferation of lateral buds, separation of buds from the other mother corms and poor growth of roots. The mottled plants produce small corms. The disease is spread through aphids.

Control measure

- Corms should be collected from heathy disease free crops for planting.
- Rouguing of infected plants will help in minimizing the secondary spread.
- Spraying of dimethoate (Rogor) @ 1.5 ml/l is effective against the vectors.

Yield

- The yield of elephant foot yam depends on the size of planting material used and planting time.
- The ratio between quantity of planting material used and corm yield is generally 1:8.
- For commercial cultivation, if 75 cm x 75 cm spacing is provided and approximately 500 g size planting material is used, corm yield of 50-60 t/ha can easily be expected.

27. Sweet potato

Scientific Name: Ipomoea batatas
Family: Convolvulaceae
Origin: South America
Chromosome No.: 2n = 6x = 90

Introduction

Sweet potato is an important tuber crop grown mostly in tropical and sub-tropical regions of Asia, Latin America, the Pacific islands, Papua New Guinea and Africa. South America has been widely recognized as the center of origin of sweet potato. It was introduced in Europe, probably in Spain at least 60 years before Irish potato. The introduction and spread in Africa and Asia closely follows the establishment of trading settlement by Spanish and Portuguese. In India it has also been introduced through Portuguese and British trade and colonization. It spread into Polynesia by Spanish expedition, which started from Peru in the 16th century. China and India are the largest sweet potato growing countries in the world. In India it is largely cultivated in Uttar Pradesh, Bihar, West Bengal, Orissa, Madhya Pradesh, Tamil Nadu, Kerala, Karnataka and Andhra Pradesh. It is commonly known as "sakarkand" in India.

Importance and use

- Edible plant part is sowllen adventitious roots in the shape of tuber.
- Sweet potato being a major source of carbohydrate is often used as a substitute for cereals.
- Yellow and orange fleshed cultivars are rich in β carotene, the precursor of vitamin A.
- Sweet potatoes are mainly utilized as human food stuff as starch and vegetable after boiling, baking or frying.
- The tender tops and leaves are used as vegetables in Africa, Indonesia and Philippines.
- It can be eaten fresh, canned, frozen or dehydrated and used in variety of food products, like pie fillings, purees, candied pieces, soufflés, baby foods etc.
- Peeled tubers are sliced and dried in the sun to produce chips.
- Dehydrated tubers are also ground into flour and often mixed with the cereal flour.
- Sweet potato flour is used as a supplement to cereal flour in the preparation of bakery products, puddings and milk jelly.
- Sweet potato is a good raw material for the fermentation production of industrial alcohol, lactic acid, acetone, butanol, vinegar, etc.
- It is a source of starch for use in textile, paper, cosmetics and food industries, and for preparation of glucose and adhesive.
- The sweetpotato starch is suitable for sizing paper and textiles and for the use in laundries.
- Tubers are also used as high carbohydrate feeding stuff for cattle, pigs and in poultry.
- Feeding raw tubers to cattle and swine can pose problems due to presence of protease inhibitors and when tubers are infested due to phytoalexines.
- The vines also serve as good source of fodder for livestock.
- The colour extracted from the peel of red-skinned varieties, rich in anthocyanin pigment, can be used as food colourant.

Nutritional value

- Sweet potatoes are rich in starch, which are small in size and resemble corn starch granules in shape.

- Sucrose and a few reducing sugars are present in the tuber.
- General quality of sweet potato is mostly dependent on its sugar content (mostly sucrose) which varies from 3-6%.
- Sugar content increases during storage and cooking.
- It also contains appreciable amount of Vitamin A, B, C and protein.
- β carotene content, the precursor of vitamin A ranges between 0.5 to 20.0 mg per 100 g fresh depending on the cultivar.
- Yellow and orange fleshed cultivars are rich in carotene.
- The tubers contain 1.5 g protein and 24.0 mg ascorbic acid (vitamin C) per 100 g fresh.
- The tubers are also rich in minerals particularly potassium (370 mg/100 g), phosphorus (49.0 mg/100 g) and calcium 946.0 mg/100 g) contents.
- Sweet potato leaves contain about 3.2% protein on fresh weight basis and also rich in carotene, vitamin C and some minerals (Ca, P and Fe).

Medicinal value

- Varieties with orange or yellow fleshed tubers are excellent source of β carotene, which apart from being precursor of Vitamin A, is ascribed with anti carcinogenic properties.
- The tuber contains very high dietary fibres that give protection against constipation, colon cancer, heart diseases and diabetics.
- It is rich in potassium with a very high K: Na ratio, and can be recommended as a part of low salt diet.

Adaptation in India

Though a tuber crop similar to sweetpotatoes was cited in the literatures of the "Vedic" period, yet generally it is considered to have been introduced in India by Portuguese in early sixteenth century. A number of cultivars have been developed by clonal selection from these germplasm resources of India. Wide variation exists among the cultivars for different morphological characters like shape, size, color of skin and flesh of tuber, flower color, flesh texture, depth and period of rooting, etc. However, huge exotic collections mainly from Japan, Taiwan, USA, Nigeria, Sudan, Peru, Puerto-Rico, USA, Nairobi and Argentina enriched the germplasm resources of India. A number of cultivars have been developed by clonal selection from these germplasm resources of India.

Breeding behaviour

- It is highly heterozygous and vegetatively propagated crop with out-breeding habit.
- Flowering in the vine is stimulated by short days, long night and day length of

11.00 hours is ideal for flowering.

- Self-incompatibility is evident in different sweet potato cultivars.
- It is cross pollinated crop and cross pollination generally done by bees.
- The flowers are solitary or cymose and vary in colour from white to purple.
- Wide variability is created in the seedlings from the seeds of even natural settings due to perpetuation of heterozygosity through vegetative means.
- Flowers open very early in the morning between 4.00 and 5.00 a.m and wilt on the same day.
- Anthers starts to dehisce before anthesis from 11-12 p.m. at night and majority of the pollens become non-viable by 7.00 a.m in the morning.
- Stigma becomes receptive in the bud before anthesis from 6.00 to 8.00p.m. at night and may be receptive for few hours after the wilting of flowers in the morning hours.
- Hybridization by hand pollination in the morning hours is very difficult.

Origin and taxonomy

Sweetpotato has been originated in northern part of South America and southern region of Central America. It was domesticated in tropical America and Polynesia. It was grown in Peru in ancient times as evidenced from the excavation of caves. It spread into Polynesia by Spanish expedition, which started from Peru in the 16th century Guatemala and Peru have been identified as secondary center of high diversity. It was introduced in Europe, probably in Spain at least 60 years before Irish potato. The introduction and spread in Africa and Asia closely follows the establishment of trading settlement by Spanish and Portuguese.

Among the 175 species of *Ipomoea*, *I. batatas* is the only tuber bearing species although a few species are known to have thickened roots. Perhaps, *Ipomoea trifida* is the closest relative of sweetpotato although sweet potato is not found in wild state. The hexaploid sweetpotato probably derived by amphidiploidy from a tetraploid (2n=60) and a diploid (2n=30) to produce a triploid (2n=45), followed by subsequent doubling of chromosome to produce hexaploid (2n=90).*I. trifida* is thought to be the direct ancestor of sweet potato and its diploid predecessors is thought to be *I. leucantha*.

Sweetpotato, *Ipomoea batatus* belongs to the section Batatas of the family Convolvulaceae. It is a natural hexaploid (2n = 2x = 90) with the basic chromosome number is x = 15.

Improved varieties

- White/cream skinned varieties: Pusa Safed, Rajendra Sakarkand-5, Sree Nandini, CO-1, CO-2, Sree Varun, Sree Gautam.
- Brown skinned varieties: Cross-4, Samrat, V-35, Kalmegh, Rajendra Sakarkand-35, Rajendra Sakarkand-43.

- Red or pink skinned varieties: H-41,H-635, H-620, CO-3,Pusa Lal, Pusa Sunheri, H-42, Varsha, Sree Vardhini, Sree Bhadra, Sree Rethna, Konkan Ashwini, Kiran, Bhuban, Gouri, Shankar, Rajendra Sakarkand-47, VL Sakarkand-6, Sree Arun, Kalinga, Sree Kishan, Sree Sourin.
- Orange fleshed varieties: Kamala Sundari, Gouri, CO-3, Kiran, Sree Vardhini, Sree Rethna.
- Light yellow fleshed varieties: VL Sakarkand-6.
- White/creamy fleshed varieties: Bhuban, CO-1, CO-2, H-41, H-42, Rajendra Sakarkand-5, Rajendra Sakarkand-43, Rajendra Sakarkand-47, Sree Bhadra, Samrat, Sree Nandini, Varsha, Sree Kishan, Sree Sourin and Sree Gautam.
- Resistant to SPFMV: Sree Vardhini.

Cultural requirements

- Sweetpotato is grown in tropical and sub-tropical climate and also in mild temperate regions.
- It is widely grown from sea level up to the elevation of 2500 m however, its performance is better in the places of low elevation.
- It is susceptible to frost and its growth and tuber development are adversely affected at temperature below 20°C.
- Warm sunny days and cool nights are favourable for better tuber development. Night temperature of 15-25°C promotes tuber formation and growth, which prevail in the rabi season in the plains.
- Day temperature between 25°C and 30°C is ideal for tuber formation and development.
- Good rains during the period of early growth followed by dry sunny weather during the period of tuber bulking and maturity is ideal.
- It can tolerate drought moderately however, tuber yield reduces if draught prevails at the initial stage of growing period.
- The tuberization is almost absent at 75% shade and severely affected at 55% and not much affected at 25% shade.
- It is grown in a wide variety of soils.
- It grows best in well drained and fertile loam, sandy-loam or clay-loam soil.
- The soil should be rich in organic matter, friable and well-drained with a pH range between 5.6 and 6.6.
- Heavy clay soil checks the development of the tubers and causes development of long cylindrical pencil-like tubers.

- Highly fertile soil favours luxuriant vine growth which eventually reduces tuber development.
- As a rainfed crop, it requires at least 500 mm rainfall during the growing season. If grown under high rainfall areas, produce vigorous vine growth and poor yield.
- A well distributed annual rainfall of 750-1000 mm is best.
- It does not withstand heavy rainfall and water logging condition.
- It can somewhat tolerate slightly acidic soil but can not tolerate alkaline and saline soils.

Land preparation

- Well prepared soil is the prerequisite for growing good sweetpotato crop.
- The field should be ploughed 3-4 times to a depth of 25-35 cm and clods are broken.
- The field should be properly leveled to provide good drainage.
- Add well rotten organic manure preferably farm yard manure or compost @ 10-15 t/ha before preparation of ridges.

Propagation and planting material

- Sweetpotato is propagated from vine cutting taken from freshly harvested vines or from medium sized weevil free sprouted tubers planted in the nursery.
- The vine cuttings from the freshly harvested crop are also planted in the nursery to get the cuttings for planting in the main field.
- Avoid using vines obtained from freshly harvested crop as it significantly increases weevil infestation in tubers.
- Produce vines in nursery to get healthy vigorous plants.
- Select cuttings from apical and middle portion in order to get higher sprouting and better yield.
- Vine cuttings for planting in the main field are about 20-35 cm long with 3-4 nodes.

Production of planting material in the nursery

Use of vine cuttings as planting materials over the years leads to reduced yield and more incidence of weevil (*Cylas formicarius*). Therefore, healthy vine cuttings are prepared through tuberlets planted in primary and secondary nurseries.

Primary nursery

- Prepare raised nursery beds of 15-20 cm height with fine tilth and good drainage facility, about 3 months prior to planting in main field.

- About 100 kg of medium sized weevil free tuberlets (100-125g each) planted in 100 square meter primary nursery bed can provide sufficient vine cuttings for planting one hectare of land. Vine cutlings from primary nursery bed are replanted in secondary nursery bed for multiplication.
- Soak the tubers in monocrotophos or fenitrothion 0.05% for 10 minutes.
- Plant the tubers at a spacing of 20cm on the ridges, 60cm apart.
- Proper irrigation on every alternate day for 10 days and thrice a week thereafter.
- Apply urea @ 1.5 kg/100 square meter 15 days after planting to ensure quick growth.
- For replanting in the secondary nursery, clip 45 days old vines to a length of 20-25 cm.

Secondary nursery

- Vines obtained from primary nursery are planted in the second nursery.
- Secondary nursery area of 500 square meter is needed to multiply the vines obtained from primary nursery for 1 hectare land.
- Prepare the beds to a fine tilth with the addition of 500 kg FYM.
- Plant the 20-25 cm long vines from first nursery at a spacing of 20 cm on ridges spaced at 60 cm.
- Top dress with 5 kg urea in two split doses, 15 and 30 days after planting.
- Irrigate the nursery whenever required.
- Vines cutting for tuber production in the main field can be taken after 45 days.

Planting time

Sweetpotato is grown throughout the country as rainfed during kharif and irrigated crop during rabi season. However, major sweetpotato area comes under early rabi and rabi seasons because of prevailing warm and sunny days, cool night and moderate rainfall which is favourable for getting high yield.

- September-October: Autumn-winter (Early rabi)
- October-December: Winter (Rabi)
- June-July: Kharif
- November: Diara land cultivation
- April-May: Hill regions

Planting of vines

- Cuttings of 20-25 cm long from apical part of the vine having at least 3-5 nodes are ideal for tuber production.

- Keep the vine cuttings with intact leaves in bundles under shade for 2-3 days prior to planting to generate more roots and better establishment.
- There are several method of planting viz. mound-method, ridge and furrow method, bed method and flat method.
- Ridge and furrow method is most common one and practice it across the slope in sloppy lands for controlling soil erosion.
- Practice mound method in areas experiencing problems of drainage.
- The vine cuttings should be planted in furrows made with the help of tyne at an angle of 45° and burry 2-3 nodes horizontally in the soil.
- The vine cuttings are also planted to a depth of 5-10 cm with only middle portion of vines buried in the soil keeping both ends exposed.
- In mound method of planting, plant 5 cuttings /mound at a spacing of 80 cm × 75 cm.

Spacing

- 60 × 20 cm: Mostly practiced (83,000 plants/ha)

Nutrient management

- Nitrogen is the primary limiting nutrient in sweetpotato production.
- Potassium helps in better tuberization and yield.
- The quantity of NPK fertilizers to be applied varies with the soil, climate and location.
- Apply 10-15 tonnes of FYM or compost at the time of field preparation.
- Apply a general fertilizer dose of 75 kg N, 50 kg P_2O_5 and 75 kg K_2O per hectare.
- Apply full dose of P and half doses of K and N mixed well at the time of planting and top dress rest half of N and K on the sides of the ridges 30 days after planting followed by earthing up.
- To get better yield, combine chemical fertilizers with biofertilizers.
- Apply 30 kg N and 2 kg *Azospirillium* per ha as vine dipping and 10 kg *Azospirillum* as soil application along with full dose of P and K.

Irrigation

- Moist bed is required for 4-5 days after planting for proper sprouting and establishment of vines.
- If sufficient moisture is not present, irrigate daily for first 3 days followed by irrigation on alternate days for one week.

- Apply water at 15 days interval in rabi season ensuring at least 5-8 irrigations during crop period supplying 112 to 150 cm of water.
- The summer crop requires irrigation at more frequent intervals.
- Apply irrigation to the most critical stages of plant growth i.e. tuber initiation, early bulking and late bulking to get higher number of tubers/plants.
- The crop can tolerate considerable periods of drought however, water shortage 50-60 days after planting when the storage roots started bulking considerably reduce yield.

Interculture and weed control

- Spraying the crop with the growth retardant Cycocel @ 500-1000 ppm concentration, twice on 30 and 45 days after planting increase tuber yield.
- Weeding at early growth stages between 7 and14 days and 30 and 45 days after planting is essential.
- Second earthing up should be done at 60 days after planting.
- It is quick growing crop and it covers the soil quickly and suppress most of the weeds.
- For chemical control, apply alachlor @ 3.5 – 6.5 kg a.i. /ha.
- The vines should be turned over after a month of planting to prevent production of adventitious roots at the nodes that are in contact with the soil which helps in better tuber development.

Crop rotation

- In eastern zone of India, Jute and upland rice is followed by sweetpotato.
- In Northern zone, sweetpotato is rotated with pea + barley.
- In Bihar and Andhra Pradesh, maize-cowpea-sweetpotato is practiced.
- In Bihar, two-tier sweetpotato planting with shallow and deep-bulking varieties in alternative row is found successful.

Harvesting the crop

- Sweetpotato generally takes about 90 to 140 days after planting depending on the variety, season and location.
- Harvesting should start at 90 days after planting (DAP) for early varieties, 110 DAP for medium duration varieties and 120 DAP onwards for late maturing varieties.
- The yield can be increased if the crop remains in the field for longer period but the tubers become fibrous, less palatable and damaged by weevil.
- Harvesting of tubers should start when older leaves turn brown and the adjacent soil cracks near the base of the plant.
- Harvest maturity of the tubers can be judged by cutting the tuber.

- In mature tuber, the latex dries up quickly without blackening while in immature tuber, the cut surface turns dark greenish in colour.
- The vines are first cut and then the tubers are dug up without causing injury.
- Harvest the crop manually with the help of spade.

Post harvest management and storage

- Sweet potato is subjected to several forms of post-harvest losses due to physio-biochemical processes in storage, physical damage, sprouting, microbial decay and weevil infestation.
- Cure the harvested tubers by spreading in sun at about 29-40°C with 80-90% relative humidity for 5-7days which results in the healing of wounds.
- For dehydration, cut the roots into pieces and spread out in the sun to dry.
- Store the healthy roots that are free from insect damage and mechanical injury.
- Store the cured tubers in well ventilated room.
- For low cost storage pile sweetpotato roots in heaps and cover them with a layer of straw. Then plaster the heaps with a mixture of clay soils and cowdung under the shade. Tubers remain fresh for few months.
- In another method, dug 50 cm deep pits, store roots in the pits and cover it with dried banana leaves and soil and make thatch over pit.
- Keeping the roots in earthen pots reeled with fine net or cloth also prevents the entry of insect pests and roots can be stored up to 3 months.
- Store the weevil free tubers in alternate rows of dry sand up to 60 days at ambient condition.
- Ideal storage temperature is 13-16°C at 85-90% relative humidity and expected storage life is 4-6 months.

Plant protection measures

Diseases

Cercospora leaf spot (*Cercospora batate, C. ipomoeae*)

This fungus cause leaf spot disease which is characterized by the appearance of yellowish-brown spots which gradually turn deep brown, with circular ovoid or irregular margin ultimately covering major portion of the leaf. Hot, wet and humid weather is favorable for this disease.

Control measures

- Maintain field sanitation to reduce its incidence
- Spray the crop with 0.25% Dithane M-45 or zineb at 15 days interval commencing from first appearance of the disease.

Alternaria leaf spot (*Alternaria sp*)

This fungal leaf spot disease appears as brown lesion mostly on old and mature leaves with concentric rings and well defined margins. In later stage infected tissue may crack and fall off. Humid weather is favourable for this disease.

Control measure

- Spray the crop with 0.3% Dithane M-45 at 15 days interval.

Collar rot (*Sclerotium rolfsii*)

It is an important fungal disease causing severe damage in the nursery. At first small, oval, straw or brown coloured lesions appear at the collar region which gradually enlarge and encircle the stem base. The tissue becomes necrotic resulting in rotting of the sprout and plant easily breaks off.

Control measures

- Select the land where collar rot was not a problem for at least 3 years.
- Remove crop debris and maintain field sanitation.
- Drench the soil around the plants with 50ppm Vitavax.

Sweet potato scab (*Sphacaeloma batatas*)

This fungal disease is most severe in places where frequent fog, rain or dew occurs. Small brown lesions appear first on the veins and leaves later become corky in texture and veins shrink causing curls on leaves. A scab like structure forms on the stem as the lesions coalesce, hence the name.

Control measures

- Adopt crop rotation.
- Use disease free planting material.
- Spray the crop with 0.25% mancozeb at 10 days interval.

Black rot (*Rhizoctonia solani*)

This fungal disease occurs both in the field and storage. When the disease occurs in the field it affects the sprouts and veins which are referred to as 'Black Shank' or 'Black root'. Small, slightly sunken black spots appear on tubers, finally covering the entire tuber. Affected tubers have an unpleasant taste and colour when cooked.

Control measures

- Adopt crop rotation.
- Select disease free planting material.
- Do not store the infected tubers with healthy ones.
- Cure the tubers at 30-35°C temperature and 85-90% relative humidity for 5-10 days.
- Treat the planting material with 0.2 % thiobendazole or benomyl.

Sweet potato Feathery Mottle Virus (SPFMV)

Sweet potato has been reported to be attacked by more than 12 different viruses belonging to seven different taxonomic groups. SPFMV is found almost wherever the crop is grown. Atleast three different strains have been reported which include Ring spot, Russet crack, and probably Internal Cork strains.

The most characteristic leaf symptom are the appearance of ringspots or chlorotic spots, irregular chlorotic patterns (feathering) along the main ribs which may or may not have purple pigmented borders. Mostly the foliar symptoms are transitional and the infected plants may not show any external sumptoms depending on the growth stage, genotype and environment. The infected roots exhibit external or internal symptoms. The external symptoms are the formation of annular necrotic lesions which gridle the fleshy fibrous roots caused by SPFMV (Russet crack strain). The tubers give a poor appearance and losses the market value. The internal symptom on tuber is caused by SPFMV-IC (Internal cork strain) which is characterized by the formation of corky brown hard tissues within the flesh. Infected tubers do not cook well and give a bitter taste. The virus is transmitted through aphids in a non-persistent manner. Under experimental conditions, the virus could be transmitted through grafting (*Ipomoea setosa*), and through sap. *Chenopodium amaranticolor and C. quinoa* are local hosts and *Nicotiana benthamiana* is a good propagative host for the virus.

Control measures

- Collection of planting materials from diseased fields should be avoided.
- Healthy plant material plots may be protected with systemic insecticides to prevent the spread through vectors.
- Plants showing virus disease symptoms amongst disease free population must be identified and destroyed.
- Weeds belonging to Convolvulaceae from the vicinity of sweetpotato fields should be eradicated.
- Virus free planting material can be obtained by meristem tip culture.

Insect pests

Sweet potato weevil (*Cylas formicarius*)

It is the most serious problem of sweet potato. Adult beetle is black with reddish-brown pro-thorax and appendages. Eggs are white, oval in shape and covered with faecal matter. Adults and grubs cause damage to the crop both in the field and in the storage. Adult weevils feed on all parts while the attack of grubs is restricted to vines and tubers. Feeding holes on the leaf, tuber and swelling of collar region of vines are some symptoms of its infestation.

Control measures

- Destroy the plant residues and alternate hosts (other *Ipomoea* species) from the field.
- Use pest free planting material.
- Disinfect the material by dipping in 0.05% monocrotophos.
- Adopt the cultural practices like, earthing up at 60 days after planting and frequent irrigation.
- Install pheromone traps @ 1trap/100 square metre to kill the male weevil.
- Collect weevils from traps and periodically change the detergent.
- Spray the crop with fenthion or fenitrothion @ 0.05% or methyle demeton @ 0.05% at monthly interval.
- Do not left the crop for long time in the field and harvest early to avoid infestation.

Yield

- 20-25 t/ha in irrigated condition.
- 10-12 t/ha in rainfed condition.
- 10-12 t/ha of vine as green fodder or planting materials.

Okra

28. Okra

Scientific Name: Abelmoschus esculentus
Family: Malvaceae
Origin: West Africa /India
Chromosome No.: 2n = 126, 130, 132, 144

Introduction

The okra or lady's finger, known as bhindi in India is an important vegetable in the tropical and sub-tropical parts of the world. It is grown in southern Europe, Turkey, Iran, Yugoskavia, Cyprus, West Africa, Ethiopia, Ghana, Afganistan, Pakintan, India, Bangladesh, Myanmar, Malaysia, Thailand, Japan, Brazil and southern USA. In India, it is commonly grown in Maharashtra, Gujarat, Karnataka, Tamil Nadu, Haryana, Punjab, Uttar Pradesh, Bihar, West Bengal and Orissa. Okra has tremendous export potential as fresh vegetable to our neighbouring countries in the Middle-East countries like, Bahrin, Quatar, Kuwait, Saudi Arabia, Muscat, Tehran and Abu Dhabi and South-East Asia particularly Singapore, Mauritious, Malayasia, Sri Lanka and Bangladesh. The major export oriented okra producing areas in India are Nasik, Ozar, Saikheda, Dindori, Kolhar, Naraingaon and Sholapur in Maharashtra.

Importance and use

- Okra is an important annual vegetable crop in tropical and subtropical regions of the world.
- In India okra has tremendous export potential as fresh vegetable which accounts for 30% exchange earnings from export of vegetables other than onion.
- Okra is mainly used for its tender green fruits as vegetable in many countries.
- Tender green fruits are used as fried vegetable or in curry and sambar.
- It is also used in soups.
- Sun dried okra are mainly used in Africa and India for year round consumption.
- Frozen and sterilized fruits are also important market product mainly in USA.
- The roots and stems ate useful for clearing sugarcane juice in preparation of jaggary.
- Ripe seeds are roasted, ground and used as a substitute for coffee in Turkey.
- Crude fibre from dry fruits and stems are used in the paper industry and making cardboard.
- Dry seeds contain 13-22% edible oil and refined oil may be used as a substitute of common edible oil particularly cotton seed oil.
- Oil from seeds is used in perfume industry.
- The seed cake is used as animal feed.
- In Far East countries like, Papua New Guinea, Fiji and Solomon Islands the species, *A manihot* is used as leafy vegetable.

Nutritional value

- It is the rich source of iodine, an essential micro-nutrient.
- It is a good source of minerals particularly calcium (66 mg/100 g fresh) and phosphorus (56 mg/100 g fresh).
- The fruits also contain good amount of sulfur and sodium.
- Tender fruits contain Vitamin A (88 IU), thiamine (0.07 mg), riboflavin (0.10 mg) and C (13 mg) per 100 g fresh.
- Tender fruits contain 1.9% protein, 6.4% carbohydrate and 1.2% fibre.
- The mucilage on hydrolysis give polysaccharides composed of galacturonic and glucuronic acids and minor contents of galactose, rhamnose, glucose and arabinose.
- Dry seeds contain 20-24% protein.

Medicinal value

- Okra contains high amount of dietary fibre (1.2%) which includes a large spectrum of molecules viz., cellulose, non-starch polysaccharides and lignin which gives protection against certain types of cancer particularly colon cancer, regulates transit and lowers blood cholesterol.
- It is said to be very useful against genito-urinary disorders, spermatorrhoea and chronic dysentery.
- Its use in regular diet is reported to increase sperm count in men.

Adaptation in India

There is not much evidence available as to when and how the cultivated okra was introduced into India. Scientists believed it to have originated from India owing to mention of 'Tindisha' and 'Gandhamula' designating okra in ancient Sanskrit literature. Significant native diversity of okra and its allied species in India points towords its secondary diversity here. However, much variability in okra cultivars is not found in India. Some of the local cultivars enlisted in the catalogue of different seed companies are Lucknow Dwarf, Long Green, Velvet Green, Smooth Velvet, Red Wonder, Samalkota, Satdhari, Panchdhari, Mukta, etc.

Diversity of wild species of Abelmoschus in India include, *A ficulneus* in wide areas mainly in Deccan plateau, Punjab and West Bengal, *A. manihot* in West Bengal, southern Karnataka and Thiruvantapuram of Kerala, *A moschatus* in eastern and western peninsular tracts, *A. angulosus* in western ghats mainly Karnataka and Nilgiri hills, *A. pungens* in north-eastern tracts, *A tuberculatus* in north and north-western plains and parts of Uttar Pradesh and *A. tetraphyllus* in north-western terai region.

Breeding behaviour

- Okra is a mainly self-pollinated crop, but 10-15% cross-pollination always occurs.
- Cross pollination is mainly accomplished by the honey bees which frequently visit the flowers due to presence of nectar, mucilage and showy petals of the flower.
- It is better called often-cross pollinated crop considering comparatively higher propensity of natural cross pollination.
- As the stem elongates the lower most flower buds open into flower.
- A flower bud takes about 22-25 days from initiation to full bloom.
- Anthesis starts from 7.00 am, continues up to 10.30 am.
- The dehiscence of anthers is transverse and occurs 15 to 20 minutes after anthesis and dehiscence completes within 5-10 minutes liberating somewhat heavy and sticky pollens.
- The pollen fertility is the maximum in the period between an hour before and an hour after opening of the flower.

- Pollens remain viable at 25°C temperature and 50% relative humidity condition for 30 days.
- Stigma is receptive at opening of the flower hence bud pollination is not successful.
- Buds to be used for cross pollination are to be collected in the afternoon hours and kept over night wrapped with wet cloth.
- Emasculated flowers are generally cross-pollinated in the morning hours of the next day by gently rubbing the staminal colomn on the stigma head.

Origin and taxonomy

Most of the cytological reports suggest that okra is a natural amphidiploid of *A. tuberculatus* with 2n= 58 and an unknown species with 2n = 72. Chromosome numbers of *Abelmoschus esculentus* as 2n=72, 108, 120, 132, 144 are the indication of a series of polyploids of x =12.

Controversy still remains on the origin of the cultivated species *Abelmoschus esculentus* although the genus *Abelmoschus* is accepted to be of Asiatic origin. In one school of thought it was originated in Ethiopia and upper Nile region of Sudan, West Africa or tropical Asia. It was probably domesticated in the Ethiopian region or in West Africa. No documentation is available for the migration of the cultivated okra from Africa to the Mediterranean region and to Asia.

The cytotaxonomy of *Abelmoschus* is yet so confused that an Asian origin of whole or part of the cultigen does not seem impossible. Indian scientists believed that either the cultigen might have originated in Asia or it was present in Africa and India as polyphyletic species. Belief of Indian origin of okra gets strong support from the mentioning of 'Tindisha' and 'Gandhamula' designating okra in ancient Sanskrit literature and availability of the wild species *Abelmoschus tuberculatus* in India which is the ancestor of cultivated okra. South-east Asia is considered to the centre of distribution.

It was apparently introduced to America during the early post- columbian period. Okra is known by many local names in different parts of the world viz., ladies finger in England, gumbo in the USA and bhindi in India.

Some species which are found both in wild and cultivated forms are *Abelmoschus esculeutus, A manihot* and *A moschatus* .These species having both wild and cultivated forms are distributed in Africa, Southeast Asia, India and other okra growing regions of the world. The species *A manihot* is considered to be a polyspecies complex. The wild species *A angulosus, A ficulneus, A crinitus* and *A tuberculatus* are confined in Asia and Australia however, *A ficulneus* is also present in West Africa. Other wild species of *Abelmoschus* include *A. tetraphyllus, A. caillei, A. pungens, A. heterophyllus*, etc. The wild species particularly *Abelmoschus manihot* and *Abelmoschus tetraphyllus* have been extensively used in breeding okra resistant to yellow vein mosaic virus.

Open pollinated improved varieties

- Improved varieties: Pusa Makhmali, Pusa Sawani, Pusa A-4, Hissar Unnat, Azad Kranti, Arka Anamika, Arka Abhay, Punjab Padmini, Co-1, MDU-1, Punjab-7 , Punjab-8, Parbhani Kranti, Gujarat Bhindi-1, Harbhajan Bhindi, Varsha Uphar, Selection 2-2, VRO-6.
- Varieties resistant to YVMV: Arka Anamika, Arka Abhay, Pusa A-4, VRO-6, Azad Krishna.
- Varieties suitable for export: Pusa Sawani, Parbhani Kranti, Punjab-7, Varsha Uphar, Panvel.
- Important local and introduced cultivar: Satdhari, Panchdhari, Vaishali Vadhu, Red Bhindi, Perkin's Long Green, Mukta, Pankaj, etc.

Cultural requirements

- Okra is basically a hot weather crop and thrives well during hot and humid climatic condition.
- It is grown in tropical and subtropical climate in the plains and in summer months in the hill.
- It is susceptible to drought and low night temperature and frost.
- It can be successfully grown under the temperature ranging between 25 and 30°C.
- The plant grows shorter under high temperature and low humidity condition.
- Day temperature more than 42°C causes flower drop.
- Optimum temperature range for seed germination is 25-35°C and seeds do not germinate below 17°C.
- Adequate sun shine is very important for growth and yield of the crop.
- It can grow well in wide range of soil, sandy to clay soil if those are well manured with enough organic matter and good drainage facility.
- Loose, friable, well manured loam soils are the best for the crop.
- Water logging is harmful for the crop.
- Ideal pH range is 6.0 to 6.8 for most of the cultivars.

Land preparation

- Prepare the field with 2-3 ploughing upto a depth of 20-25 cm and make into a good tilth before seed sowing.
- Remove all the weeds, stubbles and level the land by laddering.

- The plant has well developed root system and is a good feeder.
- Soil should be made rich in organic matter content by adding 20-25 tonnes of FYM per hectare at the time of land preparation.
- Solarization during hot summer helps to a great extent in controlling weed and soil borne micro-organisms.
- Divide the field into several plots with irrigation channels depending on irrigation source and slope of the land.

Sowing time

Okra grows successfully in tropical and subtropical climate hence, can be grown year round in the southern states due to prevailing mild winter there. However, the crop is grown in two main seasons: spring-summer and rainy season.

- Late January – early February: Spring-summer crop in eastern region.
- February-March: Spring-summer crop in north and western region.
- May-July: Rainy season crop in most areas.
- Mid August: Rainy-autumn crop.
- November: Winter crop in southern states.

Seed sowing

- Okra gives little success on transplanting and thus seeds are directly in the field.
- Seeds are sown by hand dibbling or behind the plough or by seed drill.
- Broadcasting is not recommended due to high seed rate as well as inconvenience in cultural operations and harvesting.
- 18 –22 kg seed/ha is sufficient for spring-summer crop and 8-10 kg/ha for the rainy season crop.
- Soak the seed overnight in water for better germination.
- Soaking the seeds in 0.2% Baviatin over night protect the seedlings from wilt disease.
- Soaking the seeds with 400 ppm GA or 20 ppm NAA or 20 ppm IAA enhances germination.
- Sow the seed on ridges in spring- summer season on the side facing sun and on flat beds in rainy season.
- It is better to do seed sowing in pre-irrigated moist soil than irrigate the field after sowing.
- Pre-sowing soil application of 20-22 kg Furadon 3G helps in protecting the plants from root knot nematodes and other pests during initial 4-5 weeks.
- On an average 50-70 thousand plants can be accommodated in one hectare land.

Spacing

- Spring-summer crop: 45 cm × 30 cm (branching type)
  30 cm × 15 cm (non-branching type)
- Rainy season crop: 60 cm × 45 cm (branching type)
  60 cm × 30 cm (branching type, 50000 plants/ha)
  45 × 30 cm (non-branching type, 66000 plants/ha)
- Harvesting of small fruits for export: 3 row group of 20cm × 60 cm between groups.

Nutrient management

- The plant needs to produce 5 tonnes of dry matter to produce 10 tonnes of green fruit equivalent to 1 tonnes of dry matter in fruits.
- Apply a general fertilizer dose of 125 kg N, 75 kg P_2O_5 and 60 kg K_2O per hectare for open pollinated improved varieties. 1/3 N along with other fertilizers should be given as basal and rest N should be top dressed in two split doses at 30 days after sowing and at flowering.
- For the ratoon crop the same nutrient schedule is followed and basal dressing should be done immediately after pruning followed by irrigation and plant protection measures.
- Apply 200 kg N, 100 kg P_2O_5 and 100 kg K_2O per hectare for the hybrids. Half N along with other fertilizers should be given as basal and rest N fertilizer should be top dressed in two split doses at 30 days interval after sowing.
- In spring summer application of N in splits after every 3-4 pickings increase crop duration and yield.
- Foliar spray of 0.3% zinc sulfate and 0.02% ammonium molybdate in the standing crop enhance fruit yield.
- Soil application of calcium is beneficial for the seed crop.
- Apply vermicompost @ 4-6 t/ha as broadcasting during final land preparation.
- Apply *Azotobactor* or Phosphate Solublizing Bacteria @800g/ha as seedling root dip.
- Prepare a suspension of bacterial inoculants in sufficient water.
- Dip the seedlings in the bacterial suspension in shade for 2-3 hours in the afternoon hours.
- Transplant the treated seedlings immediately.

Irrigation

- Seed should be sown when the soil has sufficient moisture.

- Apply water at the initiation of first true leaf during spring-summer and at its expansion during rainy season.
- Irrigate the field subsequently 5-6 day's intervals during summer and as and when required for the kharif crop depending on the rain.
- The crop must be irrigated during its most critical stages i.e. flowering and fruit setting stages and water stress in these stages considerably reduce yield.
- Over irrigation by flooding may cause wilting of the plant.
- Drip irrigation system increases the yield as well as saves 70-80% irrigation water.

Interculture and weed control

- Proper weed management is urgently needed to check yield loss.
- At least two weedings are required till the crop canopy covers the soil surface.
- Application of herbicides reduces the number of weeding to zero in summer and one during rainy season.
- Apply fluchloralin (Basalin 48 EC) @ 1.5 kg ai/ha as pre-sowng soil incorporation and pendimethalin (Stomp 30EC) @ 0.75 kg ai/ha as post-sowing and pre emergence soil surface spray.
- Surface application of herbicides is effective for 4-5 weeks.
- Okra generally does not require staking, training and pruning.
- Spraying the standing crop with 100-500 ppm ethrel reduce vegetative growth and weaken apical dominance.
- For ratoon crop from quick branching varieties, prune the spring-summer crop 20-25 cm above the ground after summer fruiting with the onset of rains followed by manuring the crop and plant protection cover.

Harvesting the crop

- First harvesting is generally done 40-50 days after sowing depending on the variety.
- Fruits at harvest should be tender.
- Harvest the tender fruits by bending the pedicel with a jerk or with the help of a knife at every alternate day.
- Use cotton cloth gloves to protect fingers and harvest in the morning when hairs are soft, for nearby market.
- For distant market harvest the fruits in the late evening and transport the produce in the cool condition at night.

Post harvest management and storage

- Sort out the fruits for export purpose which are green, tender, 6-8 cm long, 4-5 ridged, slender and roundish with pointed tips.
- For local markets precool the fruit by dipping in cold water and fill in jute bags or baskets, covered and sewed and then sprinkle water on them.
- For export pack the pre-cooled fruits in suitable size perforated paper cartoons (5-8 kg size) and transport in refrigerated vans.
- In air-tight containers the fruit may turn pale during transit due to heat generated through respiration of the fruits.
- The fruits can be stored at 7-9°C temperature and 80-90% relative humidity for 7-10 days.
- Post harvest treatment of fruits with cycocel (100 ppm) enhances self-life.
- Post-harvest treatment of fruits with 250 ppm ascorbic acid retains chlorophyll of the fruit and keeps them in good condition for 8-9 days in ambient condition.

Plant protection measures

Diseases

Yellow vein mosaic disease (Yellow vein mosaic virus)

This severe virus is transmitted by the insect vector white fly (*Bemicia tabaci*). This destructive disease infects at all stages of the crop. In the leaf, homogeneous interwoven network of yellow veins enclosing islands of green tissues within are developed. In extreme cases, the entire leaf turn complete yellow or cream coloured. The infected plants are stunted and bear very few, yellow coloured fruits.

Control measures

- Grow resistant or tolerant varieties like, Arka Abhay, Arka Anamika, VRO-6, NOH-303, Azad Krishna and some resistant hybrids.
- Adjust the sowing time according to white fly population. Early sowing of crop during January-February and also autumn sowing crop (October-November) reduces the intensity of disease.
- Remove diseased plants as soon as they are noticed in the field.
- Apply carbofuran (Furadon 3G @ 20-22 kg /ha) to the soil at the time of sowing.
- Control of whitefly by 4-5 foliar sprays of dimethoate (Rogor 2ml/l) or imidachloprid (3.5 ml/ 10 l or Thiomethoxam (3.5 ml / 10 l).

Powdery mildew (*Erysiphe cichoracearum*)

This fungal disease appears under prolonged dry conditions. White floury or powdery pustules appear on the lower surface of the leaf extending to upper surface causing yellowing and death of the leaves.

Control measures

- Spray the crop with Bavistin or Benlate @ 1g/l of water 3 to 4 times at 15 days interval.
- Two sprays of 0.1% Bavistin followed by two sprays of 0.1% Karathane or tridemorph effectively control the disease.

Insect pests

Jassids/ leaf hoper (*Amrasca biguttula biguttula*)

The adults are wedge shaped, about 2 mm long, pale green with a black dot on posterior portion of each forewing. Hoppers suck the sap from lower surface of leaves causing cupping and later drying and falling of the leaves. Early sown crops are highly susceptible to jassid attack and found to be very active during February to June with a peak population attained in the month of May.

Control measures

- Apply granular insecticides like carbofuran (Furadon 3 g @ 20-22 kg /ha) at the time of sowing.
- Spraying with imidaclorprid @ 0.3ml/l at the early flowering stage and if required use neem based insecticide for effective and healthy management of the pest.
- Spray the crop with 0.05% phosphamidon (Dimecron 1.5 ml/l) or 0.05% endosulphan (Thiodan 1.5ml/l) at 10 days interval.

Red spider mite (*Tetranychus* spp.)

Four spider mite species, *Tetranychus urticae* Koch, *T. macferlanei* Baker & Pritchard, *T. ludeni* Zacher and *T. neocaledonicus* Andre are commonly found to attack vegetable crops. They are tiny, microscopic and morphologically alike without visual difference among these four species. However, *T. urticae* is found everywhere to attack the crops. Okra is severely infested by these mite species and found to feed from the ventral surface of the leaf. All the active stages of the mite suck sap resulting whitish appearance of the crop. Severely infested crops are covered by their heavy webbing. Summer crops are worst affected.

Control measures

- Use of balanced fertilizer and proper irrigation keeps the plants healthy and reduce mite attack.

- Spray the crop with neem based acaricides during first occurrence of mite in the crop.
- Release of predatory mite, *Amblyseius longispinosus* is very effective against these mites.
- Heavy infestation requires acaricidal application (dicofol @ 2.5 ml/l; propergite @ 1.5 ml/l or diafenthiuron @ 1ml/l) to save the crop.

Shoot and fruit borer (*Earius vitella; E. insulana*)

The caterpillars are brownish white with a number of black and brown spots on their body. They bore into the stem, flower buds and fruits which results in drying up of shoots and dropping of flower buds and developing fruits.

Control measures

- Destroy infested plant parts along with the larvae inside and maintain sanitation in the field.
- Avoid growing of alternate host plants like, *Malva parviflora*, *Althaea officinalis*, cotton, etc, in nearby field.
- Spray the crop with Spinosad @ 0.25 ml/l at 15 days intervals.

Yield

- 7-8 tonnes/ha (open pollinated varieties in spring-summer).
- 11-12 tonnes/ha (open pollinated varieties in rainy season).
- 15-18 tonnes/ha (Hybrids in rainy season)

Pointed Gourd

29. Pointed gourd

Scientific Name: Trichosanthes dioica Roxb.
Family : Cucurbitaceae
Origin : Indo-Malayan region
Chromosome No: 2n = 22

Introduction

Pointed gourd (*Trichosanthes dioica)*, called the king of gourd, have its probable primary centre of origin in Bengal-Assam area or Indo-Malayan region. This vegetatively propagated, dioecious and perennial cucurbit is known in India from very early times because its Sanskrit equivalent "Patol" in mentioned in two medical treatises "Caraka Samhita" and Yajurveda". The pointed gourd being the highly accepted cucurbitaceous vegetable holds a coveted position in vegetable markets of India, particularly during summer and rainy season. However, the tender fruits are available in the extensive period of eight months from February to October. This perennial vine crop survives for long time through giving rise to sprouts from the tuberous roots even left uncared. This important crop is widely grown in the eastern Uttar Pradesh, Bihar, West Bengal, Assam, Orissa, Madhya Pradesh, in some parts of Maharashtra and Gujarat and some hilly tracts of Andhra Pradesh and Tamil Nadu.

Importance and use

- Pointed gourd is called the "King of gourd" because it is one of the most nutritive and wholesome vegetables.
- It is the highest dietary fibre containing vegetables.
- Great possibility exists for exploring its export to South-East Asia, Gulf countries and even to European countries.
- Pinted gourd is also called in different Indian names of "*Parwal*", "*Parmal*", "*Panal*" and "*Patal*"
- It is grown in extended period from summer to early autumn season.
- It is also cultivated in different river beds of the country.
- The tender fruits of pointed gourd are generally consumed as cooked and fried vegetable dishes, and also used in making curries.
- The immature fruits are used for preparing pickles.
- A famous sweet is prepared in India keeping the fruits filled with milk cake and immersed in sugar syrup.
- The newly emerged tender shoots with leaves are also a preferred potherb in many households in India.

Nutritional value

- It is called king of gourd because of having higher nutrient content than other cucurbitaceous vegetables.
- Protein content of the fruit is 10 times that of bottle gourd and four times that of snake gourd, ridge gourd and wax-gourd.
- Vitamin A content is about 5 times that of ridge gourd, 3 times of pumpkin and almost 100 times of bottle and wax gourd.
- Vitamin C content of pointed gourd is much higher than that of any gourd.
- It contains 2% protein, 0.3% fat, 2.2% carbohydrates, 153 IU vitamin A and 29 mg vitamin C per 100 g edible portion.
- Tender fruits also contain 3.0 gram fibre, calcium 30 mg, Potassium 83.0 mg, Phosphorus 40.0 mg per 100 g edible portion.
- Nutritive value of this crop somewhat vary from variety to variety.

Medicinal value

- It is easily digestible, diuretic, laxative and cardiatomic.
- The fruit is particularly recommended during convalescence.
- It is recommended also for bronchitis, biliousness, and high fever and nervousness.
- It also invigorates the heart and brain and is useful in the disorders of the circulatory system.
- The fruits show some prospects in the control of certain cancer like conditions.
- It is particularly beneficial for colon problem due to very high dietary fibre content.

Adaptation in India

Pointed gourd was known in India in very early times and Sanskrit language equivalents signifying Vedic period and Aryan usage is available for pointed gourd as "Patol". Cooking of this fruit "Patol" have been mentioned in two medical treatises "Caraka Samhita" and Yajurveda". Several wild forms of pointed gourd occur particularly in Assam plains and Brahmaputra valley of India. It is grown in extended period from summer to early autumn season in the field or in different riverbeds of the country.

Extensive clonal variation in this crop exists in West Bengal, Assam, Tripura, Bihar, and eastern part of Uttar Pradesh and asexual propagation perpetuate large number of diverse cultivars on farmers' field. Only fruit characters have been found contributed significantly to the separation of the female clones of pointed gourd. The clones can be grouped under four major types depending on fruit shape and size viz. spindle shaped fruits, oval fruits, nearly cylindrical fruits and plant bearing small fruits of different shape with dark, light green or pale-white colour with or with out stripes. The major fruit sizes are 10-13 cm long, 10-16 cm long thick, small 5-8 cm long roundish and small tapering at the end. Fruits of different cultivars are marketed under different local names without any standardization in nomenclature. Some of such popular local cultivars in three important pointed gourd-growing states are given below:

West Bengal	:	Damodar, Kajli, Kajli Bombai, Kajli Damodar, Sandhyamani, Hilly, Guli, Haibatkhali, Shampuria, Dhanpa, etc.
Uttar Pradesh	:	Dandali, Kalyani, Guli, Bihar Sharif, Guthaliae, etc.
Bihar	:	Dandali, Nimia, Hilly, Santokhwa, Bihar Sharif, Niria, etc.
Madhya Pradesh	:	Green Oval, Green Long Striped, White Oval

Breeding behaviour

- Unlike other cucurbits, the seed is not recommended as commercial propagating material because of poor seed viability and germinability, production of about 50% of male plants from the seeds due to segregation of male and female plants since the sex is determined by the physiological difference in one chromosome and long, even two years time for initiation of flowering.

- It is asexually propagated and cultivated commercially by vine and root cuttings.
- The sex form is typically dioecious i.e. male and female plants are separate which leads to natural cross pollination.
- One pair of chromosomes has been reported to be physiologically distinct and carrying the determining sex genes.
- Stray instances of hermaphrodite flowers are also observed which are generally non-functional and probably indicate reversion to the primitive characters.
- Fruit set in the female plant generally occur due to cross pollination with the pollens from the male plants.
- Fruit and ultimately yield in the female plants greatly depends on the percentage of male plants in the field and underdeveloped fruits due to lack of pollination is a common problem.
- 10% male plant population is enough for supplying pollens to get maximum fruit set in the female plants.
- Major pollinators are red brown beetle (*Carpophilus dimidiatus*).
- Opening of flowers from its appearance as bud requires 7-10 days in female and 9-12 days in male flowers.
- Temperature greatly influences the development of buds and low temperature can delay the full development of flowers by 8-9 days.
- Both male and female flowers open at early hours of night, between 7 and 9 p.m.
- The female flowers always open earlier than the male flowers.
- The opening of male flowers and dehiscence of anthers tend to go together.
- The stigma of the female flowers become receptive 4 hours before anthesis and the receptivity remains 12 hours after anthesis upto 8.00 a.m. in the morning.
- Maximum pollens in the male flowers germinate at anthesis i.e., at night which gradually declines.
- Hand pollination in female flowers is widely practiced now-a-days due to paucity of pollinator population which even results in a two-fold increase in fruit set over natural pollination.
- Hand pollination must be completed as early as possible in the morning preferably with in 5.30 a.m.
- Harvesting the fully opened staminate flowers in the morning hours from the male plants and simply touching the pollen laden male flower to the stigma of the female flower is sufficient to ensure effective pollination.
- Pollens of one male flower are sufficient to effectively pollinate 6-8 female flowers.

- Application of 100-500 ppm naphthalene acetic acid induces parthenocarpic fruit set.

Origin and taxonomy

Trichosanthes is a large genus, principally of Indo-Malayan distribution, with about 44 species of which 22 are found in India. Much earlier, De Candolle recorded in his "*Origin of Cultivated Plants*" the species of *Trichosanthes* as being of Old World origin, most probably India. So far, the centre of origin of *Trichosanthes* is not precisely known even though most authors agree on India or the Indo-Malayan region as its original home. The proposition of Assam-Bengal region as the primary centre of origin of pointed gourd was based on the rich diversity of this crop in this region including Bangladesh. Several wild forms of pointed gourd occur particularly in Assam plains and Brahmaputra valley of India. However, wild forms of pointed gourd are also available through out North India.

The pointed gourd, *Trichosanthes dioica* Roxb. of the family Cucurbitaceae belongs to the sub-family Cucurbitoideae, tribe Trichosantheae and sub-tribe Trichosanthinae. It is one of the two cultivated species, *T. dioica* and *T. anguina* (snake gourd) among the 22 species occurring in India.Some of the wild species are *Trichosanthes cucumerina, Trichosanthes palmate,* etc.

The basic chromosomes number is 11 (2n=22) with no apparent heteromorphy in the male and female complements. However, a heteromorphic pair of chromosome, "an incipient sex chromosome" present in all the clones is suggested to carry the determining sex genes.

Important cultivars

West Bengal : Damodar, Kajli, Kajli Bombai, Kajli Damodar, Sandhyamani, Hilly, Guli, Haibathkhali, Shampuria, Dhapa, etc.

Uttar Pradesh : Dandali, Kalyani, Guli, Bihar Sharif, Guthaliae, etc.

Bihar : Dandali, Nimia, Hilly, Santokhwa, Bihar Sharif, Niria, etc.

Madhya Pradesh: Green Oval, Green Long Striped, White Oval

Improved clonal selection: Swarna Rekha, Swarna Alaukik, FP-1, FP-3, FP-4, FP-5, Rajendra Parwal-2, BCPG-3, BCPG-4, BCPG-5.

Cultural requirements

- Pointed gourd is a warmth-loving crop, thus, thrives well under hot or moderately warm and humid climate.
- Abundance of sunshine and fairly high rainfall favour good crop yield.
- Optimum temperature for proper growth ranges from 25° to 35°C.
- Regeneration of new sprouts is generally impaired below 20°C and severe cold below 5°C is bullying for the crop.

- Vine growth becomes highly restricted during winter, which starts again along with sprouting from fleshy root at the onset of spring.
- Pointed gourd can be forced during harsh winter months (November to February) in different river-beds or river basins, familiarly called 'Diara' lands in Uttar Pradesh and Bihar, through planting of rooted vine cuttings in sand of riverbeds with required watering till they drive roots up to the water level below. Such moisture laden sandy beds get warm up quickly and wipe low temperature effect of the winter.
- High rainfall coupled with prolonged cloudiness promote high incidence of fruit rot disease and in turn drastically reduce yield.
- It can be grown in a wide variety of light textured soils having good drainage facility.
- Well-drained sandy-loam to loam soil with slightly acidic reaction (pH 6.0-7.0) is ideally suited for this crop.
- The crop can withstand water stress.
- It is susceptible to water logging hence, grown successfully in upland condition.
- Deep well drained fertile, loamy or sandy-loam soils with pH range 6.0 – 7.0 are ideal.

Land preparation

- The selected land with good drainage and plenty of sun shine are preferred for this crop.
- Prepare the field with 4-5 deep ploughing and make into a good tilth before seed sowing.
- Remove all the weeds, stubbles and level the land by laddering.
- Manures should be mixed with the soil at the time of last ploughing.
- The plant is susceptible to water logging so land should be well drained.
- Raised beds of 15-20 cm high with 1.5 m width and convenient length are prepared.
- A spacing of 60-75 cm is maintained between two beds which serves the irrigation-cum-drainage channel.
- Mounds are prepared on both sides of the bed at close proximity of the channel at 60 cm spacing.
- Mounds in the bed method or in river bed system of planting are prepared with the mixture of 15-20 tonnes of FYM per hectare and basal dose of fertilizers are mixed with the soil.

Planting time

Early planting can be done in upland situation but it is not beneficial in medium land situation as frequent heavy downpour hampers establishment of the cuttings while, planting of the vines after the onset of winter delays sprouting and establishment of the plants.

- Late August to mid September: Upland situation.
- First week to second week of October: Medium land situation in high rainfall areas.
- Late October to November: River beds.

Planting

- Unlike other cucurbits, the seed is not recommended as commercial propagating material because of poor seed viability and germinability, production of about 50% of male plants from the seeds due to segregation of male and female plants and long, even two years time for initiation of flowering.
- Vine cutting of 60-90 cm having 10-15 nodes should be taken from mature vine.
- The cuttings should be of 15-20 cm in length and of pencil thickness.
- Pointed gourd is a long duration crop of about 8 months so, a very high-density planting affects its growth and yield adversely.
- Generally 5000-6000 cuttings are required for one hectare land in 1.5 × 1.5 m spacing.
- 6000 plants per hectare in 1.5 × 0.60-0.75 m spacing also give satisfactory yield.
- Undeveloped fruit due to lack of pollination is a common problem in pointed gourd cultivation.
- In the field 8-10 percent male plant population is considered enough for getting maximum fruit set in female plants.

Spacing

Planting in bed	:	1.5 m × 1.5 m; 1.5 m × 0.60-0.75 cm
River bed system	:	1.5 m × 1.5 m
Crop on the bower	:	1.5 m × 60 cm

Planting methods

Vine cutting

The defoliated vine cuttings are planted in the field following different methods

- *Ring Method:* This planting method is widely followed where the cutting are coiled into a spiral or ring shape and planted directly on the mounds, covering one-half of the ring under the soil.

- *Lachhi method:* The cuttings are folded into a figure of "eight" and placed flat in the mound and pressed 3-5 cm deep in the middle into the soil.
- *Moist lump method:* The cuttings are encircled over a lump of moist soil leaving both ends of 15 cm free and the lump are buried 10 cm deep leaving the ends of the cuttings above the soils.
- *Straight vine method:* The cuttings are planted horizontally, 5 cm deep leaving both the ends above the soil.
- *Rooted cutting method:* Cutting previously rooted in the nursery under sand medium are planted late particularly in the diara land of Uttar Pradesh and Bihar.

Root cutting

The tuberous roots from the matured vines are collected in September-October and planted on the mounds.

Nutrient management

- Pointed gourd responds well to manuring and fertilizer application.
- Farm yard manure should be applied @ 20-25 t/ha at the time of preparation of mounds.
- In September-October planting, apply a general fertilizer dose of 150 kg N, 60 kg P_2O_5 and 60 kg K_2O per hectare .1/3 N along with full P and K fertilizers and farm yard manure should be given as basal and rest N should be top dressed in two split doses, at 80-90 days after planting and the rest, one month later.
- At late planting (December) after harvesting aman paddy, half of nitrogen and full dose of phosphatic and potassic fertilizers are applied as basal and rest half of nitrogen is side dressed 110-120 days after planting.
- Excessive nitrogen under frequent irrigation promotes excessive vegetative growth, especially in heavy soils which reduce the propensity of flowering in both male and female plants.

Irrigation

- Irrigate the field once in 5 or 6 days in summer depending upon the soil, location, temperature, etc.
- During rainy season irrigation is generally not required rather drainage is important.
- Frequent irrigation also promotes excessive vegetative growth, especially in heavy soils.

- Irrigation water should be restricted to the base of the plant or root zone without wetting the vines, especially when flowering, fruit set and fruit development are in progress.
- Frequent wetting of stems, leaves and developing fruits promote rotting disease of the vines and fruit.

Training over trellis

The crop is generally grown on the bed without any support to trail the vines. This crop can also be grown by training the vines over low trellises or bowers of 1 m height, made up with bamboos and ropes or wires. The planting distance is reduced to 1.5 m × 60 cm to accommodate more plants per unit area. Main advantages of such types of training vines are.

- Lower incidence of disease particularly vine and fruit rot.
- Fruit become attractive in appearance, shape and size.
- Longer availability of fruits.
- Increase yield per unit area due to accommodation of more plants.
- Higher return from the quality product.

Interculture and weed management

- Regular weeding should be followed to ensure proper growth of the vines.
- Mulching with straw, water hyacinth, sugarcane trash, dried grass, sawdust upto 8-10 cm thickness or black polythene helps on conserving moisture, suppressing weeds and protecting the fruits from rotting on contact with wet soil.
- Pointed gourds grown on beds are intercropped with different vegetable crops, like beet leaf, radish, coriander, fenugreek, cauliflower, pea, etc. during early stages of growth (October-January) for better land use and greater economic return however, success of the companion crop largely depends on proper weeding.
- Application of weedicides like gramoxone @ 1 litre a.i. or fernoxone @ 0.8 litre i.e. per hectare as most emergent spray along with mulching helps in checking the weeds effectively.

Ratooning

- Consecutively three successful crops can be taken from a single planting.
- In ratoon cropping, the vines are prunned 15 cm from the ground level during October after the fruiting is over.
- Recommended dose of farmyard manure and fertilizers are applied by loosening the soil around the mound by the end of winter.

- Half the recommended nitrogen is applied as top dressing one month after flowering.

Harvesting

- Harvesting generally starts 90-120 days after planting depending on the time of planting.
- In October planted crop of Gangetic alluvial zone of West Bengal, harvesting of fruits generally starts from middle of February and continues to July at frequent intervals and up to September if new flushes come with the monsoon rains.
- Harvesting should be done when the fruits are immature and tender before seeds become hard.
- Harvesting of the fruits at 12 to 15 day after fruit set depending on the cultivar is ideal for both quality consideration and yield.
- Picking should be done frequently so that maximum fruits can be harvested from a vine.
- Delay in harvesting causes formation of mature seeds in the fruit, reduces fruit quality and fruiting capacity of the vine.

Post harvest handling and storage

- After harvesting, placing of fruit in cold water (10°C) sanitized with sodium hypochlorite solution (100 mg/ litre) for 20 minutes and then dipping in Carnauba wax solution (1 part wax + 10 part water) for one minute is very effective for increasing shelf life.
- Fruits remain marketable for 3-5 days under ordinary storage conditions.
- Fruit dipping in 250 ppm sodium benzoate or 100 ppm citric acid solution for 10 minutes extends the shelf life of fruits up to eight days.
- Dipping of freshly harvested tender fruits in solution of growth substances, like kinetin (50 ppm), GA_3 (20 ppm), CCC (100 ppm) or NAA (20 ppm) for 10 minutes also increases the shelf life of the fruits upto 8 days without shrinkage and yellowing.
- The fruits have short storage life for 10 days at 8°C and 90% relative humidity.
- Fruits packed in 1 kg capacity polyethylene bag remain fresh up to 25 days at at 8°C and 90% relative humidity.
- Because of thick skinned nature, pointed gourd can be transported to distant markets.
- Fruits are packed in basket or gunny bags for sending to the market in a crude way.
- Pointed gourd must be collected in plastic crates while harvesting and packed in CFB boxes and staking properly in trucks during transit.

Plant protection measures

Physiological disorder

Unfruitfulness

Being a dioecious in nature, flowers in pistillate plants are shed due to lack of pollination. In some cases, ovary of the unfertilized flower may grow a bit due to parthenocarpic stimulation which also abscises after a few days.

Control measures

- Male plants must be grown in the field with the female plants in the ratio of 1:10 to ensure adequate pollination and fruit set.
- Plant protection chemicals should not be used in the morning hours as it hinders the visit of pollinators.
- Hand pollination may be done early in the morning (before 6.00 a.m) to achieve nearly cent per cent fruit set. Pollens of one male flower can pollinate 6-8 female flowers satisfactorily.

Disease

Rot disease (*Pythium* spp. and *Phytophthora spp.*)

The incidence of shoot and fruit rot caused by the soil borne fungi is mostly seen in rainy season, particularly when water stagnation occurs. The fungi attack the tender shoots, leaves, and fruits as a consequence white cottony growth develops on the affected plant parts. The spread of fungi is aggravated through rain splash and irrigation water.

The other soil borne fungus, *Macrophomina phaseolina*, causing collar rot attacks root and collar region of the vines and the affected portion gradually rot and finally become dry. This disease becomes rampant in the crop preceded by jute.

Control measures

- Grow the crop on raised beds with good drainage facility to avoid water stagnation.
- Spray the crop with copper oxychloride @ 3-4 g/litre of water or mancozeb @ 2.5g/litre of water or metalaxyl-mancozeb @2.5g/litre of water or cyamoxanil-mancozeb @2.5g/litre of water at 7-10 days interval.

Downy mildew (*Pseudoperonospora cubensis*)

This fungal disease is prevalent in high humidity, especially when summer rains occurs regularly. The diseases is characterized by formation of yellow, more or less angular spots on the upper surface of the leaves, white-purplish spores appear on the upper surface of the leaves. The disease spreads rapidly killing the plant quickly through rapid defoliation.

Control measures

- Management schedules as mentioned in pumpkin.

Cucumber Green Mottle Mosaic Virus (CGMMV)

The causal virus (*Tobamo virus*), an undeveloped rod shaped virus, causes irregularly distributed chlorotic spots on leaf lamina. It is transmissible by contact and mechanical inoculation. Vectors of the virus are not confirmed. The virus is carried over through seasons by infected crops, weeds and contaminated planting materials.

Cucumber Mosaic Virus (CMV)

The virus particle, an isometric cucumo virus, causes mosaic mottling, irregularly distributed yellowish patches and green vein banding with deep brown to blackish vein areas on the lamina. It is transmissible by mechanical inoculation and vectored by aphids (*Aphis gossypii* and *Myzus persicae*) but it does not persist in vectors. The inter-seasonal carry over of virus occurs through infected crops, weeds and contaminated planting materials.

Control measures

- Destroy the infected crop plants and weeds to check the carry over of virus.
- Spray the crop with imidacloprid (3.5 ml/10 litres of water) to control the aphids, which are the vectors of this virus.

Insect pests

Fruit fly (*Dacus dorsalis*)

It is the most serious and destructive insect of pointed gourd. Adult fly of this polyphagous pest having a wide host range is 4 – 5 mm long, ferruginous-brown in colour with hyaline wings. On hatching, the maggots feed inside the fruits and infested fruits can be identified by the presence of resinous liquid which oozes out of the punctures made by the flies for oviposition. The infested fruits start rotting due to secondary infection of different micro-organisms.

Control measures

- Collect and destroy the infested fruits along with the maggots to prevent the carry over of the pest.
- Stir the soil under the vines frequently and plough the infested field after harvest the crop to kill pupae.
- Spray the crop with 0.05% endosulfan.

- Apply poison bait. Bait is prepared by mixing 20 g malathion 50% WP or 50 ml diazinon 20% EC with 500 g molasses + 20 g yeast hydrolysate. This mixture is mixed with 2 litres of water for poison baiting and 20 liters of water for bait spray.
- Adopt male annihilation technique using 0.1% methyle eugenol + 0.025% malathion in a bait trap during fruiting season. Methyl eugenol is a sex attractant and males are attracted towards the bait and killed by malathion. Apply this trap solution at 7-10 days interval.

Red spider mite (*Tetranychus* spp.)

Small numerous red spider mites colonize at the ventral surface of pointed gourd leaves and in many other vegetable crops. The mite infested leaves turned to yellowish and gradually dry out. The mite also produce profuse silken webbing on the plant surface that reduce the phosynthetic activity of plants.

Control measures

- Release of predatory mite, *Amblyseius longispinosus* is very effective against red spider mite comlex.
- Prophylactic spraying of botanical pesticides is useful to minimize destructive population.
- Spraying of propergite @ 1.5 ml/l, or dicofol @ 2.5 ml/l.

Brown scale insect (*Saissetia coffeae*)

It is still not a serious insect of pointed gourd except in some parts of India. Remaining within the shelter of scale this insect sucks cell sap.

Control measures

- Spray the crop with 0.2% carbaryl or 0.05% quinalphos.
- Use of *Eublemma scitula* being an important predator controls this insect effectively.

Root knot nematode (*Meloidogyne incognita*)

Root knot nematodes that harbour pointed gourd incite profuse galls on roots, resulting hamper in the supply of food and water inside the plant system. In severe attack, the growth remains stunted and plants may even die. Suppressing nematode population through different measures effectively check the infestation.

Control measures

- Application of oil cakes of *mahua*, *neem* and mustard, *neem* cake in particular, @ 250 kg/ha can significantly suppress the incidence of *Meloidogyne incognita.*
- Growing of marigold as companion crop with pointed gourd significantly lowers gall formation and egg mass development of *M .incognita* in roots.
- Application of chopped leaves of *Ricinus communis, Calotropis procera, Leucaena leucocephala* or *Melia azedarach* in soil also helps in decreasing infection by nematode and increases sprouting.

- Vines treatment with chemical pesticide, like carbosulfan (Marshal) @ 0.05% for 6 hours followed by shade drying before planting significantly reduces nematode incidence.

Yield

The fruit yield varies widely from 10.0 to 50.0 t/ha depending on the cultivar, time of planting, crop husbandry, training on trellises, pollination management and ratooning.

Appendix

Some statistics on vegetable production in India and World

Table 1. Major vegetable growing countries in the world

Country	**Area) (000 ha**	**Production (000 t)**	**Productivity (t/ha)**
China	23936	448983	19
India	7803	125887	16
USA	1333	38075	29
Turkey	996	24454	25
Russian Ferderation	970	16516	17
Egypt	598	16041	27
Iran	641	15993	25
Italy	528	13587	26
Spain	379	12676	33
Japan	433	11938	28
Others	16957	222625	13
Total	**54573**	**946774**	

Table 2. Major vegetable producing countries in the world

Name of the crop	**Country**
Brinjal	China, India, Egypt, Turkey, Indonesia
Cabbage	China, India, Russian Federation, Korea
Cauliflower	China, India,USA, Spain, Italy
Onion	China, India,USA,Pakistan, Turkey
Peas	India, China, USA, France, UK
Potato	China, Russian Federation, India, Ukraine, USA
Tomato	China, USA,India, Turkey, Egypt, Italy

Table 3. Leading Vegetable Producing States in India (2007-08)

Name of the State	Percent share
West Bengal	17%
Uttar Pradesh	15%
Bihar	11%
Orissa	7%
Tamil Nadu	6%
Gujrat	6%
Maharashtra	5%
Karnataka	4%
Andhra Pradesh	4%
Assam	4%
Jharkhand	3%
Kerala	3%
Haryana	3%
Others	12%

Table 4. Production share of major vegetables in India (2007-08)

Name of the crop	Percent share
Potato	27.4 %
Onion	9.7 %
Tomato	8.2 %
Brinjal	7.6 %
Tapioca	7.2 %
Cabbage	4.7 %
Cauliflower	4.6 %
Okra	3.3 %
Peas	2.0 %
Sweetpotato	0.9 %
Others	24.4 %

Table 5. Area share of major vegetables in India (2007-08)

Name of the crop	Percent share
Potato	22.9 %
Onion	10.3 %
Tomato	7.3 %
Brinjal	7.3 %
Tapioca	3.5 %
Cabbage	3.4 %
Cauliflower	4.1 %
Okra	5.2 %
Peas	4.0 %
Sweetpotato	1.6 %
Others	30.4 %

Table 6. Productivity trend of vegetable production in India over the years

Year	Productivity (t/ha)	Highest productive State /Union Territory (U.T.)	States/U.T.s having higher productivity than National average
1991-92	10.5	Chandigarh (37.0 t/ha)	UP, Gujrat, Maharasthra, Karnataka, Kerala, Haryana, Punjab, HP, Uttaranchal, Chandigarh, Delhi,
2001-02	14.4	Tamil Nadu (25.5 t/ha)	W.B, UP, Tamil Nadu, Kerala, Himachal Pradesh, Podicherry, Chandigarh
2005-06	15.4	Tamil Nadu (28.0 t/ha)	W.B., UP, Bihar, Tamil Nadu, Gujarat, Andhra Pradesh, Kerala, Punjab, J&K, Himachal Pradesh, Podicherry, Chandigarh, Delhi, Lakshadweep
2006-07	15.2	Lakshadweep (33.6 t/ha)	UP, Bihar, Tamil Nadu, Gujarat, Andhra Pradesh, J&K, Kerala, Himachal Pradesh, Delhi, Podicherry, Chandigarh, Mizoram, Lakshadweep
2007-08	16.1	Lakshadweep (31.7 t/ha)	W.B., UP, Bihar, Tamil Nadu, Gujarat, Andhra Pradesh, Kerala, J&K, Himachal Pradesh, Podicherry, Chandigarh, Mizoram, Lakshadweep

Table 7. Leading vegetable producing states of some vegetables in India

Name of the crop	State (Percent Share)
Brinjal	W.B. (28.5 %); Orissa (20.1 %); Bihar (12.1 %); Gujarat (10.3 %)
Cabbage	W.B. (34.3 %), Orissa (15.9 %); Bihar (10.8 %); Gujarat (6.7%)
Cauliflower	W.B. (29.0%); Bihar (17.7 %); Orissa (11.1 %); Haryana (6.0%)
Okra	W.B. (19.4 %); Bihar (16.9 %); orissa (14.8 %); Andhra Pradesh (10.8 %)
Onion	Maharashtra (32.9 %); Gujarat (18.4 %); Karnataka (9.6 %); Bihar (8.4%)
Peas	UP (54.6 %); Madhya Pradesh (9.2%); Himachal Pradesh (7.9 %); W.B. (4.8 %)
Tomato	UP (32.2 %); W.B. (28.7 %); Bihar (17.5 %); Gujarat (4.3 %)
Potato	UP (32.2 %); W.B. (28.7 %); Bihar (17.5 %); Gujarat & Punjab (4.3 % each)
Sweetpotato	Orissa (34.0%); UP (22.0 %); W.B. (21.0 %); Andhra Pradesh (5.0 %)
Tapioca	Tamil Nadu (67.53 %); Kerala (27.53 %); Andhra Pradesh (3.92 %)

Source: Indian Horticulture Database (2008)

Table 8. Export specifications for vegetables

Quality specifications for export of vegetables	
Vegetables	**Specifications**
Onion	
Big	Light to dark red, globose, 4-6 cm size, strongly pungent (for Bangladesh, South East Asia, Gulf countries); yellow, globose and 7-10 cm size (for Europe and Japan)
Multiplier	Bigger cloved bulb (2.5-3.5 cm) of bright red colour
Small	Dark red, round, 2-3 cm size
Okra	Green, tender, 6-9 cm long and slender , 4-5 ridges
Potato	White skinned long and extra long tubers with yellow to light yellow flesh (for Iraq and Iran), both white and red types of oval shape and 4-5 cm diameter (for Bangladesh)
Bitter gourd	Green, 20-25 cm long having short neck
Chilli	Dark green, tender, 6-9 cm long, smooth without wrinkles and bends
Bottle gourd	Green, tender, cylindrical, 20-25cm long with soft hairs
French bean	Green, 10-12 cm long, straight
Tomato	Round, medium size, red in colour
Brinjal	Round, thorn less, dark green calyx, purple with good lustre.
Garlic	White, round, 3.5-4 cm having 15-20 big sized cloves.

Table 9. Vegetable varieties suitable for export specifications

Suitable vegetable varieties for export	
Vegetables	**Variety**
Onion-big	N-53, Agrifound Dark Red, Agrifound Light Red, Arka Niketan (for gulf); Arka Pitambar, Granex-33, Spanish Brown (Europe & Japan); Pusa White Flat, Pusa White Round (Dehydration)
Onion-small	Arka Bindu, Agrifound Rose
Onion-multiplier	Agrifound Red, CO-3, CO-4
Okra	Arka Anamika, Arka Abhay, Varsha Uphar
Potato	Kufri Chandramukhi, Kufri Badsah (Iran & Iraq); Kufri Jyoti (Bangladesh, Sri Lanka, Nepal); Kufri Chipsona-1 & 2, Atlantic, Fritolay Hybrid 1533 (for chips)
Bitter gourd	Arka Harit, Pusa Vishesh, MBTH-101 Meghna
Chilli (Green)	Bhagyalakshmi, Arka Lohit, Pusa Jwala, Pant C-1
Bottle gourd	Arka Bahar, Pusa Naveen
French bean	Contender, Arka Komal, Masterpiece, Premier
Tomato	Round fruited hybrids; Arka Abha, Arka Ashis, Arka Ahuti
Brinjal	Arka Navneet, Pusa Anmol, Pusa Kranti
Garlic	G-282, Agrifound Parvati

Table 10. Major importing countries of our fresh vegetables, processed products and seeds

Products	**Importing Countries**
Fresh vegetables	
Onion, potato, okra, bitter gourd, chilli, bottle gourd, beans, melons	Middle East, Bangladesh, Sri Lanka, Nepal, Mauritius, Malaysia, Singapore, UK, Germany
Capsicum, sweet corn, Gherkin, organically grown vegetables	Europe and South Asia
Processed vegetables	
Tomato paste and ketchup; Potato	Middle East, UK , Africa, Australia and Italy
Canned peas & beans; Dehydrated (onion & garlic flakes/powder) & frozen vegetables; Pickling cucumber	Egypt, Sri Lanka, UAE, USA, Turkey
Chilli oleoresin	USA, Europe, Japan
Vegetable seeds	Pakistan, Bangladesh

Table 11. Common name, trade name, recommended dose of some common fungicides

Common name	Trade name	Recommended dose
Mancozeb-75%W.P.	Dithane M-45, Uthane M-45, Indophil M-45, Sparsha, Manzate, Zebtane, Rapid, Sriram M-45, Crystal M-45, Ebik	Seed treatment: 2.5-3.0 g/kg of seed; Spray treatment: 2.5 g /litre of water
Thiram-75% W.P.	Thiram, Anuseed, Debi Thiram, Tag Thiram, Sriram Thiram, Heaxathane	Seed treatment: 2.5-3.0 g/kg of seed;
Propineb-70% W.P.	Antracal	Spray treatment: 3.0 g / litre of water
Cymoxanil 8% + Mancozeb-64% W.P.	Curzate	Spray treatment: 2.0 -3.0 g / litre of water
Metiram-70% W.P.	Polyram, Sanit, Packup	Spray treatment: 4.0 g / litre of water
Copper oxychloride-50% W.P.	Blitox, Fytolan, Trucop, Dhanucop, Sutox, Copter, Critop, Unicop, Trust	Spray treatment: 4.0 g / litre of water
Copper oxychloride-77% W.P.	Kocide-101, Hi Dice	Spray treatment: 1.5-2.0 g / litre of water
Carbendazim-50% W.P.	Bavistin, Derosal, Benphil, Dhanustin, Benguard, Tagstin, Arest, Zoom, Carzim, Superstin, Kestin, Carvistin, Anustin, Agni	Spray treatment: 1.0 g / litre of water Seed Treatment : 2.0 g / kg of seeds
Carbendazim-12% + Mancozeb-63% W.P.	Saaf, Companion, Safal, Revive, Sixer, Karma-C.M-75, Spar	Spray treatment: 2.0g / litre of water
Carbendazim-25 % + Iprodion-25 % W.P.	Quintal	Spray treatment: 1.5-2.0 g / litre of water
Chlorothalonil-75% W.P.	Kavach, Salbakh	Spray treatment: 2.0 g / litre of water
Captan-50 % W.P.	Captaf, Dhanutan, Deltan, Kohicap, Cricap, Capstan, Anucap	Seed treatment: 2.0-2.5 g/kg of seed; Spray treatment: 2.5 g / litrs of water
Thiophanate methyl-70% W.P.	Roko, Cover Himnate, Laser, Tagcin, Control, Alert, Multinet, Milduvip, Anupsim M	Seed treatment: 2.0 g/kg of seed; Spray treatment: 1.0 g / litre of water
Carboxin-75% W.P.	Vitavax, Hiltavax	Seed treatment: 2.5 g/kg of seed; Spray treatment: 1.0 g / litre of water
Oxy-carboxin-20 % W.P.	Plantavax	Seed treatment: 3.0 ml /kg of seed; Spray treatment: 1.5 ml / litre of water
Metalaxyl-8 % + Mancozeb-64% W.P.	Ridomil M Z, Sanchar, Matco, Master, Tagmil, Krilaxyl Gold, Dhanuxyl, Dithomil, Anumil-72, Visez	Spray treatment: 2.5 g / litre of water

Tridemorpf-80% E.C.	Calixyn	Spray treatment: 0.5 ml / litre of water
Micobutanil-10% W.P.	Systhane, Fiza, Index	Spray treatment: 0.5 g / litre of water
Triademephon-25 % W.P.	Beleton, Cryleton, Tagton	Spray treatment: 0.25-0.5 g / litre of water
Hexaconazol-5% E.C.	Contaf, Hexacon, Hexzol Comfort, Danzol, Topper, Sitara, Cryzol Titan,	Spray treatment: 1.5 ml / litre of water
Hexaconazol-5% S.C.	Contaf Plus	Spray treatment: 1.5 ml / litre of water
Propiconazol-25% E.C.	Tilt, Pinacol, Dhan, Result	Spray treatment: 1.5 ml / litre of water
Difenconazol-25% E.C.	Score	Spray treatment: 0.5 ml / litre of water
Streptocycline	Streptocycline, Krosin-AG, Antiback, Tagmycin, Istamycin, K-Cycline, Anucin, Super Mycin Pantomycin	Spray treatment: 1.0 g / 10 litres of water
Validamycin	Sithmar, Rhizocin, Validan	Spray treatment: 2.0 ml / litre of water

Table 12. Common name, trade name, and recommended dose of some common Insecticides

Common name	Trade name	Recommended dose
Spinosad -45% S.P.	Spinostar, Conserve	1ml / 6 litres of water
Endosulphan -35% E.C.	Thiodan, Thionel, Endohit, Endosel, Cornalux, K-Sulphan	1.5 ml /litre of water
Dicofol-18.5 % E.C.	Dicofol, Tiktok, Kelthane, Colonel-S, Commando. Nomite	3.0 ml /litre of water
Propergite – 50% E.C.	Omite	1.5 ml /litre of water
Chlorpyriphos-20% E.C.	Dursban, Force, Nuchlor, Classic, Radar, Tataban, Hilban, Dhanban, Cyphos, Blaze	2.5ml /litre of water
Imidacloprid-17.8 % S.L.	Confidor, Media, Magic, Tez, jumbo, Tatamida, Imidex, Imida	1 ml / 3 litres of water
Acetamiprid-20% S.P.	Pride, Ekka, Bismark, Trekina, dhanprit, Manik, Rapid	1 g / 5 litres of water
Cartap Hydrochloride-50% S.P.	Keldan, Fast, Cartig, Critap, Nidan, Padan, Patap, Unitap, Sunvex	1 g / litre of water
Thiomethoxam-25% W.D.G.	Eltara, Maxima, Click, Ivident	1 g / litre of water
Diafenthiuron-50% W.P.	Pegasus, Polo	0.5 g / litre of water
Fipronil-5% S.C.	Regent	1 ml / litre of water
Azadirectin-1% E.C.	Nimaric, Ultitineem, Econeem Plus,	1 ml / litre of water

	Nemazol, Blocked, Neemgold	
Carbofuran 3G	Furadon, Diafuran, Carbocil, Hammer	3 kg / bigha
Cabaryl-50% W.P.	Sevin, Dhanuvin, Carin, Madex, Cavin	2 g / litre of water
Carbosulphan -25% E.C.	Marshal, Electra, Atanka	2 ml /litre of water
Dichlorovos-76% E.C. DDVP	Nuvan, Nuvon, Divap, Suchlor, Vantof, Domar,, Dash, Dum, Vapona, Divipan	3 ml /4 litres of water
Cypermethrin-10% E.C.	Ultramethin, Ostad, Shakti, Cibil, Super Killer, Challenger, Cypersel	1 ml / litre of water

Calculation of quantity of pesticides

The following formulae are used to calculate the required quantities and strengths of pestiicides from their known quantities and strengths.

Case-I

To obtain the desired quantity and strength of a pesticide from the given quantity and strength of a formulation.

Amount of formulated pesticide required=

(Total spray fluide x percentage strength required)/given percentage strength

As for example. Sevin 50% WDP is to be diluted to 100 litres of 0.5% spray.

Amount of Sevin required= (100x0.5)/50= 1kg.

Case-II

To obtain the strength of a pesticide of known quantity and strength after diluting it to a known quantity.

Percentage concentration of diluted spray fluid =

(Quantity of emulsion to be diluted × strength of the emulsion)/Quantity of diluted product

As for example. The strength of 1 litre of 20% Chlorpyriphos if diluted in 100 litres of water

The percentage concentration of diluted produc t= (1 litre × 20 %)/100 litres = 0.2%

Case-III

To obtain the quantity of formulated pesticide, if only the quantity of actual pesticide to be applied per ha is known.

The quantity of pesticide required =

(100 % pure pesticide used x quantity of actual pesticide used)/Concentration of pesticide

As for example. 1 kg carbofuran is to be applied per ha, what will be the amount of 3 % carbofuradon (Furadon 3G) ?

The quantity of Furadon required= (100 × 1)/3= 33.3 kg.

Measurement and Conversion

Area

1 bigha	=	20 katha
1 acre	=	3 bigha
1 acre	=	100 cent
1 acre	=	43560 sqft
1 Hectare	=	2,471 acre
1 Hectare	=	7.41 bigha

Conversion

Multiply	By	To obtain
Acre	0.405	Hectare
Hectare	2.471	Acre
Square meter (m^2)	0.000247	Acre
Acre	4050	Square meter
Square meter (m^2)	10.76	Square foot (ft^2)
Square foot	0.0929	Square meter
Square centimeter (cm^2)	0.1639	Square inch (in^2)
Square inch	6.451	Square centimeter
Square millimeter (mm^2)	0.00155	Square inch
Square inch	645	Square millimeter

Length

1 foot	=	12 inch
1 foot	=	30.48 centimeter
1 yard	=	3 feet
1 inch	=	2.54 centimeter
1 inch	=	25.4 milimeter
1 kilometer	=	1000 meter
1 meter	=	100 centimer
1 centimeter	=	10 millimeter
1 millimeter	=	1000 micron

Conversion

Multiply	By	To obtain
Kilometer	0.621	Mile
Mile	1.609	Kilometer
Meter	3.28	Foot
Foot	0.304	Meter
Millimeter	0.03937	Inch
Inch	25.4	Millimeter
Mile	5280	Feet
Feet	0.0001893	miles

Weight

1 ounce (oz)	=	28.3495 gram
1 pound	=	16 ounce
1 pound	=	0.454 kilogram
1 ton	=	2000 pounds
1 ton	=	10 quintals
1 ton	=	907 kilogram
1 quintal	=	100 kilogram
1 quintal	=	2.679 mounds
1 mound	=	37.3242 kilogram

Conversion

Multiply	By	To obtain
Gram	0.0022	Pound
Pound	0.4554	Gram
Gram	0.0352	Ounce
Ounce Kilogram	28.40 2.205	Gram Pound
Pound	0.454	Kilogram
Kilogram	0.0011	Ton
Ton	907	Kilogram

Volume

Conversion

Multiply	By	To obtain
Litre	0.265	Gallon
Gallon	3.78	Litre
Cubic meter	60975	Cubic inch
Cubic inch	0.000164	Cubic meter
Cubic meter	0.00973	Acre inch
Acre inch	102.77	Cubic meter
Cubic meter	35.7	Cubic foot
Cubic foot	0.028	Cubic meter

Pressure

1 atmosphere at sea level	=	14.7 pound/sq. inch
1 atmosphere at sea level	=	29.9 inch of mercury
1 atmosphere at sea level	=	33.9 foot of water
1 foot of water	=	0.8826 inch of mercury
1 foot of water	=	0.4335 pound/sq. inch.
1 inch of mercury	=	1.133 feet of water
1 inch of mercury	=	0.4912 pound/sq. inch

Weight and volume and flow

1 acre foot of soil	=	18160000 kilogram
1 hectare furrow slice	=	2240000 kilogram of soil (15 cm deep)
1 acre foot of water	=	325851 gallons
1 acre foot of water	=	1231716.8 litre
1 acre inch of water	=	27154 gallons
1 acre inch of water	=	102642.12 litre
1 cubic foot of water	=	7.8405 gallons
1 cubic foot of water	=	28.27 litres
1 acre inch/24 hours	=	18.86 gallons/minute
1 acre inch/24 hours	=	71.29 litre/minute
1 acre foot/24 hours	=	226.3 gallons/ min ute
1 acre foot/24 hours	=	855.41 litre/minute

Abbreviations

pp. – *pro parte*	(Latin)	=	in part
e.p. – *exparte*	(-do-)	=	in part
et al. – *et alia*	(-do-)	=	and others
viz. – *videlicet*	(-do-)	=	namely
vide – *vide*	(-do-)	=	see
ibid – *ibidem*	(-do-)	=	the same in the same place
e.g. – *example grata*	(-do-)	=	for example
cf. – *confer*	(-do-)	=	compare with
ca. – *Circa*	(-do)	=	about (with ref. to dates)
i.e. – *idest*	(-do-)	=	that is in the same reference.

Temperature conversion table

The bold number in the centre column represents the temperature in centrigrate or Fahrenheit that is desired to convert from one system into the other. The equivalent temperature in Farenheit will be in right hand column and that of centigrade in the left hand column.

C⁰		F⁰	C⁰		F⁰	C⁰		F⁰
- 17.8	**0**	32	5.00	**41**	105.8	26.7	**80**	176.0
-15.0	**5**	41.0	7.22	**45**	113.0	29.4	**85**	185.0
-12.2	**10**	50.0	10.0	**50**	122.0	32.2	**90**	194.0
- 9.44	**15**	59.0	12.8	**55**	131.0	35.0	**95**	203.0
-6.67	**20**	68.0	15.0	**59**	138.2	37.8	**100**	212.0
- 3.89	**25**	77.0	18.3	**65**	149.0	71	**160**	320
-0.56	**31**	87.8	20.0	**68**	154.4	100	**212**	413
0	**32**	89.6	23.3	**74**	165.2	127	**260**	500
2.22	**36**	96.8	25.0	**77**	170.6	160	**320**	6.8

Books consulted for the preparation of the text

1. Fundamentals of Horticulture (J.B, Edmond, A.M. Musser and F.S. Andrews), The Blakiston Co., INC., Toronto, 1964.
2. Botany (J.Ben Hill, Henry. W. Popp and Alvin, R.Grove Jr.), Tata McGraw-Hill, New Delhi, 1992.
3. The Physiology of Vegetable Crops (ed. H.C. Wien), CAB International, UK, USA, 1997.
4. Plant Growth and Development (A.Carl. Leopold and Paul, E. Kriedemann), Tata McGraw-Hill, New Delhi, 1981.
5. Crop Physiology (ed. U.S. Gupta), Oxford & IBH, New Delhi, 1978.
6. Plant Propagation (M.K. Sadhu), Wiley Eastern Ltd., Calcutta, 1989.
7. Vegetative Propagation of Plants (K.K. Nanda and V.K. Kochhar), Kalyani Publishers, New Delhi, 1985.
8. Glossary of Horticulture (P.Hazra, S.K.Ghosh, T.K.Maity, M.K.Pandit and M.G.Som), Kalyani Publishers, Ludhiana, 2010 (3rd Edn.).
9. Technology for Vegetable Production and Improvement (P. Hazra and M.G.Som), Naya Prokash, Kolkata, West Bengal, 1999.
10. Vegetable Science (P. Hazra and M. G. Som), Kalyani Publishers, Ludhiana, 2006.
11. Vegetable seed production and Hybrid technology (P. Hazra and M.G. Som), Kalyani Publishers, Ludhiana, 2009.
12. Varieties of Vegetable Crops in India (P.Hazra and M.K.Banerjee), Kalyani Publishers, Ludhiana, 2005.
13. Vegetable Science and Technology in India (V. Swarup), Kalyani Publishers, Ludhiana, 2006.
14. Vegetable Crops in India (Eds. T.K.Bose and M.G.Som), Naya Prokash, 1986.
15. Vegetable Crops (H.C.Thompson and W.C.Kelley), Tata McGraw Hill, 1979.
16. Vegetables for the Tropical Region (Prem Nath), ICAR, 1976.
17. Vegetable Science (H.D. Brown and C.S.Hutchison), J.B. Lippincott Co., New York, 1949.
18. Tuber Crops (S.P.Ghosh, T.Ramanujam, J.S.Jos, S.N.Moorthy and R.G.Nair), Oxford & IBH, 1988.
19. Vegetable Growing in Homegardens of Tropical and Subtropical Areas (P.F.Konokov and V.Kiran), Mir Pub., Moscow, 1988.
20. Production Technology of Vegetable Crops (K.G. Shanmugavelu), Oxford & IBH, 1993.
21. Unexploited Tropical Vegetables (P.Indira and K.V. Peter), Publication Unit, Directorate of Extension, Kerala Agricultural University, 1988.
22. Knott's Handbook for Vegetable Growers (Donald N.Maynard and George J.Hochmuth), John Wiley & Sons, New York, 1980.
23. Textbook of Vegetables, Tubercrops and Spices (eds. S.Thamburaj and Narendra Singh), I.C.A.R., New Delhi, 2001.
24. Fertilizer Use (A.Jacob and H.V.Unexhull), Varlags Gesellschaft fur Ackerban Mbh, Hannover, 1960.
25. Potassium- A Product of nature (G. Kemmler and H.Hobt), Kali-und Salz Ag, FR Germany, 1985.
26. Soil Fertility and Fertilizer Use (S. Tandon), IFFCO, 1989.
27. Soils: An Introduction to Soils and Plant Growth (R.L.Donahue, R.W. Miller and J.C.Shickluna), Prentice-Hall of India, 1990.
28. The Nature and Properties of Soil (N.C.Brady), MacMillan Publishing House, INC., USA, 1980.

29. Irrigation Principles and Practices (E.V. Hasen, O.W.Israelsen and G.E.Stringham), John Wiley & Sons, New York, 1979.

30. Cropping Systems- Theory and Practice (B.N.Chatterjee, S.Maiti and B.K.Mandal), Oxford & IBH, 1989.

31. Post-harvest Physiology, Handling and Utilization of Tropical and Subtropical Fruits and Vegetables (Ed. E.B.Pantastico), AVI Publishing Co., USA, 1975.

32. Techniques in Seed Science and Technology (Eds. P.K.Agrawal and M.Dadlani), South Asian Publishing Co., New Delhi, 1992.

33. Seed Technology (R.L.Agrawal), Oxford & IBH, New Delhi, 1980.

34. Hawk's Physiological Chemistry (Ed. Oser. L.Bernard), Tata McGraw Hill, 1979.

35. Farm Chemicals Handbook (Ed. C. Sine), Willoughby, OH: Meister, 1998.

36. Basics of Horticulture (Ed. K. V. Peter), New India Publishing Agency, New Delhi, 2008.

Colour Plates-1

Nursery Management

Soil solarization with polythene sheets

Seed bed preparation after removing the polythene sheet

Mulching of seed bed with paddy straw after seed sowing

Field view of raised bed nursery

Permanent seed bed

Covering the seed bed and plug trays with 50 mesh nylon net against white fly

Colour Plates-2

Healthy seedling in the raised bed and plug trays

Healthy seedling with intact roots grown in plug trays

Modern vegetable nursery under polyhouse

Plug trays different cell capacity

Colour Plates-3

Diseases of Potato

Early blight of Potato

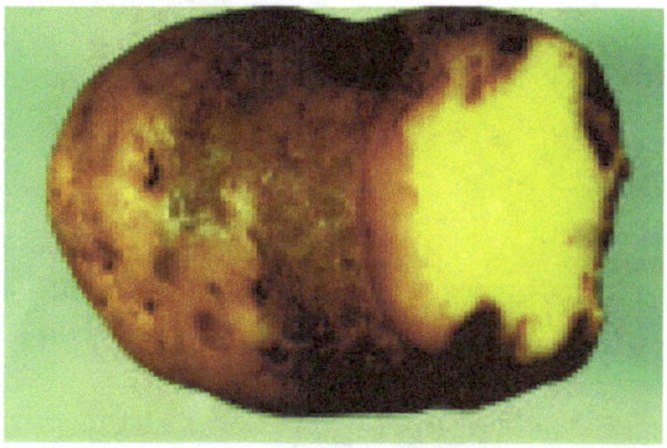

Late blight symptom in Potato tuber

Late blight of Potato stem and leaf

Soft rot of Potato

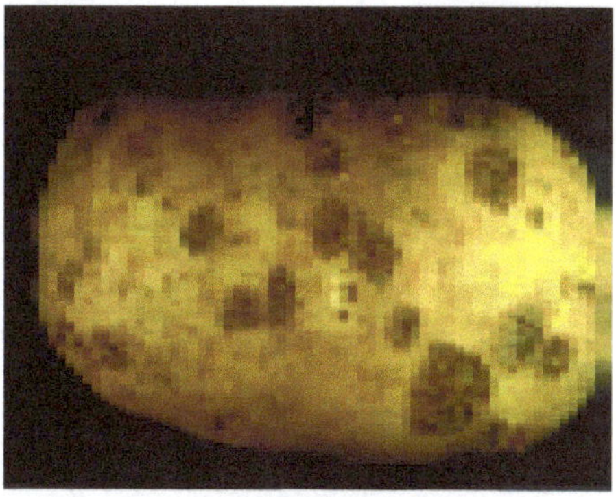

Soft disease of Potato

Colour Plates-4

Wart of potato

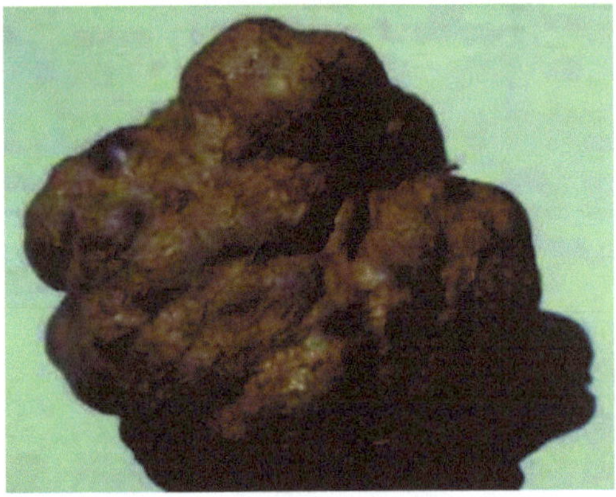
Smut of potato

Potato leaf roll virus

Potato virus Y

Insect pest of Potato

Potato leaf damaged by flea beetle

Potato tuber moth

Colour Plates-5

Potato tuber infested by the larvae of potato tuber moth

White grub of potato

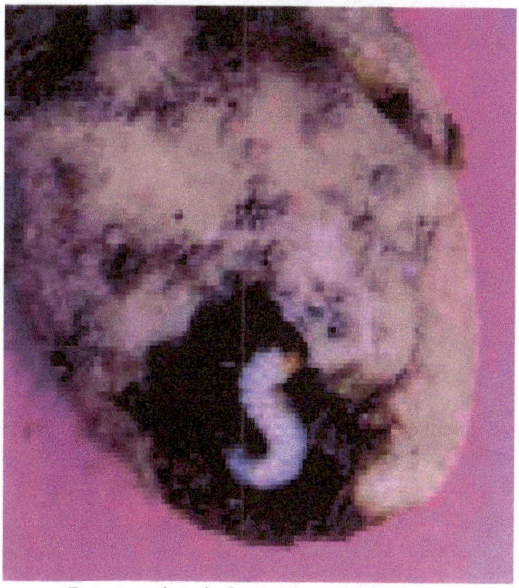

Potato tuber infested by white grub

Colour Plates-6

Physiological disorders of Tomato

Calcium deficiency symptom in tomato leaf

Blossom end rot of tomato due to calcium

Phosphorus deficiency of tomato

Nitrogen deficiency in tomato leaf

Magnesium deficiency of tomato

Potassium deficiency of tomato

Colour Plates-7

Sulfur deficiency in tomato

Zinc deficiency in tomato

Sun scalding of Tomato

Cat face of tomato

Puffiness of tomato

Colour Plates-8

Radial and concentric cracking of tomato

Concentric cracking of tomato

Iron deficiency in tomato

Molybdenum deficiency in tomato

Dumping off of Tomato

Late blight symptom in tomato fruits

Colour Plates-9

Late blight symptom in tomato leaf and stem

Early blight symptom in tomato leaves

Symptomatic diagnostics of Septoria and early blight in tomato

Early blight in tomato fruits

Septoria blight in tomato fruits

Colour Plates-10

Septoria leaf blight of tomato

Powdery mildew of tomato

Sclerotinia blight of tomato

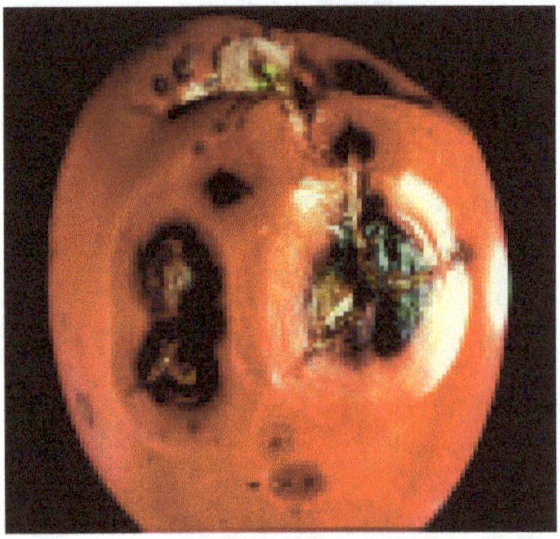

Anthracnose of tomato

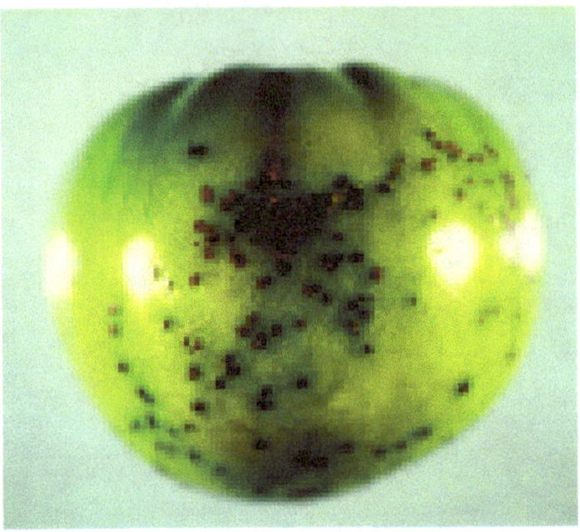

Bacterial spot in tomato fruit

Bacterial wilt of tomato

Colour Plates-11

Oozing of the wilt causing bacteria from the cut stem of tomato

Spotted wilt virus of tomato

Leaf curl virus of tomato

Yellow leaf curl virus of tomato

Insect pest of Tomato

Fruit borer of tomato

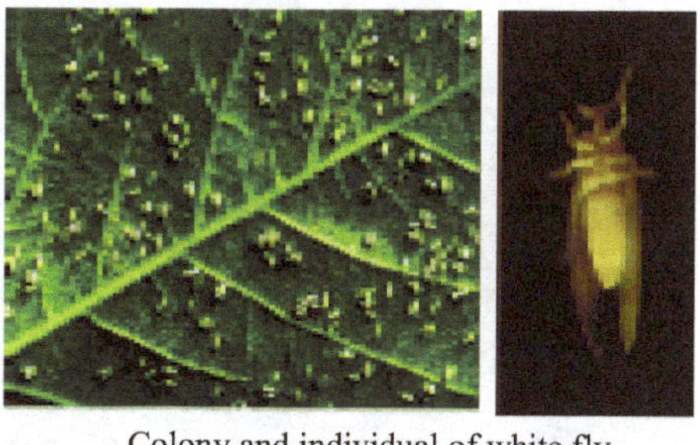

Colony and individual of white fly
Tomato leaf curl virus

Colour Plates-12

Diseases of Brinjal

Phomopsis blight of brinjal

Bacterial wilt of brinjal

Fruit infestation by brinjal fruit & shoot borer

Shoot infestation by brinjal fruit & shoot borer

Flower infestation by brinjal fruit & shoot borer

Removal of infested shoots and setting of Pheromone trap to manage fruit & shoot borer

Colour Plates-13

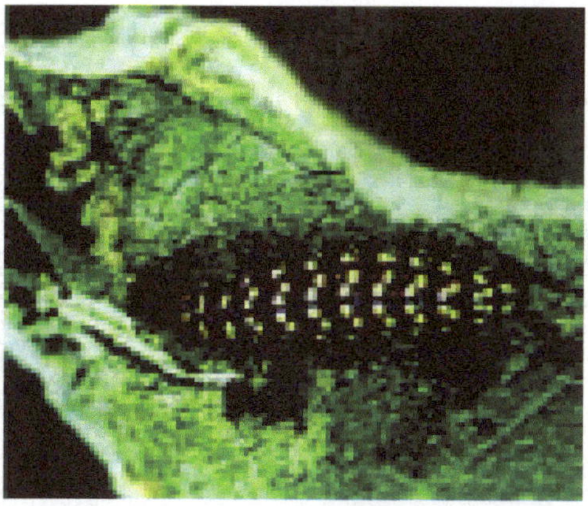

Fruit borer of Brinjal

Brinjal leaf folder

Egg mass and aadult of mealy bug infest Brinjal

Diseases of Chilli

Chilli anthracnose in fruits

Colour Plates-14

Twig blight of chilli by *Choanaephora* spp.

Fungal structure of *Choanaephora* spp.

Dieback of chilli

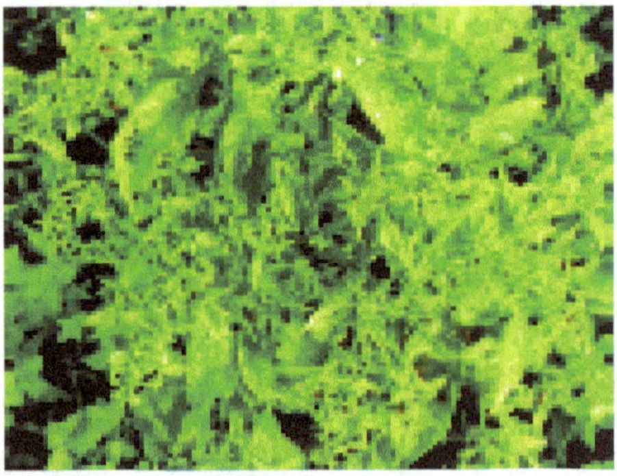

Leaf curl of chilli

Insect pest of Chilli

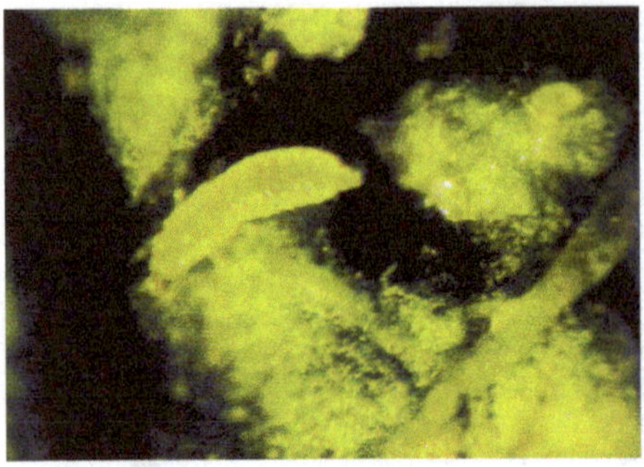

Chilli trips

Infestation of Chilli trips

Colour Plates-15

Fruit borer infestation in Chilli

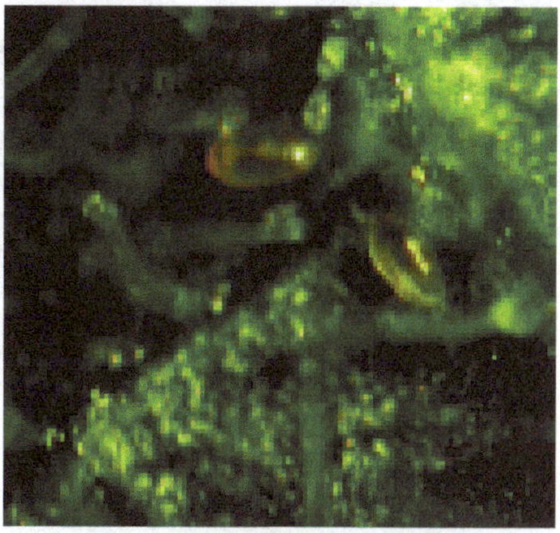

Yellow mite of Chilli

Infestation of chilli yellow mite

Dissemination of yellow mite by whitefly

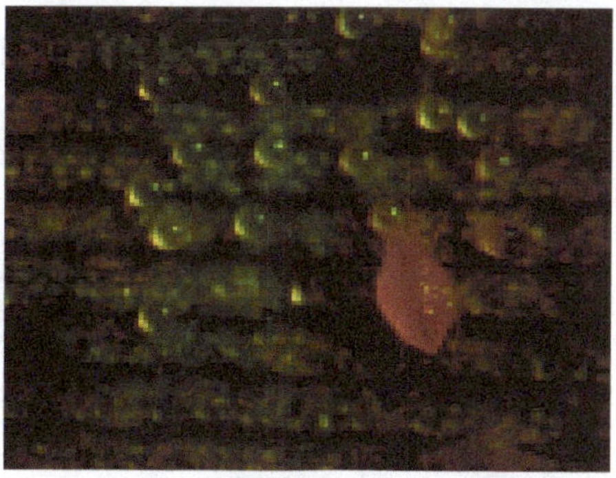

Predatory mite (*Agistemus fleschneri*) for biological control of yellow mite

Colour Plates-16

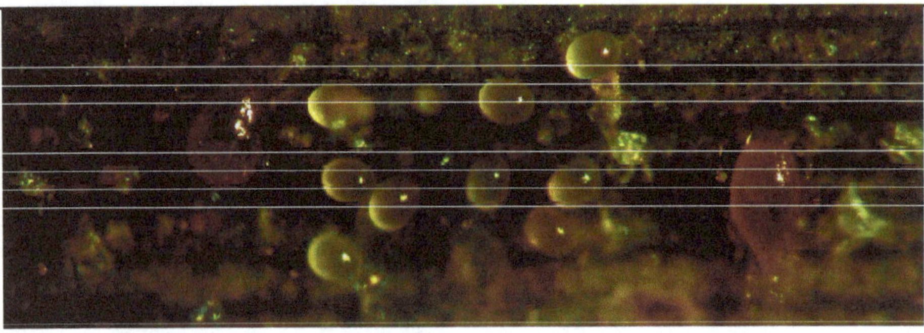

Predatory mite (*Amblyseius longispinosus*) for biological control of red spider mite

Physiological disorders of Capsicum or Sweet Pepper

Sun scalding of Sweet Pepper

Blossom end rot of Sweet Pepper due to calcium deficiency

Diseases of Sweet Pepper

Bacterial spot of capsicum in fruit and leaf

Bacterial wilt of capsicum

Index

A

ACCase inhibitors 66
Adaptation in India 72, 88. 105, 117, 131, 144, 159, 173, 183, 195, 208, 219, 232, 244, 266, 267, 276, 286, 300, 318, 324, 334, 368, 377, 390, 401
Akashin 273
Alternaria blight 273
Alternaria leaf spot 384
ALS inhibitors 66
Amaranthus 309
Angular leaf spot 253
Anthracnose 192, 203, 226, 252, 363
Aphid 127, 140, 156, 181, 242, 274, 322, 342, 345, 354
Application of bio-control agents 57
Application of herbicides 67
Ashy stem blight 352, 363

B

Bacterial blight 330, 364
Bacterial leaf spot 139
Bacterial wilt 98, 111
Basal rot 295
Bean anthracnose 340
Bean common mosaic virus 341
Bean rust 341
Biological 64
Bio-agent treatment 15
Bio-control agents 49, 56
Biological management 55
Bitter gourd 230
Black mould 307
Black rot 155, 180
Blossom end rot 95
Breeding behaviour 72, 75, 105, 118, 131, 145, 160, 173, 184, 196, 208, 232, 257, 277, 301, 311, 318, 325, 358, 368, 377, 390, 392, 401
Brinjal 103
Broccoli *or* sprouting broccoli 171
Botanical 1
Browning 167
Brown scale insect 411
Buckeye rot 98
Burning 64

C

Cabbage 157
Cabbage borer 156, 169, 181
Calcium 40
Capsicum *or* sweet pepper 129
Carrot 275
Carrot rust fly 283
Cavity spot 282
Cauliflower 142
Cercospora leaf spot 384
Characteristics of weed 59, 60
Chemica 64
Chilli 115
Classification 65
Classification of vegetable crops 1
Classification of nutrients 27
Club root 168
Cole crops 142
Collar rot 373
Common scab 83
Common weeds 63
Crop regulation 200, 212, 223, 237
Crop rotation and intercropping 372
Crop regulation 189

Crop rotation 383
Cucumber 243
Cucumber green mottle mosaic virus 409
Cucumber mosaic 254
Cucumber mosaic virus 410
Cultural requirements 107, 121, 132, 148, 162, 175, 198, 210, 221, 234, 246, 259, 269, 279, 289, 312, 319, 336, 347, 359, 370, 379, 403
Curd blight 155
Cut worms 85

D

Damping off 96, 110, 125, 138, 154, 168, 179
Damping off and fruit rot 240
Delay in fruit ripening 191, 202
Diamond back moth 156, 170
Diagnosis of the diseases 53
Die back *or* anthracnose 126, 138
Direct planting 303
Diseases 81, 96, 110, 125, 138, 154, 168, 179, 192, 203, 214, 226, 252, 261, 283, 306, 340, 351, 363, 384, 396, 409
Dispersal of weeds 61
Dormancy 80
Downy mildew 154, 168, 180, 193, 203, 214, 227, 253, 262, 307, 363, 409

E

Early blight 97
Elephant foot yam 366
Epilachna beetle 101, 241, 263
Evaluation of soil fertility 32

F

Flower thrips 354
Foliar applied group 66
French bean 332
Fruit borer 100
Fruit rot 204, 227, 228, 241, 254, 262
Fusarium wilt 226, 330

G

Garden pea 323
Gold fleck 96
Green and pot herbs 309

H

Hadda beetle 84
Hairy caterpillar 341
Hand pulling 64
Hand weeding 64
Hardening of seedlings 18
Harmful effects of weeds 62
Harvesting of crop 79, 94, 109, 124, 136, 178, 213, 238, 272, 281, 293, 305, 321, 329, 362, 373, 383
Herbicide 64
Hollow root 272
Host resistance 53
Hyacinth bean 356

I

Importance and use 70, 86, 103, 116, 129, 143, 158, 172, 182, 194, 207, 217, 230, 244, 256, 264, 275, 284, 298, 309, 317, 337, 344, 356, 366, 375, 388
Improved varieties 74, 90, 106, 120, 132, 147, 162, 186, 197, 210, 234, 258, 268, 278, 301, 319, 335, 347, 359, 378, 382, 403
Insect-pests 84, 112, 127, 140, 156, 181, 193, 215, 241, 254, 262, 283, 308, 309, 315, 322, 330, 341, 386, 397, 410
Insecticide treatment 15
Integrated disease management 53
Integrated pest management 47
Interculture and weed control 78, 94, 109, 124, 136, 151, 178, 190, 201, 213, 238, 250, 260, 271, 281, 293, 305, 321, 328, 338, 350, 362, 372, 382, 395,
Integrated nutrient management 42
IPM system in vegetables 51
Irrigation 78, 93, 109, 124, 135, 151, 178, 189, 201, 212, 237, 250, 260, 280, 292, 304, 321, 328, 337, 361, 371, 382, 394, 406
Irrigation management 54

J

Jassids 113, 397

L

Land preparation 75, 92, 108, 122, 134, 150, 164, 176, 187, 198, 211, 221, 235, 247, 259, 269, 279, 303, 312, 320, 327, 336, 348, 360, 370, 380, 392, 404,
Larval parasitoids 501
Late blight 81, 97
Latent mosaic 84
Leaf curl 99, 126, 139
Leaf eating caterpillar 315, 322
Leaf roll 84
Leaf spot 315, 322
Life cycle 60
Little leaf 111

M

Macronutrients 27
Micronutrients through chemicals 40
Macronutrients through fertilizer 36
Magnesium 40
Management of bees 47
Mechanical 63
Mechanism of action 66
Mechanisms of bio-control 57
Medicinal value 71, 88, 104, 116, 130, 144, 159, 172, 183, 195, 208, 218, 231, 244, 257, 265, 276, 285, 299, 310, 318, 324, 333, 345, 357, 367, 389, 400
Methods of planting 303, 327
Methods of weed control 63
Microbial bio-control agents 50
Monitoring pathogen 17
Mode of action 66
Mole cricket 85
Morphological 60
Mosaic 215, 227, 240, 353, 364, 374, 99
Muskmelon 194

N

Nitrogen 36
Nursery management 91, 107, 117, 133, 148, 163, 175, 290
Nursery site 121
Nursery soil 121
Nutrient management 27, 78, 93, 108, 123, 135, 150, 165, 177, 188, 200, 212, 223, 236, 249, 260, 271, 280, 292, 304, 320, 328, 337, 349, 371, 382, 394, 406
Nutrient management in soil 34
Nutrient requirement of vegetable crops 28
Nutritional value 71, 87, 104, 116, 130, 143, 158, 172, 183, 195, 218, 231, 244, 257, 265, 276, 285, 299, 310, 324, 333, 344, 357, 367, 376, 389, 400

O

Okra 388
Onion 284
Onion fly 296
Organic herbicides 67
Organic matter 34
Origin 61
Origin and taxonomy 73, 89, 106, 118, 132, 161, 174, 185, 197, 209, 220, 233, 246, 258, 268, 278, 288, 301, 311, 319, 325, 335, 346, 359, 369, 378, 391, 402

P

Peat 13, 17
Pepper maggot 141
Pepper weevil 141
Perlite 13
Phosphorus 37
Photosystem II inhibitors 67
Physiological basis 28
Plant tissue analysis 32
Planting methods 405
Planting 18, 405
Planting the seed tuber 370
Planting of vines 381
Planting time 76, 303, 370, 381, 404
Planting of tuber 77
Plant protection 14, 15, 16, 17, 18, 19, 20, 21
Pod borer 330
Pointed gourd 399
Post harvest handling and storage 79, 94, 110, 124, 136, 152, 167, 178, 190, 202, 213, 225, 239, 251, 261, 272, 281, 293, 305, 315, 321, 329, 339, 351, 362, 373, 383, 395, 408

Potato tuber moth 84
Powdery mildew 396
Predators 50
Preparation of nursery bed 16
Prevention & management methods 53
Pricking out of seedlings 17
Primary nursery 380
Principles and practices of vegetable production 18
Production technology 70
Potato 70
Propagation 302
Propagation and planting material 380
Protection of seedlings 18
Pumice 14
Pumpkin 206

Q

Quantity used by the plant 27

R

Radish 264
Ratooning 407
Red pumpkin beetle 204, 215, 228, 241, 255, 262
Red spider mite 113, 216, 315, 410
Riceyness 153
Rock wool 14
Root crops 264
Rot disease 409
Root forking 273, 282
Root knot nematode 85, 101, 216, 229, 255, 354, 365,
Rust of pea 329

S

Saw dust 13
Scheduling of irrigation 23
Secondary nursery 381
Secondary nutrients 27
Seed sowing 16, 187, 199, 211, 222, 236, 248, 259, 348, 361, 393
Seedling root dipping 41
Spacing 188, 361
Sensitivity of bio-agents with pesticides 58
Shoot and fruit borer 112, 397
Site selection and land preparation 53
Skin cracking 137
STM & ASM 25
Soft rot 83, 139, 155, 181, 307
Soil analysis 32
Soil application 40
Soil-applied group 66
Soil solarization 14
Soil treatment 14
Sowing of seeds 270, 280, 313, 320, 327, 337
Sowing time 92, 108, 122, 134, 149, 164, 176, 187, 199, 211, 222, 235, 248, 259, 270, 279, 291, 313, 320, 327, 348, 360, 393
Spacing 77, 93, 108, 123, 135, 150, 165, 177, 199, 212, 222, 236, 249, 259, 271, 280, 292, 304, 313, 320, 327, 337, 349, 382, 393, 405
Spinach beet *or* palak 31719
Splitting *or* cracking of roots 28216
Stem fly 341, 353, 364
Stemphylium blight 29518
Sulphur 39
Sun scald 96, 137
Sweet potato 375
Sweet potato scab 385
Synthetic auxin 67
Systemic 66

T

Thrips 113, 193, 205, 308
Tillage 64
Tobacco caterpillar 101
Tomato 86
Training over trellis 407
Transplant groups 11
Transplanting 164, 123, 134, 150, 177

U

Unfruitfulness 408

V

Vegetable cowpea 343
Vermicompost 13
Vermiculture 13

W

Wart 272
Water management 20, 59, 271
White fly 100
White mold 340
White rust 273

Y

Yellow vein mosaic virus 39618
Yield 85, 102, 114, 181, 193, 205, 216, 229, 255, 263, 274, 283, 297, 308, 316, 322, 331, 342, 365, 374, 387, 398, 411

FSC
www.fsc.org
MIX
Papier | Fördert gute Waldnutzung
FSC® C083411

Zeitfracht Medien GmbH
Ferdinand-Jühlke-Straße 7
99095 Erfurt, Deutschland
produktsicherheit@kolibri360.de